POLLEN ANALYSIS

POLLEN ANALYSIS
SECOND EDITION.

P.D. MOORE
BSc, PhD

J.A. WEBB
BSc, PhD

M.E. COLLINSON
BSc, PhD

OXFORD

BLACKWELL SCIENTIFIC PUBLICATIONS

LONDON EDINBURGH BOSTON

MELBOURNE PARIS BERLIN VIENNA

© 1991 by
Blackwell Scientific Publications
Editorial offices:
Osney Mead, Oxford OX2 oEL
25 John Street, London WC1N 2BL
23 Ainslie Place, Edinburgh EH3 6AJ
3 Cambridge Center, Cambridge
  Massachusetts 02142, USA
54 University Street, Carlton
  Victoria 3053, Australia

Other editorial offices:
Arnette SA
2 rue Casimir-Delavigne
75006 Paris
France

Blackwell Wissenschaft
Meinekestrasse 4
D-1000 Berlin 15
Germany

Blackwell MZV
Feldgasse 13
A-1238 Wien
Austria

First published as *An Illustrated Guide
  to Pollen Analysis* by Hodder & Stoughton,
  1978
Second edition 1991

Set by Setrite Typesetters, Hong Kong
Printed and bound in Great Britain
by William Clowes Ltd, Beccles, Suffolk

DISTRIBUTORS

Marston Book Services Ltd
PO Box 87
Oxford OX2 oDT
(*Orders*: Tel: 0865 791155
        Fax: 0865 791927
        Telex: 837515)

USA
Blackwell Scientific Publications, Inc.
3 Cambridge Center
Cambridge, MA 02142
(*Orders*: Tel: 800 759-6102)

Canada
Oxford University Press
70 Wynford Drive
Don Mills
Ontario M3C 1J9
(*Orders*: Tel: 416 441-2941)

Australia
Blackwell Scientific Publications
(Australia) Pty Ltd
54 University Street
Carlton, Victoria 3053
(*Orders*: Tel: 03 347-0300)

British Library Cataloguing in Publication Data
Moore, Peter D. (Peter Dale), *1942*−
    Pollen analysis
    I. Title II. Webb, J. A. III. Collinson, M. E.
    (Margaret E)
    561.13
    ISBN 0-632-02176-4

Library of Congress
Cataloging in Publication Data

Moore, Peter D.
    Pollen analysis / P. D. Moore, J. A. Webb,
    M. E. Collinson, − 2nd ed.
        p.   cm.
    Includes bibliographical references and index.
    ISBN 0−632−02176−4
      1. Palynology, I. Webb, J. A. II. Collinson,
    Margaret E. III. Title.
    OK658.M663 1991
    561′.13−dc20

CONTENTS

PREFACE

The value of pollen analysis as a tool for the reconstruction of past vegetation and environments, and its applications in such areas as climate change studies, archaeology, geology, honey analysis and forensic science, is now widely known. Many research workers from a variety of different backgrounds, and with a range of different aims and intentions look to pollen analysis as a means of assisting them in their research. These researchers, together with undergraduate and sixth-form students on ecology or biogeography courses, or involved in project work, require an introductory text explaining the opportunities offered by the techniques of pollen analysis and a practical guide to the operation of these techniques and the identification of the pollen grains and spores they will encounter. Those who wish to develop a deeper involvement in the subject also require key references to research papers and specialist texts that provide more detailed information. This book supplies all three of these requirements — an introductory guide, a practical manual and a bibliographic source.

The book has developed from a similar project published in 1978 (*An Illustrated Guide to Pollen Analysis* by P.D. Moore and J.A. Webb, Hodder & Stoughton, London), modified by 12 years' experience of using that guide in teaching undergraduate students in a Quaternary palaeoecology course at King's College in London. That initial attempt at producing an illustrated practical manual to pollen analysis met with some success, and the book has been widely used, often in geographical areas and in types of project for which it was not originally designed. It has been out of print for many years, but the opportunity has now arisen, largely through the enthusiasm and initiative of Susan Sternberg and Bob Campbell at Blackwell Scientific Publications, to replace it with an up-to-date and expanded version.

The new version differs from the old in several respects. It incorporates many of the advances that have been made in the past decade and guides the reader into a greatly expanded scientific bibliography. It provides a greatly extended key, fully revised by Judy Webb to supply a wider taxonomic and therefore geographical coverage. Whereas the original key was limited to the pollen flora of North West Europe, we have now introduced a number of central and southern European and North American pollen types to make the book more applicable to work in those areas. The taxonomic precision of the key has also been improved in many cases, permitting a more detailed identification and thus improved resolution in the interpretation of data. This new illustrated guide has also improved its illustrations. A new author, Margaret Collinson, has been recruited to the team with particular responsibility for pollen and spore photography. All of the original photographs have been replaced and there are now over 1000 illustrations covering about 450 pollen types. Standard magnifications have been used throughout to facilitate comparison between grains. A number of scanning electron micrographs have also been inserted, mainly with the intention of assisting in the interpretation of light microscope images of pollen grains.

It has not been our intention to supply the reader with reviews of the many fascinating results that have been obtained from the application of pollen analysis in a variety of fields (interested parties are advised where to look for such information), but rather we seek to encourage students and others to make use of pollen analytical techniques and so to expand the body of palynological information for themselves.

Many people have provided generous assistance in a variety of ways — stimulating discussion, frank criticism, the provision of material for taxonomic and photographic work, and permission to use published diagrams and information. We should especially like to mention Madeline Harley and Keith Ferguson of the Palynology Unit, The Herbarium, Royal Botanic Gardens, Kew, and Steve Blackmore and Peter Stafford of the Palynology Section, Botany Department, The Natural History Museum, London, for their help and encouragement.

Alan Howard and Tony Brain of King's College London provided much help and advice with the photography, and Andrew Evans assisted with the SEM work. Pat Wiltshire and Petra Day acted as very willing experimental animals for the testing of the keys.

M.E.C. undertook this work during the tenure of a Royal Society 1983 University Research Fellowship, which is gratefully acknowledged. J.A.W. would like to thank the Department of Plant Sciences, University of Oxford, for the use of their facilities during the revision of the key.

Finally, we must express particular gratitude to our partners and families who have endured the development of this book with unfailing (well, almost!) fortitude and patience. We trust that the sacrifices they have made in the interest of palynology will prove worth while.

P.D.M., M.E.C., London
J.A.W., Oxford

1
BASIS AND APPLICATIONS

Palynology is the study of pollen grains (produced by seed plants, angiosperms and gymnosperms) and spores (produced by pteridophytes, bryophytes, algae and fungi). The two groups differ very considerably in their function, but both pollen grains and spores (with the exception of some algae and fungi) result from cell division involving a reduction by half of the chromosome content (meiosis) and both need to be transported in order to perform their functions adequately.

The pollen grain is a container which houses the male gametophyte generation of the angiosperm or gymnosperm. The spore on the other hand is a resting and dispersal phase of the cryptogam. Both require dispersal in space, but the pollen grain can only be regarded as having served its purpose successfully if it arrives at the stigma (or micropyle) of a plant of the same species and germinates there with subsequent fertilization of an egg. Cryptogam spores require only to arrive at any site where they can germinate, but the site must be suitable for the resulting gametophyte plant to establish, survive, and produce gamete-bearing structures.

The fact that both pollen and spores need dispersal for their adequate function has led to many analogies in their form. They are of similar size, often around $20-40\,\mu m$, and they are both surrounded by tough, resistant walls which are frequently sculptured in distinctive ways. Aerodynamically, wind-transported pollen and spores behave in similar ways and their similarities have led to pollen and spores being considered together in the discipline of palynology.

Palynology is concerned with both the structure and the formation of pollen grains and spores, and also with their dispersal and their preservation under certain environmental conditions. One aspect of palynology is the study of fossil pollen grains, either in ancient or even fairly recent materials, and it is with this aspect of the science, pollen analysis, that this book is mainly concerned. The emphasis of the book rests in Quaternary studies, that is the palynology of about the last two million years, and the pollen keys refer largely to the pollen flora of temperate Europe and North America. Those whose main interest lies with the palynology of the pre-Quaternary period should consult Traverse (1988).

APPLICATIONS OF PALYNOLOGY

The examination of pollen grains, both recent and ancient, can be of value in a range of scientific studies. These include:

1 Taxonomy.
2 Genetic and evolutionary studies.
3 Honey studies (melissopalynology).
4 Forensic science.
5 Allergy studies.
6 Tracing vegetation history in:
   (a) individual species;
   (b) communities.
7 Correlating deposits and assigning tentative dates.
8 Climatic change studies.
9 The study of past human impact on vegetation.
   There are various features possessed by pollen

grains and spores which have made them useful in such a range of disciplines. In the first place they have such a tough outer coat (the exine) that they survive better and longer than many other biological materials. The chemistry of the coat renders them resistant to decay and wherever microbial activity is depressed, whether due to wetness, salinity, low oxygen availability or drought, there is a chance of pollen and spore survival. This is of particular value in the study of vegetation history and its application in such fields as dating and climatology.

The second useful feature of pollen and spores is the variation in the form and sculpture of the resistant coat. This is amply displayed in the collections of photographs, especially the scanning electron micrographs, found in this volume. The functional significance of these surface patterns is still somewhat controversial, though there is no shortage of theories concerning the value of spines, pores, grooves, and reticulations over the exine (see, for example, Blackmore & Ferguson 1986). Such variations in form provide a means of identifying pollen grains and are also a valuable character to use in taxonomic studies. Using light microscopy the identification of pollen grains is sometimes limited to family or generic level, but there are occasions when identification to species is possible (see Chapter 6). This is again a major reason for the usefulness of pollen in vegetation history studies and also in honey and in allergy investigations.

The small size of pollen grains and spores is necessary for their ease of transport either in the movement of genetic material (pollen) or in the invasion of new territory (spores). Many rely on aerial transport for their dissemination, and the study of their movement can be of interest to geneticists and to medical scientists investigating the causes of allergic reactions.

Finally, the unreliability of any particular grain finding its target means that pollen and spores must be produced in very large numbers if they are to be effective. This is particularly the case in the very chancy business of air transport, but rather less so if the more predictable habits of insects can be exploited. So pollen and spores are produced in far larger numbers than would be needed if they were more efficient in finding their targets. This excess is the material on which palaeopalynology depends. Pollen sedimenting from the atmosphere is washed out by the rain, finding its way into soils, streams, lakes and mires where it may lie for very long periods of time; a stratified information bank reflecting ancient vegetation. The very abundance of the pollen and spore populations in such deposits opens up the possibility of statistical studies which lend precision to environmental reconstruction.

Taxonomic studies

Taxonomists are concerned to establish evolutionary relationships between extant plant populations and to classify these into particular levels of organization. Fossil studies may help in this process, but often the necessary data is not available and relationships must be inferred from similarities between living individuals. In such circumstances as many characters as possible must be considered in determining similarity, and pollen grains and spores have an important part to play here.

For example, in her survey of spore diversity in the Pteridophyta, Tryton (1986) examined some 250 genera. She was able to divide them into five main spores types, based on shape, aperture, surface and wall structure and these correspond well with current classifications of these genera on whole plant morphological characters. Spores here confirm the taxonomic work already conducted.

In a study of pollen of the taxonomically complex *Acacia* genus in Australia, Guinet (1986) has found a progressive morphological series within the pollen following a north/south gradient in Australia. As in the case of pteridophyte spores, the results agree with current taxonomic views based both upon gross plant morphology and biochemistry, but serve to underline the dynamic state of the Australian acacias from a taxonomic and evolutionary point of view. The gradient of similarity in their pollen indicates a recent evolutionary divergence in which clear divisions have not yet fully emerged.

Genetic and evolutionary studies

These are closely linked to taxonomic studies, as the above example of *Acacia* taxonomy shows, but there are other extensions of palynology which can amplify the straightforward examination of pollen morphology. For example, the study of fossil pollen and spores can add a time dimension to taxonomy. In the case of certain of Tryton's pteridophytes, such as in the genus *Osmunda*, there was evidence of spore characters being conserved since Jurassic times, providing evidence of long-term evolutionary stability of spore morphology. This is in marked contrast to the rapid divergence found in *Acacia*.

An alternative application of pollen studies in genetics and evolution is the use of measurements of pollen flow in determining patterns and rates of gene flow in modern populations. Such work has been reviewed by Handel (1983) and includes the labelling of pollen grains either by dyes or by chemicals, such as rare earths, that can be rendered radioactive as a result of being bombarded by neutrons (neutron activation analysis — see Handel 1976). In this way the rate and direction of pollen movement can be followed, whether it is dispersed by air or by an animal vector without danger to the environment from direct release of radioactive compounds. Such studies indicate just how far and how fast genes can move around in natural plant populations (see also Chapter 8).

In the case of insect pollinated species, there is a distinct competition between flowers for the available pollinator and this places a selection pressure upon flowers (Ratchke 1983). This competition may even achieve economic significance, as in the competition between fruit trees and dandelions (*Taraxacum officinale*) in the grassland below for the attentions of pollinating bees (Free 1968). The evolutionary conse-

quences of such competition have been reviewed by Feinsinger (1987).

Honey studies

The analysis of the pollen loads of bees and the pollen content of honeys (sometimes termed melissopalynology) has proved of considerable economic value to apiarists and in the food industry (Cowan 1988). An average colony of bees requires large quantities of pollen for its maintenance, some estimates placing it at between 20 and 40 kg in a season. The foraging habits of bees can be followed by the analysis of their pollen loads, and sometimes it is possible to gain a fairly accurate idea of pollen load identity simply on the basis of colour (Hodges 1974). A microscopic examination of the pollen load is more satisfactory, however (Sawyer 1988), and methods by which this can be achieved are given in Chapter 4. The seasonal variation of pollen loads can be important for apiculturists (Adams & Smith 1981).

Besides the biological interest in pollination and pollen foraging, the study of pollen in honey has become important in the protection of consumers from adulteration or mislabelling of honeys.

In one British court case, for example, honey sold as 'Yorkshire Clover Honey' was found to contain significant proportions of pollen from *Helianthus*, *Onobrychis*, *Echium*, and *Salvia*, which are rarely found in true Yorkshire honey. It was concluded that the honey originated in eastern Europe and the vendor was charged and found guilty under the British Trades Description Act on the evidence of the honey analysis (Sawyer 1985).

Forensic work

The abundance of pollen in the environment, its recognizable character, and its persistence have resulted in its use in forensic science. Soil, leaf litter, even dust, contains pollen grains which may provide clues to the type of habitat or geographical area from which a sample originates. Soil from shoes, cleanings from fingernails or dust from clothing may yield enough pollen for analysis and the reconstruction of recent movements.

Erdtman (1969) describes an Austrian case history in which a murder was solved by resorting to palynological techniques. A man was arrested and charged with the murder of a male colleague while on a boat journey along the Danube near Vienna; however, no body could be found. Pollen analysis of a soil sample from the arrested man's shoes revealed much pine and alder pollen together with some spores of Tertiary origin. Fortunately, only one area was known along the Danube where pine and alder grew together on Tertiary strata, so the suspect was confronted with this fact. He was so shocked at the deduction that he admitted the crime and the precise location of the hidden body.

A particularly novel application of palynology in forensic science has been the examination of the Shroud of Turin, the remarkable and mysterious cloth upon which is an impression that was once believed to have been made by the body of Jesus Christ. As part of a scientific investigation of this material, the Roman Catholic Church permitted a forensic scientist, Dr Max Frei, to remove dust from the Shroud using sticky tape. He has published a list of the pollen types he claims to have identified (Frei 1979) and on the basis of this claims that the cloth must have been in Palestine and in the Anatolian steppes as well as in Italy. Unfortunately, many of the taxa on his list are identified to a level of precision which would leave most palynologists doubtful. For example, *Artemisia herba-alba*, *Carduus personata*, *Acacia albida*, *Silene conoidea* and *Sueda aegyptica* are not generally regarded as being capable of identification to such a precise taxonomic level. In the case of the last two, one cannot even be confident of the genus. Evidently the forensic scientist should be fully aware of the limitations as well as the potentialities of pollen analysis. It has now been demonstrated by means of radiocarbon dating that the Shroud of Turin is in fact a medieval fake (Damon *et al.* 1989). But the pollen still needs to be explained. A more cautious interpretation would be that the Shroud may well have spent some time in the Mediterranean area — not surprising since it is housed in Italy.

Pollen analysis also helped to solve a problem in South East Asia that could otherwise have become extremely serious. In the early 1980s masses of yellow spots were frequently observed on foliage in Laos and it was alleged that these provided evidence of the use of a chemical toxin, agent orange, by forces then fighting in the area. Palynological analysis of spots collected both from rocks and foliage were carried out by Nowicke and Meselson in 1984, and they found that the material was rich in pollen grains. Experimental work enabled them to demonstrate that the spots were probably caused by mass defaecation by swarms of honey bees. The small quantities of chemical toxins that had been detected in the samples were probably due to the growth of fungi on the pollen-rich debris, and claims that chemical warfare had been conducted in the area were hastily dropped.

Many illicit drugs are produced from plant products and the residue of pollen grains in these drugs can help determine their origin. In one forensic study in New Zealand by Mildenhall (1990), a sample of *Cannabis* resin was subjected to pollen analysis to discover whether it had been imported into New Zealand or whether it had been grown locally. The analysis demonstrated the presence of many pollen types exotic to New Zealand and a lack of many of the types normally found in surface pollen spectra in that country, such as *Nothofagus*. The conclusion, therefore, was that it had been imported and there was no need for police to search further for a new local source.

Allergy studies

Pollen in the atmosphere causes an allergic reaction (variously called hay fever, allergin rhinitis or pollinosis) in many people and the study of the nature of this sensitivity involves an understanding both of the chemical nature of pollen grains and of the periodicity of pollen release and its mode of transport. In a Swedish study of a whole range of allergies (Knox 1979), over 30% were caused by pollen, the remainder being due to food (about 16%), house dust, including mites (11%), and the fur or skin of various animals (about 45%). Grass pollen is one of the most serious problems, but mainly because it is so abundant in the atmosphere; other pollen types such as plantain, ragweed and birch can also cause allergic reactions. Patients sensitive to grass pollen begin to show symptoms when the pollen content of the air rises above 50 grains m^{-3}. The proportion of people sensitive to pollen varies from about 3% of the population of Britain to about 15% in the United States.

Pollen grains from the atmosphere come into contact with the cornea of the eye and with the surfaces of the trachea and bronchi, where the allergic reaction takes place. They are unlikely to penetrate into the lungs themselves because of their relatively large size and the degree of air turbulence in the upper respiratory tract. Once they have landed on the lining membranes they are removed in the surface mucous coat by the beating of cilia, probably being brought back to the pharynx within an hour of entry, but by then the irritating biochemical reactions have taken place.

The allergens which elicit a response are proteins and glycoproteins which leak from their storage sites within the complex structure of the pollen grain wall. Their likely role in nature is in the process of recognition that takes place between the pollen grain and the stigma surface (Lewis 1979; Wilson & Burley 1983). The arrival of pollen at a stigma can be a fairly random affair, especially in the case of wind pollinated species, but only compatible pollen of the correct species is permitted to germinate and develop a functional tube by the receptive stigma, and a pollen grain is recognized by the proteins it exudes. The human tissues similarly recognize and respond to foreign proteins, but some people are clearly more sensitive than others.

Monitoring the risks of hay fever at different times of year can be achieved by filtering pollen from the air using special traps (see Chapter 3) and determining the density of different types. Species composition and density of pollen in the atmosphere vary with flowering season (see, for example, Hyde 1950, 1969; Janssen 1974) and also with weather conditions, because rain washes small particles from the air. This can be seen in the studies on grass pollen in Denmark by van den Assem (1971) and summarized in Fig. 1.1. Allergic problems are thus greatest when weather conditions are warm and dry and when the offending species, such as grasses, are in full bloom.

In the United States, many sufferers have found relief by living in areas of low vegetation cover, such as the arid desert regions. But the human desire to be surrounded by plant life has resulted in an increasing extension of gardens, parks, and lawns in the desert cities, such as Tucson, Arizona, and the consequence is that airborne pollen and the incidence of hay fever in such cities is now on the increase. In the last 20 years airborne pollen from exotic species in Tucson has increased ten times.

Other examples of allergy studies can be found in Trusswell and Owen (1990).

Tracing the history of vegetation

One of the most highly exploited uses of palynology is the investigation of vegetation history (Godwin 1975; Bryant & Holloway 1985). This is of great value in palaeoecology, in archaeology and palaeoanthropology, since it provides a basic picture of the botanical setting in which past animals and human beings evolved and played their role. Most effort has certainly been expended on the reconstruction of changes during the last 10 000 years (the Holocene) during which the earth has recovered from its latest glaciation and human pressures upon global vegetation have become increasingly intense.

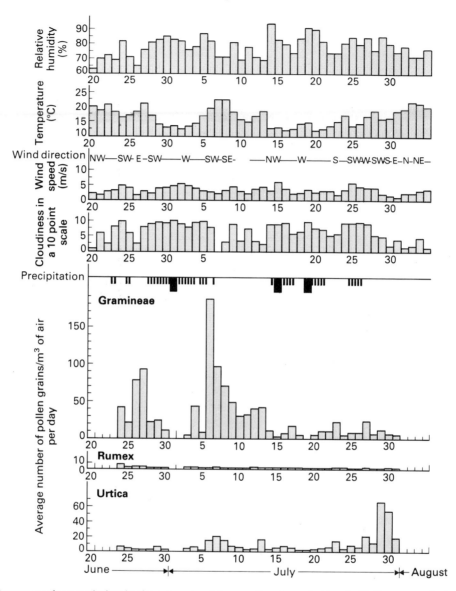

FIG. 1.1. Diagram to show variation in the amounts of three pollen types in the air (daily averages) during different weather conditions. Stinging nettle (*Urtica*), sorrel (*Rumex*) and grass (Gramineae) pollen in the air are greatest under sunny, warm conditions (from van der Assem 1971).

The applications of vegetation history studies will be discussed later, but first one must consider their value from a purely botanical point of view. Stratified sequences of pollen in lake sediments and peats provide a means of tracing the history of individual plant taxa (sometimes species, sometimes only genera, depending on the precision with which identification can be conducted). It may also be possible to document the history of plant communities by considering the whole assemblage of pollen grains and spores.

Individual taxa

Even the very earliest of pollen stratigraphic studies provided information on the changing abundance of certain plant taxa at the site of deposition. The major trees, such as pine, birch, oak and beech could be studied in terms of their arrival, expansion and sometimes contraction at that one location in space. Such studies have become more sophisticated in recent times as techniques have been developed to record the absolute accumulation rate of pollen grains from each taxon, rather than just their proportion in the whole assemblage (see Chapter 4). It is now possible to study precise curves of pollen abundance against time for a given taxon, and so reconstruct in some detail its population expansion. Bennett (1983) has carried out this type of study for forest trees in the early Holocene of eastern England and has been able to calculate the population doubling time for various tree species during their phase of exponential population growth. This ranges from about 35–46 years for hazel (*Corylus avellana*), 58–73 years for pine (*Pinus sylvestris*), to about 100 years for lime (*Tilia cordata*).

Even on the basis of percentage data, it is possible to map the spread of a taxon by observing its arrival and expansion at a number of sites over a given area. For example, Dexter *et al.* (1987) have modelled the migration of beech (*Fagus grandifolia*) in eastern North America during the last 10 000 years as this tree expanded from its glacial refugium in Tennessee and its population moved up the eastern seaboard into New England and Quebec. Tracing the changing geography of a taxon in this way demands a high density of dated pollen profiles from the area under study, but these are now available for large areas of Europe and North America.

Pollen studies concerning the origins of endemic taxa are often limited by problems of taxonomic separation of pollen grains. The work of Boyd and Dickson (1987b) on the endemic *Sorbus* species on the Isle of Arran in Scotland, for example, demanded the specific identification of pollen in a particularly difficult group (see Chapter 6).

Not only the origin and expansion, but also decline of plant taxa is of ecological interest. Allison *et al.* (1986) have studied the decline of chestnut (*Castanea dentata*) and hemlock (*Tsuga canadensis*) in the northeastern United States. The former is known to have been caused by a fungal pathogen in the early part of this century, and these authors propose that the fossil hemlock decline about 5400 years ago may also have resulted from pathogen attack. Perhaps the same is true of the crash in elm (*Ulmus*) populations in northern Europe around 5000 years ago (Perry & Moore 1987). Such studies enable us to follow the fortunes of any particular species in space and time.

Plant communities

Considerable debate surrounds the entire concept of the plant community, some considering it an interactive whole which behaves as a unit and others regarding it as a convenient abstraction on the part of plant ecologists when really it consists of an assemblage of individuals, each acting relatively independently of others in its performance and distribution. Given this disagreement among students of modern plant ecology and biogeography, it can hardly be expected that 'communities' of the past will be easy to recognize and distinguish.

Much depends upon the scale at which community detection is attempted. It is certainly possible to observe a succession of 'plant formations', or 'biomes' when following the course of vegetation change in a temperate location over the last 13 000 years. Tundra vegetation gives way to boreal birch and pine woodland and finally to deciduous forest. This is clearly demonstrated by studies of individual sites, as shown by Godwin's (1940) early work in the British Isles, or by mapping collections of species over large geographical areas and tracing the coincidence of species in assemblages, as is possible from the work of Huntley and Birks (1983) in Europe and Webb (1987) in North America. Once again, this approach is limited by the availability of a high density of well dated pollen sites in the area.

A further approach to the reconstruction of past communities is the use of current knowledge of the fidelity of certain species to certain plant assemblages. Where certain species of relatively narrow ecological amplitude are found, one usually finds an associated group of plant species with similar ecological requirements. But community reconstruction on the basis of such 'indicator species' assumes that neither the requirements of the species nor the range of possible associates has altered in the course of time. Such an assumption is rather tenuous.

The study of the current pollen rain from modern plant communities and of the processes whereby pollen reaches its site of preservation (termed *taphonomy*) can greatly facilitate palaeo-community studies, especially when combined with modern techniques of numerical analysis (see Chapter 8). With the aid of such studies, however, one may find ancient plant assemblages with no modern counterparts. One should not assume that communities remain constant in their composition through time (Moore 1990).

Correlation of deposits

The use of pollen analysis as a dating tool is dependent on prominent changes in pollen proportions being strongly correlated with a particular timing. In the British Isles, a number of distinctive pollen horizons have been regarded as reliable time indications, but such confidence has proved to be based on too limited a number of absolute datings, obtained by radiocarbon methods. As such datings have become more abundant, the unreliability of pollen horizons as time indicators has become increasingly apparent (Smith & Pilcher 1973), especially if they are extrapolated over large areas. On a local basis, the use of well authenticated datum horizons can be of value, as can recognizable sequences of changing pollen assemblages.

Some pollen horizons are reasonably synchronous over wide areas and are of broader value, such as the decline in elm at about 5000 years ago in northern Europe and the hemlock decline in eastern North America, both referred to above. But the wider use of radiocarbon dating in recent times has rendered pollen dating of less significance to those researchers who formerly relied upon it, including archaeologists and geomorphologists. In older strata, palynology remains a major stratigraphic tool (Traverse 1988).

Climatic change

Many techniques are available for the study of past climates and the study of past vegetation types through pollen analysis is clearly a useful source of evidence in this area. Much of the early Scandinavian work in pollen analysis was directed at climatic reconstruction, and for several decades the interpretation of pollen diagrams was conducted with a strong climatic emphasis. The over-riding general influence of climate on global vegetation patterns undoubtedly justifies this basic approach, but complications, such as local soil influences, the impact of human cultures, the slow response of vegetation (particularly forest) to climatic change, and the effect of chance, stochastic elements, in determining the pattern of plant arrivals at a site, must all be borne in mind when using pollen data in climatic reconstruction (see Prentice 1986a).

Webb (1980) has shown that it is possible to develop mathematical relationships between fossil pollen data (such as percentage oak pollen) and climatic variables (such as July mean temperature). On the basis of such models it becomes possible to erect reconstructions of past isotherms, but the problems relating to vegetation lagging behind climatic change and facing competitive resistance before eventually coming into equilibrium with climate, still need to be considered.

Human impact on vegetation

Understanding the environmental setting and the economy and way of life of prehistoric human cultures depends upon the collation of information from a wide range of evidence, among which pollen analysis ranks very highly (Dimbleby 1985). The detection of human modification of the environment may even provide evidence of a human presence before archaeological evidence is available. The precision with which human influence can be detected from the pollen record is greatly facilitated by the practice of agriculture on the part of the cultures being studied. In agricultural communities one expects to find elements of destruction of the natural vegetation, the introduction of crop species, the presence of weed species associated with arable and pastoral activities (Behre 1981), and the recovery of vegetation following the abandonment of the site, often in different proportions from the original vegetation cover (Delcourt 1987).

These techniques have not only been applied to the elucidation of prehistoric land management, when documentary sources are unavailable, but also to enhancing historical knowledge of more recent management of vegetation. For example, Pott (1986) has been able to construct very detailed accounts of forest management in northern Germany over the past 2000 years or so, using a phytosociological community approach to the interpretation of pollen diagrams.

But the recognition of pre-agricultural activities on the basis of pollen data is far more difficult. How, for example, can one expect to detect the impact of pre-agricultural Mesolithic cultures in the forests of northern Europe, or aboriginal cultures in semi-arid ancient Australia (Singh & Geissler 1985)? In both cases, one must place the pollen evidence alongside additional information from charcoal, inwashed soils, and evidence for nutrient flushing (such as algal blooms) in order to reconstruct local human influence.

It is also difficult to detect even agricultural or pastoral communities in environments that bear little vegetation, such as tundra or desert. Sometimes the changes induced by human activity are very subtle, such as the expansion of Chenopodiaceae pollen with more intensive farming activity in Iranian steppe-desert (Moore & Stevenson 1982).

As with all palaeoreconstruction work, the study of modern agricultural communities and their pollen rain may assist in understanding the effects of mankind on vegetation in the past, but the changes induced by modern agricultural methods, particularly herbicides, mechanization, and seed screening, have resulted in the extinction of most of the ancient assemblages of plants that once accompanied the agricultural efforts of early human cultures. Ancient weed communities very rarely have precise modern counterparts.

From these various examples it can be seen that pollen analysis has considerable potential as a tool for investigating a variety of different ecological, biogeographical, archaeological, medical and forensic problems. This book is intended to provide a practical guide to the preparation and identification of pollen grains and an outline of the problems that must be faced in the interpretation of the information they provide.

2
SOURCES OF FOSSIL POLLEN

Pollen grains (and here, for the sake of brevity, the term is used to include spores) may be preserved in a variety of materials, from which they can be recovered and identified. The interpretation of these pollen assemblages must take into account the nature of the material from which they were recovered, because the arrival, movement and preservation of pollen grains varies from one source to another. There may be occasions on which a researcher is faced with a choice of different sites for use in tackling a palaeoecological problem. It is meaningless to state that one type of site is always preferable to another, since this depends largely on the nature of the problem in hand. Different types of sites and materials have their own advantages and disadvantages. These will be discussed in this chapter, together with the behaviour of pollen in arriving at and being preserved in various sites, termed its taphonomy.

The outer part of the pollen wall (the exine) is constructed of durable polymers (see Chapter 5), the biological function of which is the protection of the gametophyte generation during its travels. Spores have similarly tough walls. Those pollen grains that fail to reach their destination (the vast majority) provide an energy resource for invertebrate animals and for microbes and are rapidly degraded. But where pollen consumers and decomposers are scarce, such as in anaerobic, saline or very dry environments, pollen exines and the coats of spores may survive. Low microbial activity is thus the basic requirement for pollen survival, and this occurs in a range of situations.

SITE INVESTIGATION

Informed interpretation of fossil pollen assemblages depends to a considerable extent on an understanding of how pollen arrives at a site of preservation and how it behaves on arrival. Where a choice is available regarding the type of site to be used for analytical work, the decision may well depend upon the precise type of question being asked by the investigator. The sites selected for regional vegetation and possibly climatic history will be different from those suitable for local history and successional studies. Models of pollen transport and the characteristics of different types of site in which pollen work can be carried out will be described below. One further point that needs to be stressed before these subjects are discussed is the variation of a particular site both in space and time.

The types of deposit and sediment described below are not uniform in their present surface, nor in their stratigraphic features. It is necessary to investigate these variations in order to establish the optimum location for recovering samples for detailed analysis and also to determine the sedimentary and hence successional history of the site. This is best achieved by field survey prior to the removal of a core for pollen analysis. The following information is valuable for interpretation of data from a site and much of this is best acquired by field investigation.

1 Geographical location, aspect and local topography in relation to prevailing winds.

2 Climatic information — location of nearest meteorological station.

3 Geology and soil types within both the hydrological catchment and the pollen catchment (which may be very different — see below).

4 Vegetation types within the hydrological and anticipated pollen catchment. Even a very preliminary survey can be of considerable value when interpreting the pollen data, though the more information available, including detailed species lists, the better. The collection of surface samples from beneath different vegetation types for subsequent analysis can also provide a valuable source of comparison, especially if numerical techniques are available (see Chapter 8).

5 The vegetation of the immediate surface of the sampling site should be surveyed in detail. The spatial variation of local vegetation is of importance, especially in small sites, and the precise location of the sampling position in relation to vegetation pattern needs to be known.

6 The immediate topography of the sampling site needs to be known in detail, especially for mires. Levelling of the surface using a dumpy

level is essential for mire sites and if possible such a survey should be correlated to precise markers of altitude through bench marks, etc.

7 The morphometry of a basin should be determined, the minimum requirement being depth measurements of water and sediments.

8 Ideally the stratigraphy of the peats or lake sediments should be investigated in as many locations as possible over the site. This should be arranged in the form of transects or grids so that a picture of the sedimentary history of the site can be achieved using two or three dimensional reconstructions (see Fig. 2.1). Obviously, the more profiles that are available, the greater detail is possible in such a reconstruction. There are several advantages in having this kind of information available: (i) the location for the choice of a pollen core can be undertaken in a more informed manner. Not only the deepest location can be selected, but also the one showing least sediment disturbance; (ii) the successional history of the location can be evaluated,

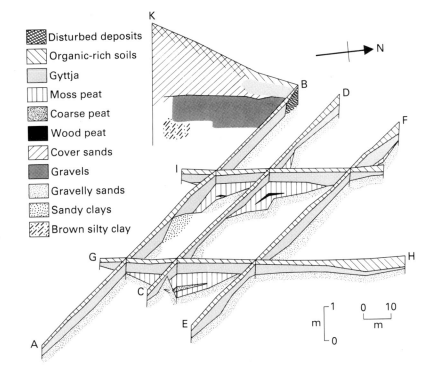

FIG. 2.1. Diagram to illustrate the stratigraphy of a peat deposit. The use of multiple sections intersecting in this way provides a clear picture of the development of the entire deposit (from Bush & Flenley 1987). Reprinted by permission from *Nature* **329**; copyright © 1987 Macmillan Magazines Ltd). (NB The symbols used in this diagram do not correspond to those recommended in Chapter 7.)

Disturbed deposits
Organic-rich soils
Gyttja
Moss peat
Coarse peat
Wood peat
Cover sands
Gravels
Gravelly sands
Sandy clays
Brown silty clay

which will not only affect the pollen assemblages of the site through the vegetation types present, but may also indicate changes in patterns of pollen influx as the nature of the site altered (e.g. lake − swamp − carr − bog successions); and (iii) the stratigraphic profile will indicate contemporaneous changes in vegetation in other parts of the site which may affect pollen assemblages but not be evident in the sediments at the site of sampling.

9 Further information concerning the geological and geomorphological history of the site, its local archaeology and land use history, and any records of local management, e.g. draining or flooding, need to be collated for pollen stratigraphic interpretation.

POLLEN DISPERSAL MODELS

In any given site containing pollen-rich sediments it is important to consider the sources of that pollen, and the means by which it arrived at its sites of preservation. Only in this way can one interpret the pollen assemblage in terms of past vegetation. Tauber (1965) has constructed a model accounting for the various mechanisms by which pollen can arrive at a site. In this model (Fig. 2.2) he considers a site surrounded by forest and regards the adventive pollen as comprising three major components.

1 *Trunk space component* (Ct). This is the pollen that falls from the tree canopy or is produced by shrubs and herbs beneath the canopy and is carried by subcanopy air movements. Some of this pollen may be transferred to air masses above the canopy by gusts of wind in the clearings, but most of it is deposited on the forest floor. Air currents beneath the canopy are greater than those within the canopy, especially once the leaf emergence has taken place in a deciduous forest, but are slower moving than those above the canopy (Fig. 2.3). Wind speeds within the canopy rarely exceed 10 m/s. Where a forest borders upon a site of pollen accumulation, such as a mire or a lake, this trunk space component is carried out for some distance onto the lake or mire surface, depending on the strength of the subcanopy air currents (Andersen 1974).

2 *Canopy component* (Cc). Some of the pollen produced within the canopy, or escaping from below, will be carried along by air currents

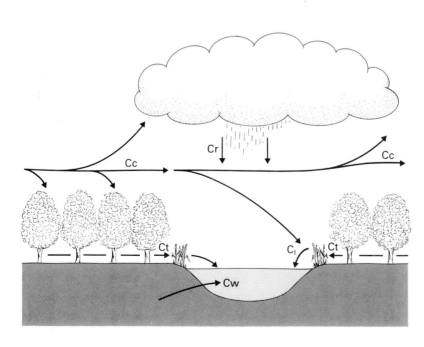

FIG. 2.2. Scheme to show the various sources of pollen in a small lake or mire within a wooded landscape (based upon Tauber 1965). Cc, canopy component; Cl, local component; Cr, rain component; Ct, trunk space component; Cw, secondary component, transported by water.

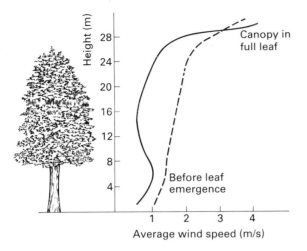

FIG. 2.3. Graph showing wind speed at different heights in a tree canopy, both when trees are in full leaf (———) and when they are bare (———). Note the slightly higher wind speeds through the trunk region when in leaf.

above the canopy itself. A proportion of this component may be transferred by thermals to high altitudes in the troposphere, in which case it may be carried considerable distances by winds in the upper troposphere (Hirst *et al.* 1967). Some of the canopy component may be trapped by eddies in the surface of the canopy and slowed down to such an extent that it sinks through the canopy and joins the trunk space component. But much of the canopy component moves in the surface winds, entering sites of pollen deposition only where these are large enough to create adequate aerodynamic interruptions in the canopy to provide downward movements in the air currents.

3 *Rain component* (Cr). Pollen grains, along with other dust particles, may act as nuclei around which water droplets form. When such droplets descend as rain they collect more dust and pollen on the way; their properties of surface tension are such that they can be said to 'forage' for pollen during their descent (McDonald 1962). This mechanism of pollen fallout probably accounts for the bulk of pollen removal from the atmosphere.

There are three other sources of pollen in many sites of pollen deposition which need to be added to Tauber's model to complete the picture of pollen input in certain types of site.

4 *Local or gravity component* (Cl). Pollen from aquatic plants growing in a lake, or from wetland species growing on the surface of a mire may be expected to deliver a large proportion of their pollen production to their immediate surroundings. This may constitute a large proportion of the total pollen input in such sites as lakes with a rich aquatic flora and mires dominated by abundant pollen producing species, such as grasses and sedges. It is also of very considerable importance in sites that are overhung by trees. This was termed a 'local component' by Moore and Webb (1978), but Jacobson and Bradshaw (1981) have referred to it as a 'gravity component'. Since gravity operates more widely than at a local level, we shall retain the term 'local'.

5 *Secondary or inwashed component* (Cw). Where a surface receives drainage water from a surrounding catchment, it is liable also to receive pollen that has been deposited elsewhere and subsequently mobilized and transported. If the transported pollen consists of recently deposited pollen grains which have never been incorporated into a sediment, then its addition to the fossil assemblage effectively increases its nonlocal component. This is all to the good if the major interest is in regional vegetation history studies. If, however, the inwashed pollen has previously been incorporated into a sediment within the catchment and has subsequently been eroded and transported in surface water, this adds reworked, older pollen to the contemporary mixture and presents problems of interpreting assemblages containing pollen of mixed ages.

This general model of Tauber forms the basis for interpreting data from any site located in a forested environment, but each site needs special consideration and it is often necessary to extend, contract, or modify the model for the specific site needs. When dealing with sediments recording a successional sequence it is even necessary to modify the model for different stages in site development. This is particularly important in mire sites which may pass through a forested stage, during which their pollen input characteristics change considerably, as in the studies of Tolonen and Tolonen (1988), who compare adjacent Finnish mire and lake sites. Evidence for rye cultivation is clearly recorded in the lake, but not in the mire until the wooded phase is ended.

One further general consideration in site selection and interpretation is its size. As mentioned above, the canopy component of the pollen rain may not reach a small site in which there is only a minor interruption in the general canopy of trees. Such a site will have its pollen input dominated by local and trunk space pollen. An extensive site area, on the other hand, allows strong representation of extra-local sources of pollen. Jacobson and Bradshaw (1981) have produced a valuable model to describe this effect (Fig. 2.4). This allows one to estimate the pollen catchment on the basis of a site's diameter. Like Tauber's original model, however, it assumes a forested environment; other types of landscape need to be considered in a different way, and will be discussed later in this chapter.

Having dealt with general models of pollen movement to sites of deposition and accumulation, it is now appropriate to consider the various types of sites individually and to examine the nature of pollen dynamics in each.

PEAT DEPOSITS

Peat deposits are accumulations of organic detritus, mainly of plant origin, that have developed in situations where the rate of production of organic matter by a plant community exceeds the combined rates of plant respiration, herbivore consumption and microbial decomposition. This usually occurs in situations where the decay rate is impaired by waterlogging (Moore 1987). The precise conditions required for peat deposition depend upon the climatic and topographic setting. In hotter and drier climates, peat may only accumulate in hollows and valleys and may never build up to levels above the winter ground water table. Peat-forming ecosystems (mires) fed by ground water flow are termed *rheotrophic mires* (Table 2.1). Where precipitation is higher, or where low temperature reduces evaporation and transpiration rates, peat may be built up to levels far exceeding the ground water table and may be fed only by rain water (*ombrotrophic mires*). Examples of the latter are the tropical forested bogs of Sarawak and the oceanic raised bogs of western Europe (Moore 1987). Such mires begin their existence with a rheotrophic

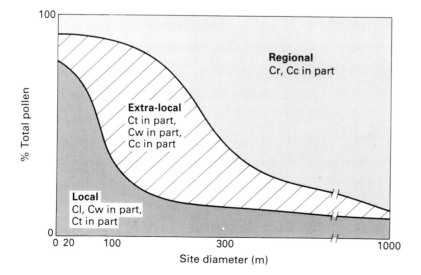

FIG. 2.4. The relationship between the size of site and the various sources of pollen entering it (from Jacobson & Bradshaw 1981). Cc, canopy component; Cl, local component; Cr, rain component; Ct, trunk space component; Cw, secondary component, transported by water.

TABLE 2.1 Classification of peat types on the basis of the hydrological conditions under which they were formed.

Peat type	Origin	Nutrient condition	Examples
Rheotrophic (minerotrophic)	Mire vegetation which receives water both from land drainage and from precipitation	Nutrient rich	Marshes, fens, swamps flushes, carrs, spring mires
Mesotrophic	Intermediate sites where ground water contributes little to the total nutrient capital	Generally nutrient poor	Poor fen, transition mires
Ombrotrophic	Mire vegetation which depends entirely on rainfall for its nutrient input	Nutrient poor	Domed mires, blanket mires

hydrologic system and develop into ombrotrophic ecosystems as peat accumulates. In some high latitude oceanic areas (such as Tierra del Fuego and the Falkland Islands in the southern hemisphere, and Norway and the British Isles in the northern hemisphere), ombrotrophic mires may form over a sloping topography, becoming relatively independent of the natural land drainage patterns. These are termed *blanket mires*.

Further information about different mire types and their hydrology and ecology can be found in Moore and Bellamy (1974), Etherington (1983), Moore (1984a), Gore (1983) and Aaby (1986). Table 2.1 provides a simple hydrological classification of mire types. Figure 2.5 provides a basic model of pollen arrival at mire sites of these two types. It can be seen that the major hydrological differentiation, namely the supply of ground water, is also of importance in supplying extra-local pollen (Cw) to the mire surface of rheotrophic mires. The importance of this pollen input will vary with the volume of water supply, the rate of flow, and the extent of the water (and pollen) catchment of the system. This will not only differ from one mire to another, but will vary between different parts of the same mire. The hydrology of mire surfaces is often complex, involving interconnected pools and runnels, interspersed with more elevated hummocks or ridges. Some of these may become effectively ombrotrophic on a small scale, and hence receive no input of pollen from drainage water. Fen pools fed by ground water, on the other hand, will behave more like lakes and ponds in their pollen dynamics (see below).

Although many mires have developed in valley sites that at some stage in their successional history have received an input of ground water flow, blanket mires may be initiated on water-shedding sites, either sloping or forming a saddle, plateau or ridge. There are peculiarities of air flow pattern in such sites which must be considered when constructing a model of their pollen input patterns. All pollen in such sites either comes from the rainful (Cr), the canopy component brought from neighbouring valleys (Cc), or from local mire plants (Cl). The first two of these are strongly influenced by the local patterns of air movement. Air masses moving up a valley become turbulent if they are forced across a break in slope (see Fig. 2.6) and create eddies in the air mass lying above the plateau. If an air mass is forced over the plateau from an adjoining valley, an interaction of eddies may result in the development of a relatively sharp divide between pollen fallout from the first and second valleys, and if these bear different vegetation there may be an abrupt break in the local deposition of certain pollen types. This has been termed a *pollen shed*. Price and Moore (1984) observed this phenomenon in their studies of blanket mire pollen stratigraphy in southern Wales. Surface deposition of pollen from coniferous woodland in an afforested valley dropped off sharply across

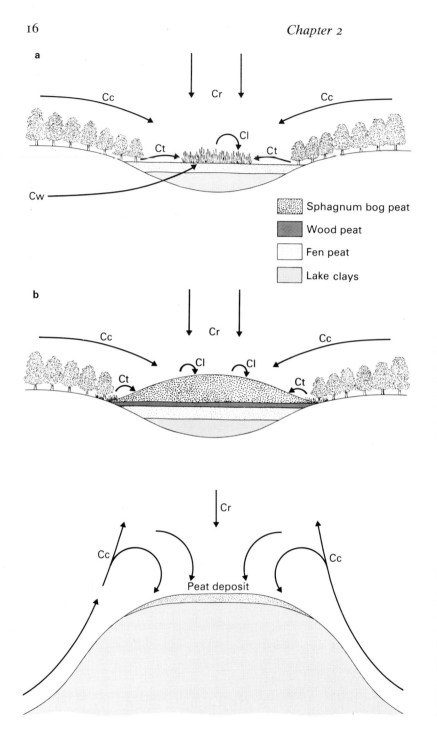

Sphagnum bog peat

Wood peat

Fen peat

Lake clays

FIG. 2.5. The input of pollen to a mire ecosystem. (a) A rheotrophic mire, receiving drainage water from its catchment. (b) An ombrotrophic mire, receiving only rain water in its central dome. Cc, canopy component; Cl, local component; Cr, rain component; Ct, trunk space component; Cw, secondary component, transported by water.

FIG. 2.6. Mires on high level plateaux may receive canopy component pollen from local up-valley winds. Eddies are created at the break of slope that may lead to considerable variation in pollen profiles across such a mire. Where a sharp transition occurs it may be termed a pollen shed (see Price & Moore 1984). Cc, canopy component; Cr, rain component.

a plateau on approaching a neighbouring valley that lacked such woodland. Pollen profiles from the plateau showed that land use in the two valleys had been different over the last 5000 years and this difference was recorded across the plateau in the series of fossil pollen profiles.

The geographical and topographical location of a mire must, therefore, be taken into consideration when selecting a site and interpreting its pollen stratigraphic record, since this will affect its pollen catchment. Over long periods of time, due to geomorphological, tectonic and successional events, the topography of the site itself may alter, so modifying the patterns of pollen transport.

From the point of view of palynology, the most important feature of peat deposits is that they develop in a stratified sequence. Pollen arrives at the surface of the vegetation, transported either by rainfall or from local sources of pollen (all mire types), or by ground water movement (rheotrophic mires). Here it suffers attack from microbes and invertebrate detritivores in the same way as other organic debris. As the surrounding vegetation continues its growth, the litter and pollen become buried by further deposits of plant detritus and are thus subjected to an increasing frequency of waterlogging. In this way the organic materials, including pollen, become stratified into horizons within the developing peat profile (Fig. 2.7).

The surface layers of peat (often to a depth of about 20 cm in the case of *Sphagnum* peats) are periodically aerated and waterlogged, depending upon the water balance of the system. During periods of aeration there is a development of decomposer activity which causes the breakdown of plant material in the aerated zone. Soluble

materials and substances which are easily degraded are removed first and are used as a source of energy by the decomposers (mainly fungi in these upper peat layers, but also bacteria). The cell contents of the pollen grains are lost quickly at this stage, together with the inner wall (intine), leaving the resistant exine. Should aeration continue, this component may also become broken down, but pollen exines, along with fungal and insect chitin, are among the last of the peat constituents to suffer decomposition under acidic conditions. As growth proceeds in the vegetation and as litter continues to accumulate, the forces of capillarity result in a gradual raising of the local water table. Thus, as more litter accumulates, the periods of aeration during dry spells at any given point in the peat profile become fewer and shorter. Decay rates, therefore, diminish gradually with depth, falling to a minimum when one enters the permanently anaerobic zone (Clymo 1965). There are some decomposer organisms which continue their activities even in oxygen-depleted habitats, such as the bacterium *Desulphovibrio*, which oxidizes its substrate by using sulphate ions, reducing them to sulphide. These organisms, together with physical reduction processes, lead to the production of hydrogen sulphide and iron (II) sulphide in wet, anaerobic deposits. Decay therefore continues in deeper peats, but at much

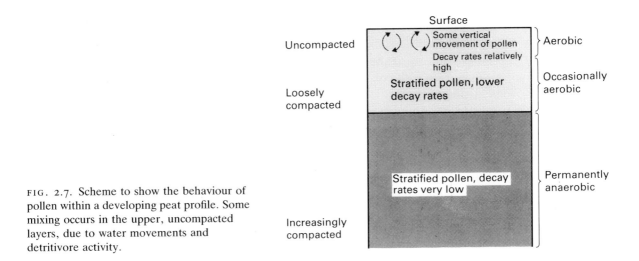

FIG. 2.7. Scheme to show the behaviour of pollen within a developing peat profile. Some mixing occurs in the upper, uncompacted layers, due to water movements and detritivore activity.

slower rates than in the surface layers. Clymo (1965) investigated the differences in decay rates between upper and lower peat layers by placing experimental bags containing *Sphagnum* moss at various depths in the peat profile. He found that bags of *Sphagnum* moss lost over 10% of their dry weight in three months in the surface peat layer, but only about 2% at a depth of 50 cm.

Hydrologists have found that lateral water flow is also very much freer in the surface peat layers and have differentiated between an upper *acrotelm* and a lower *catotelm* on this basis (Ivanov 1981; Ingram 1985). This free water movement in the acrotelm could affect pollen movement, but little information is available about pollen migration within the upper peat profile. Since the length of time during which such movement is possible in the uncompacted layers is usually less than 10 years (Clymo 1973), the influence on pollen stratigraphy is not likely to be significant. Clymo and Mackay (1987) found that some pollen movement does occur both up and down the profile of these upper layers, but Green *et al.* (1988) have nevertheless found it possible to conduct fine-resolution analyses in surface peats dating back just 30 years, so there is some remnant of pollen stratification.

In addition to the vertical stratification in peats, there is often a marked horizontal variation. Just as the vegetation on the surface of mires exhibits spatial pattern, often with a differentiation of hummocks and pools, so the peat profile shows evidence of past patterns (Stewart & Durno 1969; Moore 1977; Barber 1981). This variation is of particular concern to the stratigraphic palynologist since the conditions of pollen arrival, movement and accumulation may differ from one microsite to another (Birks & Birks 1980). Projecting vegetation, such as dwarf shrubs or tussocks of sedge and grass collect pollen by impaction and this is either washed into the substrate at the plant's base, or is deposited with the litter. Oldfield *et al.* (1979) consider that the accumulation of small particles from the atmosphere may be an order of magnitude greater on hummocks than on *Sphagnum* lawns. Flowering plants are themselves usually more abundant on the hummocks, providing a greater local or gravity pollen component (Cl). In bog pools, pollen normally arrives directly from the air (Cr), usually washed out by raindrops (McDonald 1962); local component contributions depend upon the local presence of aquatic macrophytes and the size of the pool.

LAKES

Lakes receive water draining from their catchments and with it eroded sediments and inwashed pollen. Much of the material sedimented within lakes is derived from outside the confines of the lake itself (*allochthonous* material). To this is added *autochthonous* matter derived from plants and animals living within the lake. This contrasts with peat sites, where the bulk of the material accreting is autochthonous and, in the case of ombrotrophic mires, may be almost entirely autochthonous (the additional material comprising simply the pollen, dust, etc. settling on the surface from the air).

Lake sediments contain inorganic and organic allochthonous materials from the catchment, together with autochthonous organic matter from its resident organisms. Pollen will reach the lake by water transport of contemporary pollen from the surfaces and soils of the catchment, and by the erosion of older pollen from soils, peats and other materials suspended in the drainage waters. It is probable that for most lakes this represents the major source of sediment (Pennington 1981a). In addition, some pollen will arrive from the atmosphere by direct rain washout and some will be produced by locally growing aquatic plants. These transport processes are summarized in Fig. 2.8.

Lakes vary in a number of features, such as catchment size, topography and geology, the shape of their basin and in their nutrient status, and these factors influence their patterns of pollen influx and sedimentation. Table 2.2 provides a simple classification system based upon nutrient status and productivity.

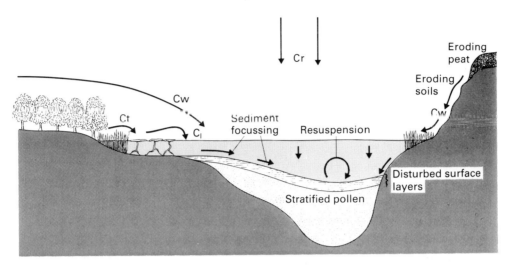

FIG. 2.8. The sources of pollen at a lake site and its subsequent behaviour. Secondary sources (Cw) may be important, depending on the catchment. Within the lake, sediment transport and seasonal mixing have an influence on pollen stratification. Cl, local component; Cr, rain component; Ct, trunk space component, Cw, secondary component, transported by water.

TABLE 2.2 Classification of lake types based upon their nutrient status.

Lake type	Nutrient availability	Sediment composition	Rate of accumulation
Eutrophic	High, leading to high productivity	Large proportion of organic material termed *gyttja* or *neckron mud*. Precise nature varies with producer organisms involved, e.g. algae, broad-leaved aquatics. Highly calcareous deposits are termed *marl*	Generally rapid
Oligotrophic	Low, therefore low productivity	Largely minerogenic, i.e. low organic matter content	Slow, depending on erosion rates in the catchment
Dystrophic	Low nutrient status and low mineral input, therefore low productivity	Colloidal precipitates called *gel-mud* or *dy*, rich in pollen but poor in other types of organic detritus	Slow

Like peat deposits, lake sediments develop in a stratified sequence, hence depth is related to age, though the relationship is not necessarily a linear one. As Table 2.2 indicates, different types of lake sediment accumulate at different rates. Lake sediments do not suffer from all the disadvantages found in peat deposits. They do not experience the same downward movement of water through the sediment profile which occurs in peat deposits, hence there is far less likelihood of downward movement of pollen through the profile. On the other hand, there are detritus-feeding animals within the upper layers of sediments which cause mixing of the surface layers of sediments (bioturbation). One further feature which results in such mixing is

physical turbulence at the sediment/water interface. This has been examined in some detail by
Davis (1968) in Michigan in the United States.
She measured sedimentation rates in lakes by
two methods. In one she took cores from the
surface layers of the lake sediments and analysed
these for pollen. The types of pollen present
changed very abruptly at a depth of about 25 cm,
trees giving way to grasses and weeds. This
corresponded to forest clearance and land settlement in the area in 1830. From this it was
possible to work out the sedimentation rate over
the past century and a half and also (from
measurements of pollen density within the
sediment) the rate of pollen sedimentation per
unit surface area. The second line of investigation
was the use of sediment traps (see Chapter 3).
These were open bottles which were suspended,
open neck upwards, 2 m above the bottom of the
lakes and in which sedimentation rates and pollen
accumulation rates were measured. The rate of
pollen accumulation in these sediment traps was
between two and four times faster than indicated
by the sediment cores. Detailed analyses of
the pollen input into these traps revealed that
pollen accumulated most rapidly in the periods
October–November and April–June. In the
case of the October–November and the April
periods, the vast bulk of the pollen sedimenting
into traps was of species which were not flowering
at that time of year. This pollen must have been
resuspended from the surface sediments and was
being deposited for a second time. Only during
May and June was the pollen deposition made
up largely of newly arrived grains from the
flowering plants of surrounding areas.

The periods during which pollen resuspension
and redeposition occurred corresponded with
times when the lake was not frozen, but was not
thermally stratified and stable. During the
summer months the surface waters become
warm and lie in a stable fashion above the cooler
water below. In this condition there is little or no
disturbance of surface sediments. When this
stratification is broken down by falling temperatures and rising winds in the autumn, there is a

resuspension of these sediments. With the freezing of the lake some stability returns to the
sediments, but this is lost again when the lake
thaws in spring until eventually the summer
stability sets in again.

An additional complication is provided by the
fact that erosion of the surface sediments tends
to be greatest in shallow water near the lake
margin, but redeposition takes place over the
whole basin. This results in the actual accretion
rate of sediment being lower in shallow water
than in deep water. Sediment effectively becomes focused in the deeper parts of the basin
(Davis & Ford 1982).

These lateral and vertical mixing effects may
result in a greater uniformity of sediment than is
the case with peat deposits. It can be argued,
therefore, that there is greater justification in
taking a single core as representative of the
whole site. But successive mixing of the pollen
input from sequential years reduces the resolution that can be achieved, even by very close
sampling (Moore 1981).

Obviously, when coring lakes, areas of distinct
erosion (such as along the course of entry
streams) or of excessive alluvial deposition (as in
the deltaic fans of entry streams) provide different sedimentary environments and the pollen
stratigraphy in such profiles need careful interpretation. If a single core is to be taken from a
site for analysis and if it is hoped that it will
prove representative of the site's sedimentary
history, then such locations are better avoided.

In deep lakes, where the basal waters are
never disturbed by winds or by changes in
thermal stratification (termed *meromictic* lakes),
there is no resuspension and the sediment may
even develop a laminated structure, associated
with the seasonal variation in catchment erosion
and biological productivity. These laminations
take the form of light and dark couplets, the
dark being associated with higher organic matter
from late summer and autumn blooms of phytoplankton (Simola *et al.* 1981; Green 1983;
O'Sullivan 1983; Peglar *et al.* 1984). Studies in
England (Peglar *et al.* 1984) have demonstrated

that the pale layers were mainly calcium carbonate deposited in spring. The annual nature of these couplets can be of considerable value in determining time scales and sedimentation rates (Craig 1972; Saarnisto 1979, 1986).

It is of value to the palaeoecologist to know the proportion of a pollen assemblage derived from inwashed pollen from the catchment, and that derived from aerial transport. Bonny (1978) has conducted experiments in which sediment traps were placed inside large (45 m diameter), open-topped tubes, extending down to the sediment/water interface. The pollen input to these tubes (derived entirely from the air) was compared with the pollen collections from traps left in the unenclosed lake waters (derived from the air and input streams). From the difference between them it was possible to calculate the contribution of stream inputs. This amounted to about 85% of the total pollen (see also Bonny & Allen 1984).

Pennington (1979) has compared two lakes in northern England, one receiving inflow streams and the other in an enclosed basin, and has shown that the difference between the sites varies with the vegetation of the catchments. Prior to human deforestation of the region, the stream-borne pollen amounted to less than 50% of the total pollen input to the unenclosed site. But after deforestation this rose to over 80%, thus confirming the figure previously arrived at by Bonny. The quantitative pollen input to a lake is thus dependent both on the catchment size (Bonny 1976) and the vegetation it bears.

The transport of pollen by rivers and streams into lakes varies with the nature of the stream and with the pollen type involved. Experimental studies have been carried out by Peck (1972) and by Holmes (1990). In the latter study laboratory flume experiments were carried out and it was found that pollen movement and differential sorting were dependent on flow velocity, the nature of the stream bed and also on the pollen size. Field studies have confirmed the validity of these findings.

As in the case of mire sites, the size of the lake will influence the pattern of pollen arrival. A small site, less than about 100 m in diameter (Jacobson & Bradshaw 1981), will be dominated by local pollen from the trunk space and gravity components. It could be argued that this restriction will not be as severe as that experienced by a mire site of similar dimensions because of the additional influence of the stream input bringing pollen from a wider catchment.

It may be easier to identify the local components in a lake than in a mire, for many submerged and floating aquatics are distinctive pollen types, whereas many mire species, such as Cyperaceae, Ericaceae and Gramineae, may also derive from wider sources. Emergent aquatics may also belong to Cyperaceae and Gramineae, however, so discerning the local component is still a problem. Stratigraphic profiles of the site and detailed macrofossil studies are of great value in elucidating such difficulties.

DITCHES AND MOATS

Small waterlogged locations, such as those provided in ditches and moats, are often of value in environmental archaeological studies because of their capacity to preserve a range of useful materials, including pollen and spores. But their study and interpretation present many problems to the palynologist.

In some respects they may behave like small ponds, having a limited pollen catchment and therefore representing essentially only the local conditions of vegetation. But they are more complicated than most ponds in their intensive human management. Their hydrology may be closely controlled, especially in the case of moats. Some of these are fed by streams, or even rivers, thus having a larger than expected catchment. Since moats are usually artificial structures, their excavation may have produced unstable banks and ramparts that may subsequently provide a source of material for erosion into the moat. In this case, the reworked pollen component will be high and will dilute the input of contemporaneous pollen, sometimes to the complete masking of pollen stratigraphic patterns. There is also the very real possibility that the moat has been excavated or dredged subsequent to its original construction, which is difficult to ascertain except by documentary studies. One further problem is the use of the moat as a means of refuse disposal. This practice may provide some interesting material for the archaeologist, but may again disturb pollen stratigraphy.

In view of the problems associated with the study of moats, it is not surprising that there have been relatively few investigations of them. Some unpublished work by Moseley and others on the moats of southeast England showed little variation in the pollen stratigraphy of their muds. Greig (1986) has described one pollen study, but again does not indicate whether a stratigraphic sequence could be detected. He does provide some informative graphic representations of pollen and macrofossil movements around and into moats.

Several other types of waterlogged habitat from archaeological settings may prove suitable for pollen preservation, such as ditches, wells, lynchets, post holes and sewers, but many of them suffer from the same kind of problem as those described above. Often, however, they are the only available source of pollen evidence and must therefore be exploited. More details on the interpretation of the pollen record of such sites are found in the account of Dimbleby (1985).

SOILS

Soils have received less attention from palynologists than waterlogged materials. This is both because of the poorer preservation in aerobic environments and also because of the vertical mixing of the profile to be expected in soils with high populations of invertebrate detritivores (Andersen 1986). Long ago, however, Erdtman (1943a) demonstrated that certain soils were relatively rich in pollen with densities of up to 0.5 million grains/g, and Dimbleby (1957) has found values as high as 1.5 million grains/g of

dry soil. This level of preservation, however, is achieved only in soils of pH less than 5. Above a pH level of 6, there is virtually no preservation of pollen. Whether the soil is a podsol or a brown earth, or whether it is light or heavy in texture, does not appear to affect the level of preservation. Temperature also appears to have little effect, since pollen is even preserved in acid tropical soils.

Havinga (1964) considered that pollen degradation begins with the oxidation of the exine (pollen coat) and has shown that certain pollen types are much more readily destroyed than others (Havinga 1971, 1985). This is a serious problem because it will lead to the under-representation of some pollen types and the over-representation of others. *Quercus*, for example, is considerably more sensitive to oxidation than is *Tilia*, which could lead to changed proportions in the course of time.

Where pollen grains are not preserved, phytoliths (minerals deposited in plant cell lumina) may be used in the study of soils. A combination of approaches (pollen and phytolith) is particularly valuable in the study of grasslands (Kurmann 1985) as grass phytoliths may be species or group specific, whereas grass pollen is generally not (Rovner 1986).

Soil fauna, particularly earthworms, are responsible for considerable movements of pollen in soils, and experiments have shown that they are capable of producing an even distribution of pollen throughout a soil when supplied with a layer of pollen, either on or some distance below the surface (Walch *et al.* 1970). Stratified pollen zones can still develop, however, if earthworms burrow to progressively shallower depths during soil profile development (Keatinge 1983a).

But even in the absence of the larger soil animals, such as earthworms, pollen still percolates downwards through soils, though this is not regarded as significant by Havinga (1974). Dimbleby (1985) suggests a figure of about 10 cm in 300 years as a reasonable estimate of downward movement for pollen, but this varies both with soil types and with different types of pollen grain. Old root channels are often rich in pollen and it is likely that pollen descends along such routes.

Despite, or perhaps because of, these downward movements of pollen in soil profiles, it is often possible to demonstrate pollen stratification and, if interpreted with care, these can yield useful environmental information.

MOR HUMUS

Under certain vegetation types, such as coniferous forest and ericaceous heathland, the plant litter falling on the soil surface is incorporated into the mineral soil only very slowly. This appears to be associated with a high polyphenol content in the litter and a low rate of breakdown by micro-organisms (Davies *et al.* 1964). Such litter accumulates on the surface of a soil in a stratified manner similar to that of peat deposits. Indeed, the initiation of certain types of blanket peat may be preceded by a phase of humus accretion.

Aaby (1983) has made detailed studies of soil humus in Danish forests and he differentiates a series of humus types developed on soil surfaces. These relate closely to the activity of soil fauna. A lumbricid humus stage is characterized by a high level of humification and intensive faunal activity (sometimes it is termed *mull humus*). Pollen deterioration is consequently high, with corroded and thinned grains common (see Chapter 7 for further information on pollen deterioration). Above this is an arthropod humus stage (transitional between mull and mor, sometimes termed *amphimull*) which has undergone only moderate humification and in which pollen preservation is considerably better. The small invertebrates responsible for physical breakdown of the humus, such as collembola and enchytraids, ingest pollen and this does cause some degradation, particularly exine thinning and crumpling. This humus type grades into true raw (or *mor*) humus in which humification is low and pollen preservation good. Arthropod activity is less intense than in preceding humus types,

though enchytraids persist, as do bacteria and fungi. Often the pollen represents the most persistent form of organic matter, hence the mor humus can be very rich in pollen.

Stockmarr (1975) distinguishes between two types of mor humus, *copromor* which has a crumbly structure, having passed through the guts of arthropods, and *mycomor* which has a tough structure and an abundance of fungal hyphae. This latter is usually developed only under certain types of canopy, such as *Fagus*, *Calluna* and possibly *Ilex*. Figure 2.9 summarizes these horizons and demonstrates their sequence in time.

The processes leading to this succession of humus types have often been associated with human activity, either by forest clearance (Iversen 1964) or by alteration in the tree canopy, such as replacement of *Tilia* by *Quercus* (Heath *et al.* 1965; Aaby 1983).

Within the mor humus layers the stratification of pollen can be interpreted as representative of a time sequence, as in the case of peat deposits. Mor humus profiles developed beneath a woodland canopy must be regarded as reflections merely of the immediately local sequence of woodland history, possibly only within a 30 m radius. It is advisable, therefore, to analyse several profiles from a site in order to perceive general forest history at that location. On the whole, however, such studies have provided consistent records of woodland history and can be extremely valuable tools in tracing past human management practices (Bradshaw 1988).

Mor humus profiles from sites lacking a tree canopy, such as heathlands, may reflect a wider pollen catchment, but still provide a record of immediate changes in the vegetation and environment.

TUFA

The constant trickling of spring water rich in calcium bicarbonate can lead to a rapid accretion of calcium carbonate as carbon dioxide is given off (Ford 1989). Algae and cyanobacteria also

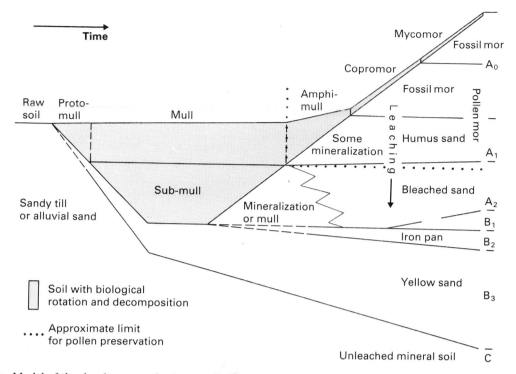

FIG. 2.9. Model of the development of a forest soil through time (redrawn with permission from Stockmarr 1975). It demonstrates the changes in biological activity during the development of a podsol profile and the consequences for organic matter stratification on the soil surface.

contribute to the calcium carbonate deposition (Pentecost 1978). Tufa is thus a lime-rich deposit that accumulates around springs in calcareous habitats and it may contain embedded pollen, though usually at very low concentrations because of its rapid rate of build up. Other organic matter, such as leaves, twigs, and especially portions of bryophytes, are often important components, as may be other minerals and siliceous material and the shells of gastropods and ostracods. The development of tufa deposits is sometimes so extensive that lakes develop behind tufa dams. Pollen probably survives in tufa (but not in high pH soils) because of the constant waterlogging and anaerobic conditions.

Because of their stratified sequence of sedimentation, tufas have been used in pollen stratigraphic work (e.g. Kerney *et al.* 1964, 1980; Preece *et al.* 1986) and have been of considerable value in the provision of a record of past veg-

etation in sites where other sources of evidence may be scarce or absent. They have also proved particularly useful at some archaeological sites (e.g. Preece 1980).

The pollen catchment of a tufa deposit is not easy to define, however. By the nature of its origin (in spring waters) the site of tufa formation is usually small and often situated on valley sides or depressions. For these reasons, there is likely to be a strong input of local trunk space and local pollen (Ct and Cl). In some tufa diagrams this seems to dominate the pollen profile, as in the work of Preece *et al.* (1986) near Dublin, where the early post-glacial pollen assemblage becomes dominated by hazel (*Corylus avellana*), which is likely to have been abundant in the immediate vicinity of the tufa-producing site.

In addition to the immediately local sources of pollen, there is also the pollen transported in the percolating water to the site of fossilization (Cw), the influence of which will depend upon the catchment from which it is derived. In most sites its origins and movements may also be local. So tufa deposits, like most of the soil mor humus layers described below, usually reflect only the very local sequence of vegetation history.

ALLUVIAL AND FLOODPLAIN SEDIMENTS

Alluvial materials form a very important source of a range of palaeoecological evidence on the basis of which reconstructions of past conditions can be attempted (Gregory 1983). Generally, alluvial sediments are predominantly inorganic, although this depends on the nature of the substrate being eroded in the catchment. The rate of erosion and hence the rate of sedimentation depends upon a range of factors, but particularly upon the climate and the vegetation of the catchment. An increase in climatic wetness or a destruction of vegetation cover can cause an increase in sedimentation rate (Burrin & Scaife 1984; Moore 1984b).

Alluvial material often, though not invariably, contains fossil pollen. Some of the pollen is derived from secondary deposition of eroded substrates (Cw) and this leads to some problems both in the quality of preservation and in the interpretation of alluvial pollen assemblages. The local pollen input depends on the pollen productivity of the floodplain vegetation. Martin and Mehringer (1965), working in the United States, found that most of the pollen in their alluvial samples was derived from floodplain vegetation. The physical transport of pollen in an inorganic matrix often leads to mechanical damage, particularly crumpling and folding. If the origin of the pollen is the soil cover of the river catchment, then oxidation and microbial degradation is also likely to have led to further pollen deterioration (see Chapter 7). In the lower reaches of rivers, marine influence may be felt, particularly in the erosion and reworking of the alluvial sediments themselves.

Many alluvial deposits are stratified in a chronological sequence and hence one can determine pollen stratigraphical sequences as in the case of lake sediments (see, for example, the work of Planchais 1987 in southern France; Burrin & Scaife 1984 in southern England; Solomon *et al.* 1982 in the United States). Such diagrams must be interpreted with care, for the secondary nature of much of the pollen means that it cannot be regarded as an immediate reflection of contemporaneous vegetation. But gross trends can provide a record of vegetation history. Planchais (1987), for example, provides a pollen record covering the past 10 000 years in 34 m of deposit and from it the course of vegetation history in this part of the Mediterranean coast can be determined in some detail.

Alluvial sediments are not only associated with temperate river systems, but also with downwashed soils of semi-arid sites where winter snows melt in spring or seasonal rain provides a means of transport before summer heat results in its evaporation. Such conditions are found in present day Iran (Moore & Stevenson 1982) and have existed in the past in the Rajasthan Desert of North West India (Singh *et al.* 1974). In both of these areas it has been possible to determine pollen stratigraphy and vegetation history from alluvial sediments.

Figure 2.10 depicts a model of pollen transport by alluvial processes in a semi-arid area such as Iran. In this situation, where local pollen production by sparse, anemophilous vegetation is low, the influence of long-distance pollen movement washed out by rain becomes proportionally greater. Tree pollen carried many hundreds of kilometres may account for 10% or more of the total pollen assemblage. In some arid areas, such as northern Australia, organic deposits develop around artesian spring outflows and these can be a source of fossil pollen (W.E. Boyd, unpublished).

LOESS

Aeolian, wind-blown materials are found particularly in steppic and arid environments. In

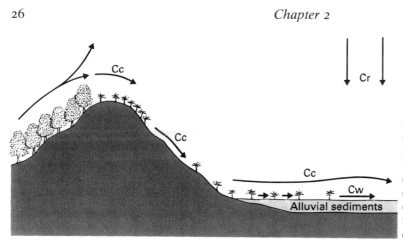

FIG. 2.10. Alluvial sediments are strongly influenced by secondary materials deposited by floodwaters (Cw). In semi-arid situations, the canopy component (Cc) may be small due to low plant density and the long-distance transport (Cr) then accounts for an unexpectedly large proportion of the pollen input.

central Europe, for example, loess sands have often been used as a source of information concerning palaeoenvironments (Fink & Kukla 1977). Such deposits are often found in currently oceanic temperate areas, like western Europe, as a consequence of the existence of suitable conditions for their formation in glacial and late-glacial times (Lowe & Walker 1984). Pollen is found in these deposits but in very low concentrations (see, for example, Couteaux & de Beaulieu 1976). Where they have been exploited for pollen analysis, conventional methods of pollen concentration have required modification to cope with the high quantity of silica (see Chapter 4).

CAVE DEPOSITS

The accumulation of rock debris in caves results in a stratified material that sometimes contains fossil pollen, though it is usually present in very low concentration, as might be expected given that there is no local production. Perhaps it is not surprising, therefore, that in some areas these deposits have been somewhat neglected by palynologists (Hunt & Gale 1986). The origin of pollen in cave sediments is discussed by Dimbleby (1985), who points out that in addition to the input of airborne pollen brought by air movements, there is the likelihood of animal, including human, activity in transporting pollen. This could be brought into a cave as contaminant

mud from feet, or as herbaceous material brought by human agencies as bedding. In this case the pollen spectrum can only provide a biased indication of local vegetation, but it may be of considerable interest to the archaeologist. Cave deposits in Australia have been used to document vegetation changes resulting from aboriginal occupation phases (Martin 1973).

Where water enters a cave, both the preservation of the pollen and the catchment area of the fossil assemblage is improved. Even stalactites have been found to contain pollen grains trapped in mineralized deposits and these could have been brought some distance by underground water movement.

In addition, Dimbleby (1985) points out the possible importance of faecal material from bats. If bats consume moths that have visited flowers, one would expect to find pollen in their faeces, but little information is available on this. Any other animal using a cave could also be responsible for the introduction of pollen in faeces. Many of the caves and rock shelters of North Africa, for example, are rich in sheep faeces and the accumulation of cave debris contains pollen that may well come from this source.

ICE

Compacted snow in high, mountainous regions eventually becomes consolidated as ice, forming a stratified deposit within which dust particles and pollen grains are incorporated and stratified. The first studies of the pollen record of glaciers was conducted by Vareschi (1934) on the Alpine glaciers of Austria and Switzerland and his work was discussed by Godwin (1949). Subsequent work in the Alps by Ambach *et al* (1966) and Bortenschlager (1969) demonstrated the full potential of this source of material. Seasonal variation in pollen input could be detected and freak events, such as dust storms bringing pollen from the Sahara, were recorded. High latitude ice studies, such as those at Camp Century, Greenland, by Fredskild and Wagner (1974), have shown that the pollen concentration in such areas is very low and large samples, up to a litre of ice, need to be used for extraction.

FAECAL MATERIAL

Faeces have often been used as a means of analysing the dietary preferences of animals, particularly herbivores (see Bhadresa 1986), but most studies have concentrated upon the plant macrofossil content. Pollen is present in faeces, however, and may provide additional information about diet and feeding habits. The most thorough study is that of Moe (1983) who investigated the pollen content of sheep faeces in Norway. He found that there was a seasonal variation in the faecal pollen content, but that this did not correspond with the pollen rain. The variation was more likely to have been connected with diet, such as the ingestion of Cichorioideae pollen (probably *Taraxacum*), and Umbelliferae pollen (*Conopodium*) when these species came into flower. Van der Knaap (1989) has used the pollen in fossil reindeer faecal pellets to study the past vegetation of Spitsbergen.

The value of pollen in faeces as a possible indicator of diet has been exploited to only a limited extent. It could provide some interesting information about prehistoric peoples where other dietary information, especially relating to plant intake, is scarce. Schoenwetter (1974) has made use of this technique in investigating human palaeofaeces in dry cave deposits

in North America and found a preponderance of wind pollinated types, such as Chenopodiaceae, Amaranthaceae and Ambrosiaeae (Compositae). In interpreting his results he considers four possible sources of pollen in the human gut:

1 Pollen ingested as part of food or drink.
2 Pollen transferred between foods as a result of storage.
3 Pollen settling from the atmosphere onto food.
4 Inhaled pollen, swallowed with mucus. (This source could be contemporaneous or derived from older material in dust).

It is difficult to determine which of these sources is the most important, but macrofossil analysis of the faeces can be of assistance. If the pollen intake is largely from insect pollinated flowers, then food ingestion is the most likely source, but with wind pollinated types, interpretations are more difficult. In the case of Schoenwetter's data, the relatively low levels of grass pollen suggest that the pollen found in the faeces is unlikely to have been substantially derived by inhalation or food contamination. Direct ingestion is again the most likely source.

Several archaeological investigations have been carried out in which the entire contents of latrines have been examined rather than just the faecal material (see, for example, Greig 1981). This complicates the interpretation very considerably because a far greater diversity of potential pollen sources is now present. These include atmospheric sources after deposition, waste food materials (latrines were often used for general garbage), plant material used for personal hygiene, etc. Work on this type of material should always be accompanied by macrofossil analysis (as well as analysis of gut parasites) so that the origins of the pollen can be more effectively determined.

HONEY

Honey is normally rich in pollen grains, which may enter the hive as contaminants of collected nectar (normally from the same flower), or may in some species be the main reward offered by a flower to its pollinator. In either case, the pollen content is usually regarded as a sound guide to the floral sources of the honey (Cowan 1988). There is, however, the possibility of pollen loads falling accidentally into cells used for nectar storage, and this can cause some confusion in the interpretation of honey pollen. Some honeys are derived from honeydew, the sugary secretions of aphids and other plant-sucking insects. Honey derived from these sources will usually appear dark because of the abundance of fungal material (hyphae, spores, yeasts, etc.) that is normally present. The incidence of extraneous pollen derived from the atmosphere is also likely to be higher in honey of this type.

Fossil honey has been recorded from archaeological sites. An example is a Bronze Age burial in Scotland where Dickson (1978) analysed a black material in an overturned beaker accompanying a skeleton, and found high densities of pollen from lime (*Tilia*) and meadowsweet (*Filipendula*). The site was north of the natural limit of lime trees and Dickson considered it likely that the material consisted of the remains of honey, or mead (an alcoholic drink made from fermented honey), which had been imported into the area from further south.

3

THE COLLECTION OF SAMPLES

Many of the applications of pollen analysis depend upon the sampling of stratified sequences of peat, lake sediment or soil. Under ideal conditions, these samples can be taken from exposed surfaces of deposits, either as a result of natural erosion processes, such as on cliffs and peat erosion faces, or by excavating a section of the deposit. When excavation of the site is not feasible, specialized equipment is necessary for the recovery of samples at known depths. Such sampling should be carried out with the least possible disturbance to the sequence and the least risk of contaminating the samples obtained. For this reason, a large range of sampling instruments has been developed, each of which may prove appropriate for specific tasks and sediment types. General accounts of the various sampling equipment available have been given by Aaby and Digerfeldt (1986), Moore (1986), West (1977) and Barber (1976).

EXPOSED SECTIONS

There are many advantages in the exposure of a complete profile of a deposit, be it peat, lake muds, soil or tufa. They provide a very direct and simple means of assessing the degree of lateral variation in a deposit and hence avoid the difficulties of conducting transects of cores (see Chapter 2). They also allow field collection of samples without the risks inherent in coring equipment of sediment distortion, compression and contamination.

Barber (1976) has discussed at some length the various techniques that can be used for investigating exposed faces of peats. He advises some initial investigation prior to thorough cleaning of a weathered face because some peat features, such as tussocks of *Eriophorum*, may be more prominent at this stage than after the cleaning process. He also recommends the use of gridded quadrats for detailed mapping of the face and also photography as means of accurate recording of lateral variations.

The taking of samples for laboratory analysis, however, should be preceded by careful cleaning, initially by spade or trowel, but also using a sharp knife or scalpel. Final cleaning should be conducted using horizontal movements of the knife to prevent accidental contamination in a vertical plane. It is often useful to record detailed structural features and colours at this stage, since the exposure of fresh surfaces to oxygen causes rapid darkening and features which may be apparent e.g. banding, yellow layers of algae, unhumified *Sphagnum* or black lines of charcoal, can be lost by the time the samples are taken to the laboratory. Sometimes such stratified features are visually lost within minutes of exposure. Standard colour charts can be used, as for soils (Ball 1986). The recording of sediment types and humification is of considerable value and a full description of techniques and conventions is given in Troels-Smith (1955), later modified by Aaby and Berglund (1986) and summarized by Moore (1986), see Fig. 3.1.

The removal of samples for pollen analysis can be conducted from such faces in the field. Perhaps it is wise to add that this can be carried out efficiently in the field only under good

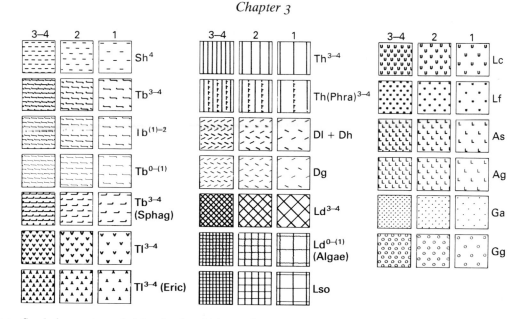

FIG. 3.1. Symbols recommended for the description and representation of sediment types, as modified by Aaby from the original scheme of Troels-Smith. Numbers at the heads of columns refer to density (greatest on the left) and the letters and figures in the right hand column refer to the nature of the material and its humification respectively. Sh, structureless organic matter; Tb, moss peat; Tl, wood peat; Th, herbaceous peat; D, detritus muds (l, lignaceous; h, humic; g, granular); Ld, fine fragments of organic matter; Lso, siliceous diatomite; Lc, marl; Lf, iron-rich material; As, clay; Ag, silt; Ga, sand; Gg, gravel. For more details see Aaby and Berglund (1986).

weather conditions. Collecting in the rain can lead to contamination from the raindrops themselves and collecting from such sites as cliffs in high winds may also lead to inefficiency and contamination. The other main problem with field collection of individual samples is that the vertical interval between them needs to be determined in advance of the analysis. The interpolation of samples at a later stage in order to expand part of the sequence is not possible.

This final difficulty can be overcome by the use of monolith tins in which whole sections of the exposed face can be recovered and transported to the laboratory. Stainless steel, 1 mm in thickness, can be used for constructing monolith tins. A useful dimension is 50 × 15 × 15 cm, formed by bending a steel sheet to form three sides of column, the fourth side and the ends remaining open. These can be hammered into a peat face with a mallet and then dug away from behind and below.

CORE RECOVERY FROM PEAT AND LAKE SEDIMENTS

Where exposed sections of a deposit are not available, cores must be extracted from the surface of the site. The collection of cores from intact peat sites is generally simpler than from lakes simply because equipment can be operated from the peat surface. It is usually possible to remove cores by hand, but if certain layers are stiff one can employ portable extractors, such as that described by Blyth (1984) and used by Hulme and Shirriffs (1985), or that of Cushing and Wright (1965), a photograph of which appears in Smith *et al.* (1968).

The coring of lakes is most effectively conducted by using two small boats roped together with a sampling platform between them. Alternatively, a lake may be cored when ice covered (Colinvaux 1964), though this can be dangerous in the lower latitudes. It has been successfully carried out as far south as southern England.

Various types of coring equipment have been devised to suit different situations and types of sediment.

Hiller sampler

This is a chamber sampler (Fries & Hafsten 1965) fitted with an auger head, allowing it to be twisted as it penetrates the sediment (Fig. 3.2). This provides it with a considerable capacity for penetrating even fairly stiff or fibrous materials. The sample is obtained by twisting in the opposite direction when the inner rotating flanged chamber opens and scours a sample from the adjacent sediment. Although a robust instrument, it suffers from a number of disadvantages:

1 The projecting flange of the rotating chamber, as well as the auger head, are liable to trap root and other plant material during descent and thus cause contamination. It is possible to replace the auger head with a steel cone, but this reduces its penetrating power and does not solve the problems with the flange.

2 The material sampled has been partially disturbed by the penetration of the auger head.

3 The auger also disturbs material to a depth of 10−20 cm below the level of the sampling chamber, thus affecting the next sample in sequence. This can be overcome by using adjacent boreholes and alternating samples, but this results in problems of core correlation if the sediments vary laterally (see Deevey & Potzger 1951).

4 The most serious disadvantage with the conventional Hiller sampler is the fact that intact cores cannot be removed from the sampling chamber; samples from specific depths must be removed in the field. It is impossible to maintain an adequate degree of hygiene in the field to perform this task efficiently, especially in summer when aerial pollen may be abundant, or in wet weather when rain drops carry pollen into

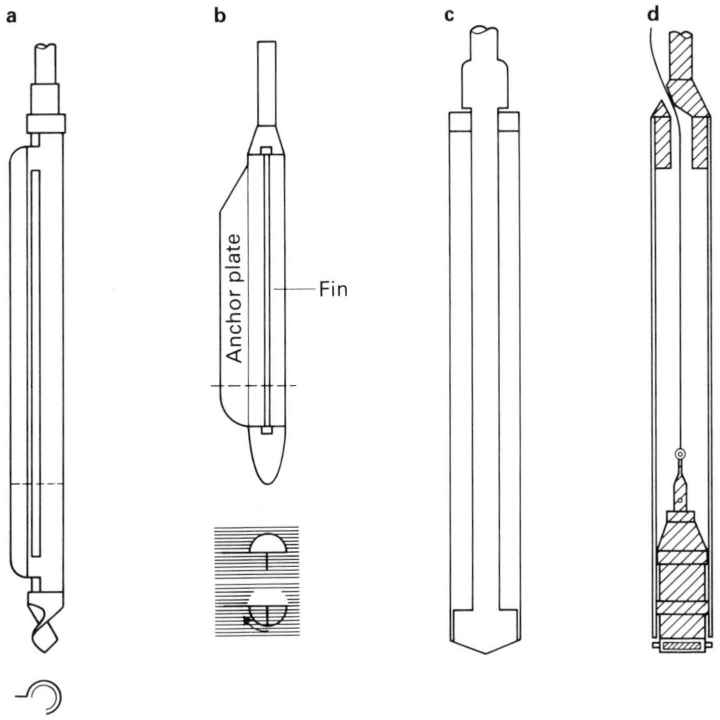

FIG. 3.2. Sediment samplers in common use: (a) Hiller, with cross-section showing the projecting flange and inner rotating chamber; (b) Russian, with cross-sections showing the 180° rotation of the movable chamber; (c) Dachnowski; and (d) Livingstone (from West 1977).

the samples. Field sampling also precludes the possibility of taking intermediate samples at a subsequent time if a pollen sequence proves particularly interesting and more detailed analysis is necessary. It also renders impossible the detailed laboratory analysis of sediment stratigraphy and the precise location of sediment boundaries.

The removal of samples from the chamber and the cleaning of the chamber before the next core is taken is rendered even more difficult by the narrow chamber opening of most Hiller corers. The modified Hiller sampler described by Thomas (1964) overcomes some of these problems by having a detachable auger head so that the inner chamber can be removed intact from the outer one. If zinc or plastic liners are inserted into the inner chamber, these can be wrapped in polythene and transported to the laboratory for later sampling. In dismantling the head in this way an opportunity is also provided for the careful cleaning of the chambers and a reduction in the risk of contamination.

One further problem with Hiller-type samplers is their limited capacity. If radiocarbon dating is required for a sample, then a 2 cm diameter core can provide enough material only if samples are pooled over a considerable depth, perhaps 10 cm or more. This reduces the precision of the dating very considerably.

Generally, the Hiller sampler is not to be recommended even for preliminary survey work, except where the stiffness of the sediments permit no alternative (Burrin & Scaife 1984).

Russian sampler

This sampler, which was designed by Russian research workers, has been described by Jowsey (1966). It is very widely used for peat stratigraphic work because of its clean action and its speed of operation and cleaning.

It differs from the Hiller corer in that it lacks an auger head and must therefore be pushed vertically into sediments (Fig. 3.2). This limits its use to fairly soft materials, but has the great advantage that the sediment through which it passes is not disturbed by churning action. The main blade of the sampler passes the material to be sampled as it descends, and the 180° rotation of the movable chamber cuts a semicylinder of peat or lake sediment and withdraws it intact upon the surface of the fin. When the sampler is withdrawn, the opening of the chamber reveals the entire sample in a clean and undisturbed condition. This is ideal for field measurement and description and can also provide excellent material for intact removal, wrapping and transport to the laboratory. Plastic drainpipes cut longitudinally in half and into 50 cm lengths supply ideal containers for the removal and storage of samples from a Russian corer.

The advantages of this sampler over the Hiller are several:
1 Projecting flanges are more effectively designed and smoothed to avoid catching fibrous material and carrying it down the profile.
2 Sediment disturbance is minimized.
3 Cleaning of the surfaces is facilitated by their complete exposure between samples.
4 No dismantling is necessary, so the entire process of sampling can be fast and efficient, keeping exposure of the samples to the air and ambient pollen rain to a minimum.
5 For stratigraphic work, the full exposure of the core provides an excellent opportunity for the recording of sediment features such as colour, banding, charcoal layers, humification changes, etc., which are quickly lost on oxidation and can only be achieved in the laboratory with much greater difficulty. Field description of cores at the time when they are collected is therefore strongly recommended.

This type of sampler is particularly useful for preliminary survey work and for the study of general stratigraphy of peatland sites. It can also be used for core recovery for pollen analysis in many types of sediment. It is not very suitable for work in stiff mineral sediments, very soft unconsolidated muds, under water, where there is abundant wood in the profile or for shallow profiles through peaty soils. It suffers from the same lack of capacity as the Hiller, but larger capacity versions have been designed.

Piston samplers

Both of the chamber samplers described above are of very limited value in soft sediments or under water. In this type of location a piston sampler is more appropriate (see the reviews by Wright *et al.* 1965; Wright 1980; Aaby and Digerfeldt 1986). The piston sampler can be used in peats, as long as they are not too coarse or fibrous and it is also very appropriate for the recovery of limnic sediments from beneath a peat deposit.

The operation principle of a piston corer is that a hollow tube is driven vertically down into the sediment, but at the same time a plunger is withdrawn up the tube creating the negative pressure which prevents the compression and distortion of the sediment column (see Fig. 3.2). Even when a piston is employed, however, there comes a point when the tube with its sediment content act effectively as a solid mass and further insertion into the sediment does not result in the collection of any more material, but simply the compression of the layers beneath the sampler. Before this point is reached the corer must be withdrawn and the sediment core extruded before taking the next sample. Generally a depth sample of 1 m at a time is maximal. It is also wise to extract cores alternately from separate boreholes and to allow an overlap between cores to permit some estimate of compression and distortion to be made.

The capacity of the corer depends upon its diameter. Although a wide diameter tube provides larger samples for additional analyses, such as chemical or radiocarbon, it leads to problems of sediment loss as the core is withdrawn from the mud. The effects of sediment type and water depth on this process are described by Aaby and Digerfeldt (1986). Most piston samplers are designed with a diameter of about 5 cm, though they range up to 10 cm or above. Smith *et al.* (1968) have described a simple, large capacity device consisting essentially of a tube, with a split in one side, which is driven into the peat deposit mechanically. The split side avoids problems of compression of the core. Many other variations on the basic design of piston samplers are to be found, but the best known is probably the Livingstone modification of the Dachnowski sampler (Rowley & Dahl 1956). The compressed air sampler of Mackereth (1958) is also well known and can be used to recover long cores in deep water.

Piston corers have not been used very widely in peat deposits, mainly because of the problems experienced with coarse or fibrous plant debris, or with wood layers. Wright *et al.* (1984) have designed a modified piston corer in which the cutting edge of the sample cylinder is serrated to permit easier cutting through such material. Their corer is also modified so that the chamber can be rotated back and forth, and hence enhance the cutting motion.

Some operators of piston corers in lake sites recommend the use of casing tubes for relocation of boreholes and for the prevention of bending in the extension poles. When sampling from boats, it may be found useful to use two vessels, bound together by a sampling platform and securely anchored. If coring is conducted on an ice layer, then care must be taken to ensure that the thickness of ice is adequate to support the equipment and crew, and that it is not weakened by penetration of the ice.

The recovery of undisturbed sequences from the upper layers of lake sediment and from the sediment/water interface is particularly difficult and has led to the design of a range of modified piston corers which are described and illustrated by Aaby and Digerfeldt (1986). Such cores are of particular value in the study of environmental changes which have taken place in the last few centuries, or even decades.

The surface layers of peat deposits are particularly difficult to sample for a number of reasons. They are loosely compacted and therefore easily deformed by a corer, or even the approach of a human operator. They are often very wet, and the draining of superfluous water distorts the peat matrix. The instrument designed by Smith *et al.* (1968), mentioned above, is liable to such distortion as water drains from it during core recovery. Clymo (1988) has designed a cylindrical sampler, 75 cm long and 20 cm in diameter, with a sinuous cutting edge. It has no piston, and can be rotated as it enters the peat. The cutting implement is then withdrawn and replaced by an open-ended plastic cylinder. A steel plate is used to seal off the base of the cylinder, being inserted by a system of rods and wires. The complete sample is then withdrawn. The instrument is sufficiently well sealed to permit even the sampling of the water content of the peat core.

Freezing samplers

Just as the surface layers of peat deposits are difficult to recover intact and undistorted, so are lake sediments at the interface with the water body. An alternative to piston corers for the recovery of unconsolidated surface sediments from lakes and ponds has been the use of samplers that freeze the material *in situ* and recover samples while they are still frozen. The technique was pioneered by Shapiro (1958) and has undergone a range of evolutionary developments. One of the best known of these is described by Saarnisto (1986). This consists of a metal tube up to 4 m in length and 8 cm in diameter with a plastic, pointed plug at the lower end (Fig. 3.3). It is filled alternately with dry ice and trichlorethylene, butanol or ethanol which produces a very low temperature (below −80°C). A polythene tube is attached to the top to allow gas to escape and the entire tube (with a cable attached for recovery) is dropped into the water from a boat and is allowed to penetrate the surface sediments under its own weight. Within 15 min a layer of sediment about 3 cm thick has frozen onto the outside of the tube and this remains intact when the sampler is recovered. The entire apparatus may be transported frozen to the laboratory, wrapped in foil

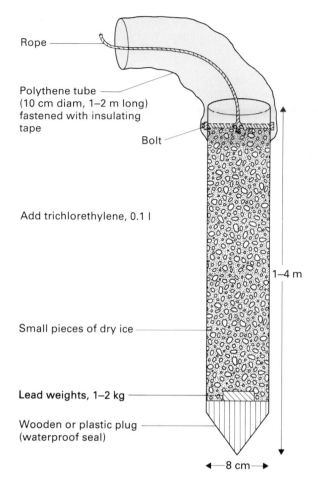

Rope

Polythene tube (10 cm diam, 1–2 m long) fastened with insulating tape

Bolt

Add trichlorethylene, 0.1 l

1–4 m

Small pieces of dry ice

Lead weights, 1–2 kg

Wooden or plastic plug (waterproof seal)

←— 8 cm —→

FIG. 3.3. Freezer sampler for lake sediments. The tube is packed with dry ice and is dropped vertically into the sediments from a boat. Trichloroethylene is then added to effect rapid freezing.

to prevent drying, or may be emptied of its dry ice and treated with hot water, when the outer skin of frozen sediment can be slid off.

This method has proved particularly valuable for the study of annually laminated lake sediments in regions such as Canada (Cwynar 1978) and Finland (Huttunen 1980), where very detailed studies of land use history in recent times have proved possible.

SAMPLING THE POLLEN RAIN

The interpretation of fossil pollen assemblages

relies upon the relationship between vegetation and the pollen which accumulates in adjacent sediments. It is possible to study this relationship directly by conducting analyses of surface sediments from ponds, lakes and mires, or by using traps in which the current pollen rain can be collected.

Surface samples

The type of surface material used for sampling must be selected with care, for the composition of pollen assemblages has been shown to vary with sedimentation conditions. For example, Potter (1967) examined the surface sediments in water storage tanks for stock (about 2 m × 15 m and 2 m high) in New Mexico and compared them with nearby surface soil samples and found some consistent differences between them. In general, the soil samples were richer in grasses, chenopods and *Ambrosia* than the tank samples, but the latter contained greater quantities of tree pollen. The most appropriate type of sample to use for a particular study is best decided by the comparisons which are to be made. If pond or lake sites are being interpreted, then clearly one should select surface samples from comparable sites. If a peat core is being analysed, then surface moss samples would be more appropriate.

Aquatic surface samples

The distribution of pollen grains laterally over sediment surfaces in lakes is not even. Davis *et al.* (1971) have shown that *Pinus* and *Ambrosia* pollen have higher abundance relative to *Quercus* in shallow water than in deep water, suggesting a differential movement laterally through the water. The implication of this finding for surface pollen studies is that single observations will differ from one another and that several samples should be taken from various parts of the lake. Even pools of a few metres in diameter show lateral variation in their surface

pollen samples, as shown by the studies of Salmi (1962) on a bog pool in Finland. The size of the lake will also affect the relative contribution of various pollen components (see Chapter 2); Wright (1967) advises the use of large lakes for such studies, but this depends on the type of fossil site with which comparisons are to be made. Dried mud samples from seasonal ponds have proved acceptable as a source of data on pollen rain in arid areas (Ritchie 1986b).

Terrestrial surface samples

The location of terrestrial surface samples depends upon the pattern of the vegetation under investigation. In reasonably uniform vegetation, such as that found in west Greenland by Pennington (1980), random samples can be taken. Where there is a gradient of vegetation, a transect of samples is recommended by Wright (1967). The spacing of samples along such a transect will depend on the scale of vegetation change. Tinsley and Smith (1974), in their study of a woodland/heath transition, selected samples every 10 m, whereas in their study of *Calluna* pollen deposition in moorland, Evans and Moore (1985) sampled every 2 m.

In these heathland and moorland studies the vegetation was described within a 2 m and 1 m radius, respectively, of the sample point. In forests a larger radius is generally favoured, of the order of 20 m (Bradshaw 1981) to 30 m (Andersen 1970). In maquis vegetation in Spain (Stevenson 1985) and in desert shrub vegetation in Iran (Moore & Stevenson 1982), 10 m × 10 m plots have been used, the vegetation recorded, and ten surface samples removed at random from within the plot. The necessity for multiple sampling within plots has been demonstrated by Adam and Mehringer (1975) who found considerable lateral variability in terrestrial surface samples. An absolute minimum of five samples is needed to overcome this variability. These samples may be combined, mixed and subsampled (Birks 1973; Stevenson 1985), or treated separately (Andrews 1985), in which case further information on lateral heterogeneity of samples is obtained.

In terrestrial surface sampling, the most commonly used material is the moss polster, though where these are not available (such as in arid land studies) surface soil samples have been used. In the latter case there are dangers of incorporation of older pollen by mixing, selective pollen decay, and long-term pollen accumulation in the soil.

Where moss polsters are used, the general practice is to collect only green parts of the polster (Bradshaw 1981; Prentice 1986b), though Andersen (1967) has used the humic material at the base of the polster as well. He recommends, however, that in woodland studies the moss polsters should ideally be selected from fallen branches to avoid the risk of mixing with soils. Bradshaw, on the basis of the work of Pitkin (1975) on the growth rates of mosses, considers that the green parts of the moss represent the last 5 years of growth, and hence pollen accumulation. He has also demonstrated that there is no alteration of the pollen assemblage by selecting pleurocarpous or acrocarpous mosses.

In meadow sites, where mosses may not be available, whole leaf tissues of herbs have been used (Hall 1989) but, when sampled in autumn, they were found to underestimate early spring flowering species, such as most trees. The duration of pollen retention on leaf surfaces is evidently fairly short. Parallel studies using both moss polsters and pollen traps have been carried out by Vuorela (1973).

POLLEN TRAPS

The use of a pollen trap rather than a surface sample of vegetation helps to overcome the problem of extraneous material, such as soil, contaminating the contemporary pollen assemblage. But the assumption that pollen trapped in this way is comparable to that landing on a mire surface, or becoming incorporated into lake sediment, depends upon the efficiency of the trap. This efficiency is particularly affected by the aerodynamic characteristics of the trap.

A variety of pollen traps have been used; some of the early ones are discussed by Lewis and Ogden (1965). Some samplers act simply as rain collectors, but because of their aerodynamic properties they are unlikely to sample adequately the particles suspended in the air. The collection efficiency of such traps varies with meteorological conditions, particularly wind (Ranson & Leopold 1962). Davis (1967) has used similar, open-necked traps for the determination of pollen deposition within lakes and she found that the number of grains accumulating in these traps was linearly correlated with the area of aperture, suggesting that the efficiency of the trap is not altered if different sized traps are used. She recommends a 4.5 l bottle with a mouth diameter of about 8−10 cm, suspended in a wire cage from a float and anchored 2 m above

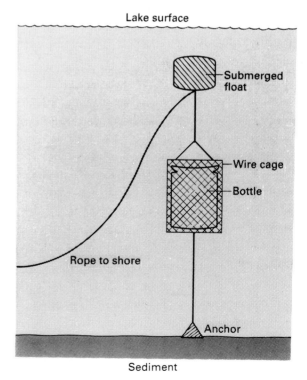

FIG. 3.4. Sediment trap for use in lakes.

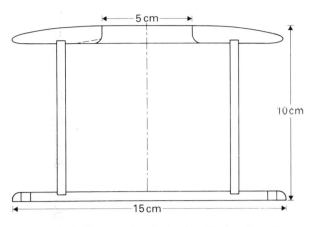

FIG. 3.5. Pollen sampler devised by Tauber (1974).

the lake bed (Fig. 3.4). Similar traps have been used in the studies of Bonny (1976, 1980).

Tauber (1974) designed the trap which has been most widely used in the study of pollen sedimentation from the air (Fig. 3.5). It differs from a conventional cylinder in its cap, which is curved and is of greater diameter than the cylinder it covers. The central hole, on the other hand, is of smaller diameter. These features are designed to avoid the turbulence associated with the lip of a cylinder. It was tested by Peck (1972) in underwater situations and she found that

smaller pollen grains were trapped less efficiently than larger ones, especially at low water flow. When the speed of water flow was increased, the size of grains was of less influence, but the overall efficiency of the trap was lowered. Bonny and Allen (1983) have compared the efficiencies of Tauber traps with cylinders in aerial conditions and found that the Tauber trap picks up more pollen than the cylinder in open conditions, but less under a tree canopy. Cundill (1986) has proposed an alternative type of trap (Fig. 3.6) in which pollen is filtered from rainwater by passing through acetate wool which can later be dissolved in acetone. Tests of this trap in eastern Scotland have proved generally satisfactory. By filtering the rain, the problem of insects (with their pollen loads) being caught up in the trap is easier to overcome. Also the sinking of the structure to ground level solves some of the aerodynamic problems associated with free standing pollen traps.

Sticky traps, held both horizontally and vertically, have been used for pollen sampling, but they usually need protection from rain, which obviously modifies the pollen input. Glycerine jelly on glass plates can be used in this way, but it is subject to removal by water. The orientation of slides is important, since lateral air movement

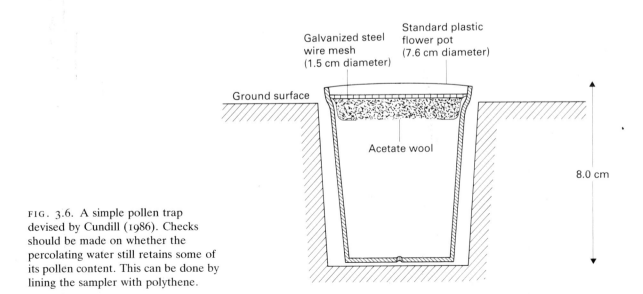

FIG. 3.6. A simple pollen trap devised by Cundill (1986). Checks should be made on whether the percolating water still retains some of its pollen content. This can be done by lining the sampler with polythene.

results in the impaction of pollen grains (Chamberlain & Chadwick 1972) and vertical surfaces thus collect much higher densities of pollen than horizontal ones. The efficiency of pollen capture by impaction is found to increase with increasing windspeed. Such results imply that vertical objects like stems of trees, or even reeds, could act as a filter, collecting pollen from the air which could then be washed down into the soil or sediment beneath.

An alternative to the use of these passive, receptor traps is the active, filter trap which draws air through an aperture and directs it at a sticky plate with a known rate of volume flow. One widely used design for such a trap is the Hirst trap (Hirst 1952), which is particularly valuable in aeropalynology and allergy studies. The trap has a continuously moving receptor slide, so is ideal for long-term trapping experiments in which diurnal or seasonal fluctuations in the aerospora are to be observed. Some other traps are described by Hyde (1969).

FORENSIC SAMPLING TECHNIQUES

The materials submitted for forensic analysis are varied, but usually consist of mud, soil or dust samples which require testing for their source of origin. Most such samples can be analysed in precisely the manner described for other types of samples in the next chapter.

Occasionally pollen or spores need to be recovered from clothing or other materials onto which they have become impacted. The most appropriate method in this case is to use clear sticky tape, attaching it to the cloth and then peeling it off, when loosely attached fibres, dust and pollen grains will be removed with it. This may be examined directly (in which case the use of double-sided sticky tape is an advantage) or, preferably, can be removed and prepared

for microscopy by processing in the manner described in the next chapter.

TRANSPORT AND STORAGE OF SAMPLES

Cores should be transported in robust, rigid containers, such as plastic drainpipes or guttering. Monoliths are usually collected in steel cases, so do not normally present problems, unless they require packing to prevent movement in transit. This may create difficulties in the case of friable mor humus layers especially if the total depth of the monolith is less than the size of the collection tin. Careful packing is required at the ends.

Efforts must be made to ensure that samples do not dry during transport to the laboratory. Individual samples from exposed faces or from surface material can be kept in polythene tubes with lids or in polythene bags. Cores and monoliths should be wrapped very thoroughly in polythene sheets and should be labelled on the outside as well as the inside with water-resistant ink. Cores do not appear to suffer for moderate periods in higher temperatures as long as they are not allowed to dry. Samples taken and retained for some weeks in the arid subtropics, then carried back to the laboratory by air have not suffered loss of pollen as a result.

Long-term storage is best achieved in a refrigerator or cold room between 2°C and 5°C. Freezing can result in expansion and distortion of cores and above these temperatures there is a very real possibility of microbial activity and consequent pollen degradation. Even in the cold, the efficient sealing of samples in plastic containers must be maintained to avoid drying. Under these conditions, samples can be stored for several years if necessary without deterioration of the pollen content.

4

THE TREATMENT OF SAMPLES

Having collected and, if necessary, stored cores or monoliths of material, the next stage of treatment consists of subsampling. This is usually necessary since only small samples (of the order of 1 cm^3 or less) are required for pollen extraction, so these must be removed from the original material. The greatest danger at this stage is that the subsample may become contaminated by atmospheric pollen in the process. Ideally, the pollen preparation laboratory should be fitted with an air filtration system to avoid the intrusion of modern pollen. Where such air filtration facilities are not available, a simple check on the fallout of atmospheric pollen can be achieved by placing microscope slides covered with glycerine jelly in strategic positions on the work bench to register the level of pollen input. If this is unacceptably high, then the laboratory must be regarded as unsuitable for pollen extraction. It is always sensible to prepare type material from pollen sources in the vicinity of the laboratory, so that potential contaminants are known.

Contaminant pollen grains can sometimes be recognized in a preparation, for if they have arrived after processing they may still contain their cytoplasmic contents and will not have been stained. But contaminants arriving before processing will be indistinguishable from fossil grains.

In the absence of pollen filtration systems it is still possible to obtain acceptably low levels of pollen fallout if several simple steps are taken to reduce the aerial contamination of the laboratory.

1 Windows should always be kept closed.

2 House plants should never be kept in the laboratory.

3 Herbarium or dried plant material should never be taken into the laboratory where pollen extractions are to take place.

4 If possible, extractions should be conducted in the winter period (October–March in north temperate areas), when atmospheric pollen is at a minimum.

5 The highest levels of pollen in the atmosphere generally occur in the afternoon, so this is the least suitable time for carrying out extractions.

6 Separate glassware should always be used for fossil and type material.

SUBSAMPLING

In the case of unstratified material, subsampling can be carried out in a random manner, having thoroughly mixed the original bulk. This can be achieved by spreading the mixed bulk in a tray and selecting random points on a grid. Such a method is useful for the subsampling of bulked surface samples and for some kinds of archaeological material.

More frequently, palaeoecological materials are stratified in a temporal sequence and the sampling process must be designed to elucidate changes in sediment pollen with time. Sequential sampling with depth is therefore the most appropriate scheme to adopt. The intervals at which samples should be taken depends upon a variety of factors, such as the rate of growth of the deposit, the rate of change in pollen assemblages with depth, the degree of precision with which it is desired to detect such changes and the time available to the investigator for the project. Given that all projects have some time limitation, the investigator often has to face the choice between more frequent sampling (i.e. smaller sampling intervals) and the number of pollen grains counted within each sample. In such a case, one must ensure that the number of grains counted at each sampling point should be an adequate representation of the pollen components that are of particular interest (see Chapter 7 for details on this process).

A commonly adopted scheme of sampling is to begin by using fairly wide intervals, such as 20 or 10 cm and then to fill in gaps where more detail is required at a later stage. If there is little change in pollen assemblages between such wide sampling intervals, either because of rapid sedimentation or slow rates of vegetation change, it may not be necessary to extract samples at closer intervals. But where marked changes with depth occur, further samples can be analysed. This approach is economical on time and effort, but is only possible when entire cores or monoliths are available in the laboratory. This is a major consideration when deciding to sample in the field or to take cores back to the laboratory.

There is one minor problem with the use of 20, 10 and 5 cm sampling intervals, that is the problem of placing further intermediate samples. The recommended quantity of material for pollen extraction is approximately 1 cm^3 and this is normally extracted from the core in such a way that the sample occupies 1 cm depth of the core. It is this fact that makes close sampling awkward if one has begun sampling on a decimal system. A simple alternative is to begin 'skeleton' sampling at 16 cm intervals, which allows closer interval sampling to be used at 8, 4, 2 or even 1 cm intervals. This means that if a high degree of resolution is required from a diagram one can insert intermediate samples at closer and closer intervals until eventually contiguous samples of material are being analysed.

The extraction of the sample from the core or monolith should be preceded by a careful cleaning of the core surface. This involves the cutting away of the superficial material that may have suffered some contamination, either from adjacent sediments or the coring equipment, or from the atmosphere during the short period of inevitable exposure in the process of taking cores in the field. Cleaning should be undertaken with a clean, sharp scalpel and all cuts should be made in a direction horizontal to the axis of the core. In other words, movement of the scalpel should always be sideways to avoid contamination of any given level of sediment with material from a position above or below its own.

Unless sampling is to begin at close intervals, there is little merit in cleaning an entire core initially. It is better to clean only the area around the position of sampling, since rewrapping and storing cores is bound to cause further superficial contamination.

The 1 cm sample can be cut out with a scalpel or spatula. Unless the absolute density of pollen per unit volume is to be calculated (see below), there is no need for a high degree of precision in determining the exact size of the sample. The pollen content will eventually be expressed in relative percentage terms and will thus not be affected by sample size.

The simple use of a ruler and scalpel for sampling has been made for samples of 0.5 cm and 0.25 cm depth without difficulty, but if thinner samples are required a more sophisticated system is called for. One technique, recommended by Sturludottir and Turner (1985) is to freeze the core and use a microtome for thin slicing. They placed the peat core in a freezer at $-20°C$ for 72 h, cut it into 2.5 cm lengths and then sectioned these lengths in a cold room using a hand-operated microtome. In this way they obtained peat samples 1 mm in thickness and 0.5 cm^3 in volume. These were reckoned to represent only one or two seasons of growth, thus providing a very high level of resolution in the pollen profile. One problem with freezing is that it may cause expansion and distortion in the core and make it difficult to determine the relationship between sample thickness and the original thickness in the core. Cloutman (1987) has designed a monolith cutter which pushes the core from its container in carefully regulated volumes.

An alternative technique has been devised by Wiltshire (1988) (Fig. 4.1), who has used a series of mounted razor blades which can be pressed into a core, monolith, or even an exposed peat face to recover adjacent thin sections. Its main limitation is its inefficiency in fibrous, coarse and also in unconsolidated peaty materials, but these present a problem for any system of close sampling. Wiltshire has been able to obtain samples of 2 mm thickness using this device.

For piston core samples, a close interval fractionator has been described by Fast and Wetzel (1974). This slices the sediment core as it is extruded. A similar extruder for unconsolidated sediment has been described by Glew (1988).

Woodland and moorland litter and mor humus layers are particularly difficult materials for close sampling. They are normally taken to the laboratory in three-sided monolith tins (see above) which can be used as a support for friable organic detritus while it is sliced into contiguous samples. A stainless steel square of the same dimensions

as the cross section of the tin and with one sharpened side is useful for slicing the monolith, but it has not been found very effective for slices less than 1 cm in thickness.

Thinner samples have been obtained directly from the field (Perry & Moore 1987) by placing steel pins into the ground at the four corners of a 10 cm × 10 cm square, each pin being marked at 5 mm height intervals. Successive litter layers, each of 5 mm thickness, could then be removed using scalpel and forceps. The main problem is the abundance of very fine rootlets which must be carefully dissected and cut. Such operations are time consuming and can be performed in the field only under the most equable of weather conditions.

POLLEN CONCENTRATION PROCEDURES

Obtaining pollen in a countable form from lake, peat and soil materials does not so much involve *extraction* as *concentration*. The techniques used in the process are aimed largely at the disintegration and dissolution or otherwise removal

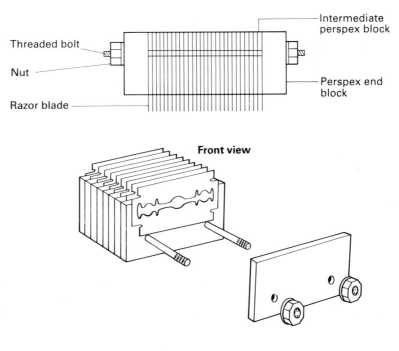

Front view

FIG. 4.1. A simple device for taking serial samples of small volume from a core or from an exposed peat face (Wiltshire 1988). Razor blades are set at 2 mm intervals in a perspex block.

of the non-pollen matrix in the sediment. The various ways in which this can be achieved are based upon the small size of pollen grains and spores, coupled with their extreme resistance to many corrosive chemicals.

The complexity of the pollen concentration process depends upon the original concentration of grains and the nature of the matrix. In most ombrotrophic peats (see Chapter 2) the matrix is almost entirely organic, whereas in rheotrophic peats and most lacustrine sediments (and also in soils) there are varying quantities of mineral material, either siliceous or calcareous. If the concentration of mineral material is high, then it becomes necessary to remove it, otherwise the pollen contained within the sample will be obscured.

The ultimate aim of the concentration process is that pollen-rich samples shall be produced, mounted on microscope slides, which will permit: (i) the accurate identification of as many as possible of the grains encountered, and (ii) the counting of an adequate sample of pollen grains and spores to provide a representation of the total population in the original sample. Time spent in the treatment of samples that results in the removal of potentially obscuring particles and cleans the pollen and spores of detritus, is well invested, for it not only saves time during counting, but also permits more accurate identification.

The various chemical processes developed for the treatment of pollen samples relate to the different matrix materials in which the pollen may be embedded. Here, those processes will be described individually and the combination required for any given sample can then be determined by the precise nature of the sample under investigation.

Potassium hydroxide digestion

This treatment on its own can produce a reasonable concentration of pollen from certain peats. The quality of the pollen and spores extracted in this way, however, leaves much to be desired. It should normally be accompanied by acetolysis except in the most cursory of analyses.

1 Place approximately 1 cm of the sample into a boiling tube and add 10 ml of 10% potassium hydroxide (KOH). The precise quantities of sample and reagents are not critical unless pollen densities are to be determined (see below). Place for 10–15 min in a boiling water bath, stirring to break up the material with a glass stirring rod. Ensure that the concentration of KOH does not rise above 10% by occasional addition of distilled water. Prolonged boiling should be avoided as it may cause swelling of the pollen grains. This digestion process not only breaks up the matrix, but also dissolves humic materials, producing a dark brown solution.

2 Pass the sample through a fine sieve (either metal or plastic); an aperture of 100–120 μm is probably the most suitable as this allows all pollen and spores to pass through, but retains larger particulate matter, such as sand grains and plant fragments. This debris may well prove valuable for macrofossil surveys and for elucidating the sedimentation processes, so it should be washed in water, and collected and stored in tubes for subsequent examination using a binocular dissecting microscope (Grosse-Brauckmann 1986; Wasylikowa 1986). The suspension passing through the sieve should be collected in a polypropylene centrifuge tube.

3 Balance the tubes with distilled water and centrifuge for 3 min at 3000 r.p.m. This should result in the deposition of a small pellet of material at the base of the tube, though the dark colour of the humic solution may obscure its presence. It is preferable that a swing-out head centrifuge be used rather than the solid head type with inclined tubes. The latter produces a pellet in the form of a smear up the side of the tube, and such material can easily be lost in decantation, especially when its precise position may not be immediately apparent because of the darkness of the supernatant.

4 Decant off the liquid in the tube. This should be done with a combination of care and confidence, so that the pellet is not disturbed and lost. The decanted liquid consists largely of organic colloids and can be discarded. Those lacking confidence may prefer to decant into a beaker until assured of the presence of the pellet remaining in the tube.

5 Resuspend the pellet in distilled water using a mechanical 'Whirlimix' stirrer. If clumping of the fine detritus fraction should occur, this can usually be overcome by the addition of a few drops of a detergent, such as 5% sodium lauryl sulphate solution.

6 Centrifuge as previously, then repeat the washing process in distilled water as in steps 4 and 5 above.

Hydrochloric acid treatment

If there is an abundance of calcium carbonate in the sediment, as in the case of tufa or lake marls, then it may prove preferable to begin with this treatment. Analysis of pre-Quaternary sediments with a largely inorganic matrix is also better commenced with hydrochloric acid (HCl) treatment, followed by hydrofluoric acid (HF) and then KOH digestion. Hydrochloric acid treatment should certainly precede any digestion with HF and all mineral sediments should be tested with a few spots of dilute HCl to check for the fizzing and production of carbon dioxide (CO_2) that results from the presence of carbonate.

Treatment consists simply of adding cold 10% HCl in excess (i.e. until all obvious reaction ceases). The reagent may be used warm, but this is not usually necessary and there is a danger that hot HCl will cause corrosion of the pollen wall.

Hydrofluoric acid treatment

When there is an abundance of silica present in the sample, then removal is essential to avoid pollen being obscured when mounted. One way of doing this is to dissolve the siliceous material in hot HF. This is an extremely corrosive reagent, however, and a number of precautions are absolutely essential.

Always ensure that the sediment is free from carbonates (use HCl), and that any KOH has been thoroughly removed from the sample before adding HF. Always use protective clothing, including gloves (disposable ones are preferable as they are less likely to become punctured) and eyeshield as any contact with the skin can be very dangerous and the after-effects are persistent. Ensure that HF burn cream (1 part magnesium oxide to 1.5 parts glycerine jelly) is available as a first aid treatment. Any contact with HF should be treated by thorough washing in water (20 min under a tap), application of the burn cream, followed by immediate hospital attention. Only use HF in specially equipped fume cupboards with perspex fittings. Check that the drainage system is not made of glass and that the extraction system is efficient and HF resistant. Always inform other workers that you are using HF and label tubes clearly, especially if you leave them for prolonged treatment.

Processing of the sample (in the fume cupboard) consists of the following steps:

1 Add about 6 ml of concentrated (30–40%) HF to the washed pellet in a polypropylene centrifuge tube (NB HF dissolves glass and some plastics). The liquid should be added directly to the tube from the special dispensers in which the acid is supplied. There is no need to measure out a precise volume.

2 Resuspend the pellet by mixing carefully with a polypropylene stirring rod. Do *not* use a 'Whirlimix' shaker.

3 Place in a boiling water bath or, preferably, in an electrically heated thermostatic block at 100°C for 15 min and check for grittiness in the tube using the stirring rod. If necessary, continue boiling for an hour or more; no damage will result to the pollen as a result of prolonged boiling in HF or even cold treatment for several days. When all silica has been removed, proceed to the next stage.

4 Centrifuge while hot, inside the fume cupboard and with caps over the centrifuge tubes.

Any contact with the fumes should be carefully avoided. It is useful to check the atmosphere with damp litmus paper to ensure that acid fumes are not leaking.

5 Decant the hot supernatant into a clearly labelled polypropylene vessel. When cool it should then be carefully neutralized with a suspension of sodium carbonate. Test with litmus until neutral before the resulting fluorides are discharged down the drain.

6 Resuspend the pellet in 10% HCl and warm, but do not boil. This removes any silicofluorides produced during HF treatment. Centrifuge in the fume cupboard and dispose of the supernatant in the same way as the previous stage.

7 Wash twice in distilled water to remove all remaining traces of HF. Ensure that the polypropylene tubes are free from any remaining spots of HF which might contaminate other equipment or come into contact with the hands.

Acetolysis

Cellulose is a polysaccharide and can be removed most effectively by acid hydrolysis. The technique described here is basically that of Erdtman (1960). For a peat lacking any siliceous content, the treatment should follow on immediately after KOH digestion and washing. If HF treatment is necessary then acetolysis should be used after the final HCl and washing stages.

The reagents used in acetolysis, concentrated sulphuric acid and acetic anhydride, are not only corrosive, but react vigorously with water. Care should therefore be taken in their handling and especially in ensuring that materials with which they come into contact are dry. To this end, the pollen sample should be dehydrated before being subjected to acetolysis. The procedure (carried out entirely within the fume cupboard) is as follows:

1 Resuspend the washed pellet in glacial acetic acid; this dehydrates the organic material. Centrifuge and decant; discard the supernatant. One word of caution here, many of the waterproof inks from felt pens are acetone based and, although safe when boiled in water, they dissolve in glacial acetic acid and in acetic anhydride. Care should be taken, therefore, to ensure that labels are not lost from tubes at this and subsequent stages. Engraved tubes are the safest.

2 Add about 6 ml of acetolysis mixture to the pellet and resuspend by stirring with a rod (mechanical whirlmixing of these corrosive materials is not to be recommended). Acetolysis mixture consists of acetic anhydride mixed with concentrated sulphuric acid in the ratio 9:1 by volume. The mixture should be made up freshly each day. Although the mixing of these two liquids does not generate the same degree of heat as that produced when concentrated sulphuric acid is mixed with water, there is some heat liberated. The sulphuric acid should be added drop by drop to the acetic anhydride and stirred. Sometimes a brown colour results from charring. This operation should be performed in a fume cupboard with gloves and eye protection.

Disposal of surplus acetolysis mixture can be achieved by pouring it very carefully into cold running water.

3 Place the suspension in a boiling water bath for 3 min. Longer periods may prove harmful to certain pollen and spore types. *Sphagnum* and *Pteridium* spores seem particularly prone to destruction during prolonged acetolysis.

4 Centrifuge as above, and decant the supernatant carefully into running water.

5 Resuspend in glacial acetic acid, centrifuge and decant. The pellet is now ready for rehydration without any danger of exothermic reactions.

6 Resuspend the pellet in distilled water, centrifuge and decant the supernatant. Repeat. It is very necessary to be thorough in the washing out of residual acetic acid, otherwise crystals will form later if glycerine jelly is used for mounting.

Staining

The observation, identification and, if required, microphotography of pollen grains and spores can be conducted on unstained material. Acetolysis usually leaves grains with a slightly yellow colour which may prove adequate for the tastes of some palynologists. Others, however, prefer a stronger enhancement of the contrasts in the optical image of the pollen wall and this is particularly helpful, in some cases essential, when photography is intended. The most widely used stain for this purpose is aqueous safranine, but fuchsin may also be used.

1 Subsequent to acetolysis, it is helpful to add a few drops of 10% KOH solution to the final wash. This will neutralize any residual acid in the sample and ensure that the stain will take more effectively.

2 Add about 2 ml of distilled water to the clean, neutral pellet, plus two drops of 5% aqueous safranine solution. If there is a large quantity of organic debris remaining at this stage, then more than two drops may be required to achieve adequate staining, but over-staining should be avoided as this obscures wall structure. Re-suspend the pellet thoroughly.

3 Top up with distilled water, stir, centrifuge and decant. The resulting pellet should be red in colour and should comprise only fine organic material. There should be no mineral layer at the base. If there is, then the HF treatment needs to be repeated.

Glycerine jelly mounting

Of the various mounting media available (see Praglowski 1970; Chapman 1985; Collinson 1987), glycerine jelly has one major advantage, it is easily handled, being molten when placed in a boiling water bath, yet solid at room temperature (in temperate climates). It also has excellent optical properties (Batten & Morrison 1983). Its main disadvantage is that it absorbs water from the atmosphere and this causes pollen grains to swell when mounted in it (Faegri & Deuse 1960), often increasing their size by 1.25–1.5 times. Over very long periods, i.e. decades, grains deteriorate and their wall structure becomes unclear. But room temperature solidity does mean that when slides are stored the grains

remain in position. It is possible, therefore, to locate grains from mechanical stage map references and return to them for further observation and comparison at a later date. This is not easy to accomplish with silicone oil (see below). Mounting in glycerine jelly is carried out as follows:

1 Melt glycerine jelly in a beaker suspended in a boiling water bath. Add to the pellet a volume of jelly about twice the volume of the pellet; this usually means about 1–1.5 ml of jelly if the original sample was 1 ml of sediment, but it obviously depends upon the density of pollen and other fine fraction organics.

2 Keep the tube warm in a water bath and stir thoroughly to produce a uniform suspension of red-stained solids in the clear jelly.

3 Heat a clean microscope slide and coverslip on a slide-drying hotplate and dispense a portion of the suspension onto the slide. This may be carried out by creating an even smear of preparation over the entire area to be covered by the coverslip, or it may involve the positioning of a central drop which spreads out under the pressure of the coverslip. If the latter method is used, it must be remembered that there may be differential movement of the pollen grains from the centre out towards the edges (Brooks & Thomas 1967). Smaller grains will move out faster than larger grains, creating a concentrically zoned pattern. Problems arising from this, however, can easily be overcome by arranging the counting of the slide in a radial series of scans.

A useful technique to ensure that the pollen grains are in one focal plane and are close to the coverslip (helpful in the use of oil immersion microscopy) is to invert the slide, supported at its ends, so that grains sediment to the coverslip as the jelly solidifies.

Large coverslips (22 × 50 mm) are often useful, especially if the pollen density is not very high. It is also valuable to make up at least two slides from each level. This can provide a statistical check on variation within the pollen sample which can subsequently be compared with variation between samples.

4 Sealing of slides with clear cellulose in acetone (clear nail varnish is often used for this purpose, but Hill (1983) considers it unsatisfactory) is sensible if slides are to be stored for long periods (i.e. more than a year). A useful slide sealant is BDH gluceal. After cooling, the slides should be labelled and stored horizontally in trays. Vertical storage in racks is less satisfactory, especially if there is a danger of raised temperature or if grid references to particular grains are being used. Boxes can be housed on edge to ensure that slides are horizontal. There is no danger of jelly melting at room temperature in temperate climates, as long as the slides are housed away from direct sunlight. Glycerine jelly is not very suitable for tropical climates.

POLLEN SIZE PROBLEMS

One of the criteria which may be useful in the identification of pollen grains is their size. But it has been found that certain of the processing and mounting procedures described above may cause alterations in the size of pollen grains and also some changes in their form, for example, the width of colpi. Many of these problems are difficult to avoid, but can be circumvented to some extent by ensuring that type material is treated in precisely the same way, so that it is entirely comparable with the fossil material.

The following techniques have been reported as having some influence on pollen size or form.
1 Potassium hydroxide treatment. There are suggestions that prolonged treatment in 10% KOH can cause the swelling of pollen.
2 Acetolysis is also reported as causing swelling (Reitsma 1969; Martin 1973). Martin found it necessary to determine correction factors for use when comparing the sizes of acetolysed and non-acetolysed grains. Acetolysis certainly has considerable influence on the final form of the colpus and porus, depending upon the time for which the mixture was boiled and the precise composition of the mixture. Even slight departures from the 9:1 ratio of acetic anhydride to sulphuric acid can result in structural changes in the pollen grain. There is a clear need here for consistency between the treatment of fossil and type material.

3 Hydrofluoric acid treatment. Pollen treated with HF often appears smaller than untreated material. This has been regarded as a direct effect on pollen size by the treatment (see Martin 1973), but it may be a consequence of the conditions of pollen preservation, since HF treatment is usually reserved for silty materials. Andersen (1978) has pointed out that the sediment type does have an effect upon pollen size. The precise effect of HF needs closer attention, since it may interfere with some critical size distinctions of pollen types at junctions between silty and organic sediments. For example, the identification of *Betula nana* at the commencement of the Holocene may be affected by HF treatment. In such situations there is much to be said for treating all samples in the same way, whether HF is strictly required or not.
4 Glycerine jelly mounting. As has been mentioned above, glycerine jelly does cause swelling of pollen grains, and alternative mounting media have been sought as a consequence of this. Silicone oil has proved a most popular alternative medium (Andersen 1960). It has a low refractive index (see Berglund *et al.* 1960), a low viscosity (2000 plus centistrokes) and does not appear to cause swelling for periods up to 17 years. Praglowski (1970) recommends the use of low viscosity mounting media as this avoids the collapse of pollen grains and maintains their spheroidal structure. Andersen (1978) has examined in some detail the effects of long-term pollen storage in silicone oil and has found that changes do occur, but are of a random nature. One important source of error is the degree to which the solvent (especially benzene — see below) has completely evaporated from the preparation. If solvent remains in the sealed slide, then pollen grains may shrink by as much as 10% over the first 3 weeks of storage.

ADDITIONAL TECHNIQUES

The techniques described above will normally serve to concentrate any pollen contained in a sediment and should usually be applied first. If the results are not satisfactory, then a range of additional and alternative techniques for pollen preparation and mounting have been devised and these will now be described.

Oxidation

In some cases it may prove necessary to remove further materials from a pollen preparation before counting can take place, especially if there is an abundance of lignin in a deposit which resists acetolysis. In such circumstances, oxidation is appropriate. When used it should follow acetolysis. Erdtman and Erdtman (1933) have described the use of chloric oxides as oxidizing agents for the process.

1 After acetolysis, resuspend the pellet in glacial acetic acid. Add 5−6 drops of sodium chlorate solution and then (carefully) 1 ml of concentrated HCl. The reaction is violent and rapid, resulting in the oxidation and bleaching of material in the tube. More than a few seconds' treatment is liable to result in the destruction of the pollen grains.

2 Centrifuge, decant and wash several times.

3 Stain and mount as before.

The use of various other oxidizing agents, such as nitric acid, hydrogen peroxide, etc. has been described by Gray (1965). Oxidation is a particularly effective way of removing lignin and may also serve to clear microfossils that have become darkened.

Bromoform flotation

The principle of differential flotation was first used by Knox (1942) for the recovery of microfossils. The method has been described by Frey (1955) and the following is a summary of his account. The process should be carried out after KOH digestion and washing. The method entails the use of bromoform, which is an extremely unpleasant and potentially dangerous liquid to handle and this factor greatly reduces the value of this technique.

1 Resuspend the pellet in acetone in order to dehydrate the material, then centrifuge and decant.

2 Add 5 ml of a mixture of bromoform and acetone with a specific gravity of 2.3. The latter must be checked with a hydrometer and the density of the mixture adjusted by adding acetone (less dense) or bromoform (more dense) as necessary. The contents of the centrifuge tube should be mixed thoroughly and then centrifuged.

3 The supernatant now contains all particles with a low density, including all pollen grains, spores, sponge spicules, diatom frustules, chitinous remains of arthropods, etc. This is decanted off and the pellet should be treated with a further 5 ml of bromoform and acetone. If necessary, several extractions can be made (three are usually sufficient) to ensure the removal of all microfossils.

4 Microfossils can be recovered from the bromoform/acetone mixture by reducing its specific gravity. This is achieved by the addition of acetone; roughly twice the mixture volume should be added. On centrifuging, the microfossils remain in the pellet and the supernatant can be discarded into a waste bottle.

5 Wash the sediment in acetone and centrifuge once more.

The material may now be stained, acetolysed and mounted as described previously.

Practical difficulties are sometimes encountered with this method, particularly that of obtaining a free suspension of fine organic particles in the bromoform/acetone mixture and the avoidance of clumping. Such clumping is especially to be avoided if dense mineral particles are involved, since it can lead to the loss of microfossils with the heavier material.

One very obvious advantage of this method when compared with HF treatment is that siliceous microfossils, such as diatom frustules and phytoliths, can also be recovered, whereas they are dissolved by HF.

Guillet and Planchais (1969) have used a solution of zinc chloride at a density of 1.8 as an alternative to bromoform. This has the advantage that it is not such an intensely toxic compound.

Density gradient separations

Flotation principles can be operated using less dangerous liquids than bromoform, such as sucrose solution. Continuous density gradients of this type have been used, for example, in the separation of red blood cells and of protozoan parasites (Cox 1970). Since pollen grains come in a variety of sizes, the use of discontinuous gradients is perhaps preferable as this permits the extraction of a range of particle sizes at the same level, if the sucrose densities are correctly chosen.

Precise instructions concerning this technique are impossible, for the choice of densities will depend upon the nature of the sediment matrix as well as upon the speed and time of centrifugation. The following account should therefore be taken simply as a starting point for experimentation and adaptation to the needs of the particular sediment under investigation.

1 Carefully pipette the following aliquots into a centrifuge tube, permitting no mixing between each of the layers:

    4 ml sucrose of specific gravity (SG) 1.28;
    2 ml sucrose SG 1.10;
    4 ml sucrose SG 1.05;
    2 ml suspension of pollen preparation (after KOH and washing).

2 Centrifuge at about 1000 r.p.m for 2 min. At the end of this process, most of the pollen will be concentrated at the base of the 1.10 SG sucrose, together with small fragments of siliceous material and other organic detritus. Larger mineral fragments will reside in the basal aliquot and the uppermost layer (the 1.05 SG sucrose and water suspension will probably have mixed) will contain fine mineral particles.

3 Carefully withdraw the dark, pollen-rich layer using a pipette, dilute with distilled water and centrifuge to recover the washed pellet.

This technique, or adaptations of it, provides an alternative to HF treatment, but it does have certain disadvantages. If there are large mineral particles present in the sediment matrix, these may carry pollen and other material with them into the denser sucrose layers when they are first introduced into the gradient. The use of detergents prior to introduction can result in a better dispersion of particles. Even so, some larger microfossils, such as *Polypodium* spores may sink into the lowermost layer of sucrose.

Silicone oil mounting

Unlike glycerine jelly, silicone oil requires the dehydration of material before mounting. This is carried out as follows:

1 Wash with 96% alcohol, centrifuge and decant.

2 Wash with 99% alcohol, centrifuge and decant.

3 Wash with benzene (Andersen 1965) or preferably with tertiary butyl alcohol (Davis 1966), centrifuge and decant.

4 Add 1 ml tertiary butyl alcohol (or benzene) and transfer to an open dish in a fume cupboard. Add silicone oil (precise quantity depends upon the volume of the sample remaining), mix and allow to stand for 24 h, during which time the volatile component (tertiary butyl alcohol or benzene) evaporates, leaving the silicone oil with its suspended sample. Warming to 50°C is advantageous since the retention of any benzene can lead to pollen shrinkage (Andersen 1978).

5 Finally, dilute with silicone oil to the required microfossil density and mount. Sealing of slides is to be recommended and they should always be stored horizontally as silicone oil is far more fluid than glycerine jelly.

The use of tertiary butyl alcohol as a volatile medium miscible with silicone oil is preferred to benzene because of the carcinogenic properties of the latter.

Coverslip support

Cushing (1961) has pointed out that the weight of a coverslip can distort pollen grains if they are subjected to its weight. This can be avoided by providing some support to the coverslip, either by introducing small particles, such as polythene beads, into the mounting medium, or by placing very small quantities of plasticine under the corners of the coverslip. This precaution is particularly worthwhile if size criteria are to be used for identification or analysis.

Extraction of pollen from honey

Honey is normally rich in pollen, unless it has been treated commercially to remove the pollen component, such as by filtering through diatomaceous earth. It may also contain fungal material, especially if it is derived from honeydew rather than nectar (Lieux 1972). Various methods have been described for the concentration of pollen from honey (e.g. Louveaux *et al.* 1978; Adams *et al.* 1979; Sawyer 1988), and some neglect the use of acetolysis. For precise work, however, involving critical identification of taxa, acetolysis is essential, for most published keys and descriptions of grains are based on acetolysed material. Photographs of frequently encountered pollen grains prepared without resorting to acetolysis are given by Sawyer (1988).

The main additional technique required in dealing with honey is the removal of honey sugars. A sample of about 10–20 g honey should be dissolved in 20 ml of warm (about 40°C) distilled water and then centrifuged. If the honey is derived from honeydew, Sawyer (1988) recommends that dilute sulphuric acid should be used instead of water. The pellet should then be resuspended in 10% KOH, warmed and centrifuged once more. After washing, the sample can be dehydrated with glacial acetic acid and acetolysed as described above.

Pollen can also be analysed in pollen loads collected directly from worker bees. Collection from bees may involve killing the individual workers, or can be achieved by stunning them with a short exposure to coal gas. Their loads may be removed safely while the bee is unconscious. It is also possible to obtain pollen loads by restricting the aperture to the hive temporarily in such a way that the pollen loads are lost by the incoming workers and can be collected in a receptor tray. Nectar from combs can be obtained simply by shaking over a tray (Adams & Smith 1981).

Extracting from clay

The recovery of pollen grains from sediments rich in clay can be particularly troublesome and even the use of prolonged HF treatment or bromoform flotation may prove inadequate for the removal of very small particles in the clay fraction. These can be a serious problem as they obscure pollen grains, especially when phase contrast microscopy is being employed. Several techniques have been developed to improve the removal of these clays during processing.

Pyrophosphate treatment

Bates *et al.* (1978) recommend the use of sodium pyrophosphate as a deflocculant in the preparation of clay-rich materials. After treatment with KOH or HCl, the samples are washed thoroughly in water until neutral in reaction. 0.1 M sodium pyrophosphate is then added, shaken vigorously and placed in a hot water bath for 10–20 min. The supernatant becomes cloudy because of the suspension of clay particles. The samples are then centrifuged at 3000 r.p.m for 5 min and the supernatant (including its load of clay) is decanted and rejected. If larger particles of siliceous material are also present, HF treatment can follow, otherwise the samples may then be acetolysed.

Ultrasonic dispersion

Gray (1965) proposed the use of ultrasonic vibration to disperse clays and Dodson (1983) describes a method in which samples are washed with ethanol, then treated with bromoform of SG 2.0 and exposed to ultrasonic vibration in a 390 W vibrator at 80 kc/s for 5 min. Marceau (1969) recommends a series of short, strong bursts rather than prolonged treatment. The samples were then dehydrated in ethanol and mounted in silicone oil. But Dodson found that samples treated in this way had more pollen fragments in them than those treated simply with HF. The conclusion is that ultrasonic vibration at this level causes some damage to pollen grains, especially small grains with thin walls.

Fine sieving

The sieving processes described so far employ relatively large aperture sieves (over 100 μm) and their function is to retain coarse material and allow pollen through. An alternative, especially when attempting to discriminate against very small particles (less than 5 μm) is to use sieves so fine that they retain pollen along with coarser materials, but allow finer particles to pass through. Cwynar *et al.* (1979) have used nylon screens of various dimensions; 5, 7, 10 and 15 μm. They found that 7 μm apertures proved most successful in pollen recovery from clays. In their method, they combined the use of sodium pyrophosphate, HF, acetolysis and finally sieving with 7 μm and this involved a negligible loss of pollen (determined by absolute techniques). Larger aperture sieves led to an increasing loss of pollen in the filtrate. Smaller aperture sieves (5 μm) were unsuitable because they become clogged with fine sediment. Counting time for preparations was reduced four to five times as a result of this processing.

Tomlinson (1984) has reported pollen losses as a result of using 10 μm mesh sieves, especially of *Filipendula* grains, and she recommends the use of a 5 μm mesh in conjunction with ultrasonic vibration and a tap-operated suction filter.

Jemmett and Owen (1990) have investigated the effect of different mesh weaves used in sieve construction on the loss of pollen. They have found that some types of weave, such as those using a 7 over 1 under system, can lose pollen grains even at an aperture size of 1 μm. *Alnus* grains (used as an exotic additive in this Australian work) were found to be depleted after using this type of sieve and it is possible that they were becoming wedged under the weaves. Clearly fine sieves should be checked for this type of problem before being used in routine work.

Extracting from tufa

Calcareous sediments, such as tufa, are normally not rich in pollen and the material contained in them is often degraded. The extraction process described here is essentially that of Preece *et al.* (1986).

1 Dissolve the calcium carbonate by four or five treatments with 10% HCl, interspersed with centrifugation.
2 Sieve through 160 μm sieve.
3 Wash with water.
4 Boil with 10% sodium hydroxide.
5 Wash with water.
6 Hydrofluoric acid treatment.
7 Acetolysis.
8 Wash, stain and mount.

Since the concentration of pollen is likely to be low, it is advisable to begin with a large sample and to make up several slides in case the final density is poor.

Pollen concentration from surface moss polsters

Surface samples of bryophyte material or litter are often low in their pollen concentration. This may be improved by immersing the sample in a flask of water and shaking for 10–15 min on an automatic shaker, followed by centrifugation. Alternatively, the sample may be homogenized in a blender for a few seconds, then sieved through a 250 μm sieve and centrifuged. This technique is faster, but care needs to be taken to avoid pollen damage. Two or three short (1 s) bursts with the blender have proved more effective than longer treatment (Andrews 1985).

Extracting from ice

Ice from glaciers is generally low in pollen concentration. Ambach *et al.* (1966) found concentrations of between 600 and 60 000 grains per litre. Fredskild and Wagner (1974) have found similarly low concentrations in the ice of the Greenland ice cap. This means that samples may need to be up to one litre in volume if sufficient pollen is to be recovered for counting.

The procedure recommended by Fredskild and Wagner is to melt a sample of between 200 and 1000 ml and centrifuge. The material recovered can then be subjected to standard preparation techniques, commencing with KOH treatment.

ABSOLUTE EXTRACTION TECHNIQUES

None of the techniques described in this chapter so far provide the necessary data for determining the density of pollen grains in the sediment. The sampling systems described have not been volumetric and the subsampling used in counting pollen grains means that the total number of grains of any particular type found within a given volume of sample is never known. The outcome of such analysis is that one must express findings in relative terms; each pollen taxon can only be expressed as a proportion of a selected pollen sum, such as total pollen and spores, total land pollen, total arboreal pollen, etc., and this can lead to many problems of interpretation (see Chapter 8). Ideally, one would like to have information on the abundance of each pollen taxon in such a form that it was not dependent upon fluctuations in other taxa. This could best be expressed as number of pollen grains settling on a given area of surface sediment (say 1 cm^2) in a given time (say 1 year). If this data were available, then one could derive precise infor-

mation concerning absolute fluctuations in pollen abundance with time, irrespective of alterations in the status of other pollen taxa.

There are two prerequisites for this information to be acquired. The first is a knowledge of the density of each pollen type within the sediment, and the second is an appreciation of the rate of sedimentation. In the case of the latter requirement, it must be borne in mind that sedimentation rates in lakes and mires can vary considerably with time, and therefore a simple knowledge of a deposit's age is an inadequate basis for determining sedimentation rate. A series of independent dates is needed, usually supplied in recent (Holocene) deposits by radiocarbon dating techniques (see, for example, Pennington 1981b). Alternatively, and preferably, there may be annual laminations in a deposit which enable sedimentation rates to be determined very precisely. In general, the use of absolute techniques is more difficult in peats than in lake sediments (see, for example, Beckett 1979) because of their more irregular rates of accretion, their variable degrees of compaction and problems with volumetric sampling (see below).

As far as extraction of pollen is concerned, the main problem is the development of a technique which will allow the determination of the absolute density of pollen grains in a sample of known volume, which occupies a known vertical depth of sediment. This latter requirement is necessary for the calculation of how long the sample took to accumulate once the sedimentation rate is known.

Several techniques have been used to determine pollen densities. Very early in the history of pollen analysis, von Post (1916) used an absolute method by treating and counting completely a known volume of sediment. At that stage, however, no methods were available for the determination of sedimentation rates and major variations in absolute pollen densities with depth could often be accounted for by changes in sediment deposition rate. For this reason von Post's absolute methods were abandoned in favour of conventional relative counts in which density estimates are not required.

With the availability of radiocarbon dating as a means of estimating sedimentation rates, interest has been revived in density counts. The techniques employed in the determination of density (the relative merits of which have been tested by Peck 1974) can be divided into three basic types:

1 Volumetric methods (Davis & Deevey 1964; Davis 1965, 1966).
2 Weighing methods (Jorgensen 1967).
3 Exotic marker grain methods (Benninghoff 1962; Matthews 1969; Bonny 1972).

In the first and third types of method it is necessary to determine the sample volume prior to treatment and this can prove a difficult process. Several techniques for determining the precise volume of sediment samples have been developed. Davis (1969) used a spatula of exactly 1 cm capacity for sampling her cores. This method suffers from the disadvantage that the core of sediment is easily distorted or compressed, thus changing its density. A boring device which produces samples that are apparently free from differential compaction has been designed and described by Engstrom and Maher (1972).

Middeldorp (1982) took monolith samples of peats in the field, using a container 50 × 15 × 10 cm in size, and then cut the peat monolith into 1 cm thick slices and froze them in the laboratory. It was then possible to subsample the frozen slices volumetrically using a cork borer of known diameter.

A relatively simple system is the determination of volume by displacement in water, as recommended by Bonny (1972). Here, a 10 ml measuring cylinder containing 5 ml of water is used to receive the sample and the change in meniscus height is recorded. This method has the advantage that it can be used with coarse, fibrous material, such as peats, as well as with fine-grained lake sediments.

Volumetric methods

Having taken a sample of known volume, the extraction treatments described above can be carried out. At the final stage, it is necessary to suspend the pellet in a known quantity of mounting fluid, so that subsamples can be taken and the total number of pollen grains of each taxon determined. This is not easily achieved for two reasons. First, the subsampling process must be conducted with extreme accuracy and any portion of an aliquot remaining in a pipette represents a serious error. This can be overcome to some extent by using a fluid of low viscosity for the suspension. Glycerine jelly would clearly be unacceptable for this purpose. Davis (1966), for example, uses tertiary butyl alcohol as a dehydrating agent prior to mounting in silicone oil, and recommends that aliquots for density determinations be taken at this stage rather than from the silicone oil, because of the lower viscosity of this medium. Subsamples can then be evaporated and mounted in silicone oil.

Weighing methods

In these methods a known weight of sample is treated (see Jorgensen 1967) and is finally suspended in a known weight of glycerol. A weighed subsample of this suspension is then mounted and counted; all pollen grains in the subsample must be counted. It is then possible to determine the number of pollen grains of each taxon present/unit weight of the original material. There are several problems associated with such a method, for example the use of wet or dry weights for the original sample. The final information acquired is also of less immediate value than the volumetric data, for it does not relate simply to sedimentation rates and hence to the calculation of pollen influx.

One of the major disadvantages of both the volumetric and weighing methods, as described here, is that all pollen grains must be counted in the final subsample. This means that assumptions must be made initially concerning the approximate pollen density so that a reasonable volume of subsample can be estimated. This difficulty can be overcome most effectively by use of marker grains.

Exotic marker grain methods

The idea behind this technique is that a known number of exotic marker pollen grains (or other appropriate particles) is added to a known volume of sample at the commencement of processing. All subsequent abundance determinations can then be carried out in relation to counts of the marker grains. The technique was first employed by Benninghoff (1962) and has subsequently been modified by Matthews (1969), Pennington and Bonny (1970) and Bonny (1972).

The first requirement for this technique is a suspension of known concentration of a pollen grain (or other marker particle) of an appropriate size (i.e. about 20−30 μm, similar to that of pollen grains), which disperses easily in the suspension medium and does not agglomerate into clumps, which is easily recognizable to the observer and which does not occur naturally in the pollen sample. It is not easy to satisfy all of these requirements and the most appropriate grain is likely to differ from one site to another. Many workers have used *Eucalyptus*, but clearly this is not suitable in Australia (but a Northern Hemisphere type, such as *Alnus* can be used as an alternative − Dodson 1983), or for work in the Northern Hemisphere for documenting recent vegetation changes where *Eucalyptus* has become a common introduced tree. Bonny (1972) used *Ailanthus glandulosa* and Webb (unpublished) has used *Jasminium* stained green. Both of these satisfied the requirements for their particular sites of study. Preparations of marker grains should be carefully checked for contaminants. Craig (1972) used small plastic spheres of pollen grain size as markers in his work, following an idea first proposed by Cushing.

The preparation of exotic marker suspensions can be conducted by acetolysing fresh material (staining, if required, in a distinctive way) and suspending in glycerine jelly. The use of acetolysis in preparing such a suspension is recommended since it avoids problems of differential flotation in grains still coated in waxes or fats (Matthews 1969). The suspension must be mixed very thoroughly, stirring for several days using a magnetic stirrer (Tipping 1985) and the precise density determined using a haemocytometer. Once made up, such a suspension can be retained for re-use, but care must be taken to check its density and to ensure that it is well mixed before use.

Stockmarr (1973) has devised a method of producing tablets made up from *Lycopodium* spores and containing uniform numbers of spores per tablet. This has proved an extremely valuable and labour saving facility since a known number of marker grains can be added to a sample simply by adding a known number of tablets. The necessity for making up one's own exotic suspension is thus avoided. The main disadvantage of this technique is that the researcher is constrained to use *Lycopodium* as a marker and this taxon may be present and of interest in the fossil assemblages under investigation. If present in any quantity as a fossil it can interfere with the determination of density. Obviously, the fossil record of this taxon is lost by dilution with exotic spores, which represents a loss of information that may be of interest, especially in sediments deposited under tundra conditions.

These tablets were originally available directly from Stockmarr, but are now supplied by B.E. Berglund of the University of Lund, Sweden. There has been a change in their formulation since the reorganization of their supply and some incidents of spore clumping have been reported (see Francis & Hall 1985).

The precise timing of marker grain addition varies with different workers. Maher (1972) proposed the addition of exotic pollen suspension before processing of the sediment sample, and this is logical, since any pollen lost during processing will be accompanied by an equivalent loss in exotic pollen, so the overall proportions of fossil taxa to exotic pollen will remain unchanged.

The quantity of marker grains to be added depends upon the expected pollen density in the fossil sample. If too much is added, then an inordinately large proportion of counting time may be used in recording marker grains. Ideally one should aim at a final ratio of exotic to fossil grains of between 1:5 and 2:5. The process of counting is carried out as for a conventional analysis except that marker grains are recorded in the course of the count. Edwards and Gunson (1978) recommend mechanical counting of the marker grain using a coulter counter. The pollen sum chosen does not, of course, include the marker element, but all fossil taxa can be expressed in terms relative to the marker grain. This use of proportions means that it is not necessary to count all pollen grains in the subsample (as was the case for the other absolute methods described above). Since the original number of marker grains added to the sample is known, one can calculate from the proportions observed the original number of grains of each taxon in the sediment sample, and hence the density of grains in that sample. Given a knowledge of sedimentation rates in the deposit, it then becomes possible to calculate the rate of influx of each taxon as grains/cm^2 per year.

TYPE SLIDES

Although keys and photographs are very helpful aids to the identification of pollen grains, there is no substitute for a collection of well documented modern samples of pollen grains for comparison with fossil material. Palynologists working with Quaternary sediments have the great advantage over those dealing with older samples that the majority of types are still extant. Species still living can therefore be used for direct comparison with the fossils and can provide confirmation of identifications provisionally made from keys such as the one in this book. Using type materials may also permit the researcher to identify with greater precision than is indicated in keys. Small

differences in sculpturing, shape, etc. may be difficult to describe verbally in a key, but may be fairly evident when observing the actual grains. It is also possible that numerical separation of populations of pollen types is feasible, whereas identification of single grains is not. So measurements taken on large populations of modern pollen samples may provide a basis for the determination of fossil populations.

A type collection of pollen grains is therefore an essential tool of the practising palynologist, and it is inadvisable to undertake serious work without at the same time developing a full type collection. It must also be stressed that single collections of a particular species are of limited value; they provide no indication of the degree of variation found within that species. The number of collections required to provide such information depends upon the variability of the pollen grains of the species in question. Three collections may prove adequate to confirm some uniform species, but many more may be required for variable grains, especially if their features overlap with those of other species. Generally the more collections available from as wide a geographic range as possible, the better.

Several points need to be stressed with respect to pollen type collections. The first and obvious one is that the quality of identification of fossil material will depend upon the quality of the identification of the type material. Samples used for the preparation of type slides should be submitted to thorough taxonomic scrutiny, otherwise severe confusion will be encountered later on. Ideally, the sample should be accompanied by a herbarium specimen, so that if doubt arises concerning the pollen sample, one can refer back to the specimen. The type slide should be labelled with the reference number of the herbarium sample.

When selecting flowers from herbarium sheets (or from living specimens) care must be taken to remove only mature flowers. If young flower buds are included in the sample then some of the pollen grains on the resulting slides may be immature and show incomplete sculpture as well as size abnormalities (usually sizes that are far larger than expected due to over-expansion during acetolysis). It is also worthwhile remembering that open flowers may have been cross pollinated by insects and may contain grains from other individuals of the same species, or even from completely different species.

A further point needing emphasis is that the type sample should undergo the same treatment as the fossil material with which it is to be compared. It is advisable, therefore, to process type material with KOH and acetolysis and to use the same staining and mounting procedure as is normally adopted in the laboratory for dealing with fossils. Most type material is not submitted to HF treatment, but there may be a case for this, especially if size criteria are to be used, as HF treatment is said to cause shrinkage of grains. If type samples are mounted in glycerine jelly, their swelling must be taken into account. There is also a danger of degeneration of the pollen wall after long-term storage in glycerine jelly (say more than 7 or 8 years — see Hill, 1983) and it is advisable to replenish the type collection with fresh material from time to time. Supporting the coverslip with small lumps of clay during the slide-making process may prevent coverslip compression. Sealing the coverslips with wax or glyceal (see above) is required to prevent drying, as this helps to reduce the degeneration process.

MICROSCOPY

The precision with which pollen grains can be identified depends in part upon the quality of the microscope used for observing them. Many of the features used in the key require sophisticated optical equipment, though experience with undergraduate students using relatively simple microscopes suggests that reasonable results can be obtained without recourse to expensive equipment, provided that the instrument is properly set up and operated. There are several features in a microscope, however, which greatly assist the speed, comfort and precision of pollen analysis.

One requirement is a binocular head. It is entirely possible to identify pollen grains with a single eyepiece microscope, but this causes considerable discomfort and eyestrain when used for routine counting. In addition, the microscope should be fitted with a mechanical stage so that traverses across the slide can be conducted in a systematic manner. This stage should preferably be marked with scales and a vernier gauge so that the coordinates of any position on the slide can be taken and grains relocated for subsequent observation, checking, comparison with type material, photography, etc.

An additional item of equipment, particularly useful for the location of pollen grains when the slide is transferred from one microscope to another, is the England Finder. This is a slide with an etched grid that provides coordinates for locating a position that is completely independent of individual mechanical stages on microscopes (Collinson 1987; Traverse 1988).

The most appropriate magnification for routine scanning and counting is ×400, i.e. a ×40 objective and ×10 eyepiece. This provides sufficient magnification for the identification of many pollen grains, and an adequate field of view for comfortable counting in all but very low pollen concentrations. Wide-angle eyepieces are extremely helpful and may permit more rapid counting. Some grains will require more detailed perusal before they can be identified and it is advisable to have available a ×100 oil immersion objective to use in these cases. The extra time taken to put a grain under oil immersion when in any doubt is well worthwhile in terms of the additional precision resulting.

Phase contrast microscopy is an extremely valuable asset in palynology, indeed it is essential for high quality work. It is useful to have this facility available with both ×40 and ×100 objectives. Many people use phase contrast even for routine scanning and counting since it provides considerable extra information about the structure of the pollen wall and sculpture type. Phase contrast microscopy enhances the differences in refractive index of the materials observed, displaying denser objects as dark. Thus it is possible to distinguish very simply between solid elements such as columellae and echinae, and perforations in the wall, both of which can be confused at first glance in bright field illumination (see Chapter 5). This facility is even more essential for high magnification observation using oil immersion objectives. Several of the features used in the key (Chapter 6) require the use of phase contrast microscopy.

LIGHT MICROPHOTOGRAPHY

The documentation of pollen grains is best achieved by microphotography and this can be used both to illustrate type material as an aid to rapid identification, or to record important fossil grains permitting easier comparison between grains and allowing more critical assessment by other workers. The following account is based on the equipment and methods used in the compilation of the photographs for this book. It should provide useful guidance for those undertaking pollen photography.

Equipment

The majority of illustrations in this book (Plates 1–58) have been produced using transmitted light and a Zeiss standard laboratory microscope fitted with a trinocular head to which a Zeiss MC63 photographic system was attached. A variable 6 V 20 W illuminator was used. Variable high intensity illumination is essential in order to accommodate the variation in grain thickness and to allow phase contrast photography with exposure times sufficiently short to minimize risk of movement (see below). A standard (achromat) ×100 phase oil immersion objective was used with a ×10 photo eyepiece. The resultant 35 mm negatives magnified the grains ×250. Rarely, very large grains were photographed without oil immersion using the ×40

objective (final negative magnification ×100). The negative magnification should always be checked by photographing part of a stage micrometer slide with each objective. One should not rely on manufacturers' statements nor on proportional extrapolations from one objective, or microscope, to another. In some cases, with very poorly stained, thin-walled grains a green filter can be used to enhance contrast between the grain and background. Routine use of the green filter, often recommended for black and white photography (e.g. Traverse 1988), was not employed in the present work as it produced no significant general improvement and sometimes worse results especially with heavily stained grains. Ilford FP4 35 mm roll film was used in this work and developed in ID 11 with slight enhancement of contrast in the development time.

The photographic system used here is comparatively inexpensive and the optical system is similar to that which will be available routinely to most pollen analysts. We thus feel that the resolution and information content shown on our illustrations should be comparable with that obtained by the average user of this volume when observing pollen grains under the microscope. The user, however, has the added advantage of the focusing facility on the microscope giving the ability to construct images effectively in three dimensions, which is lost when taking transmitted light photomicrographs.

Recording and negative storage

A complete record should be kept of all photographs, ideally in one record book or on sheets filed with the negatives. This record should include the full identification, source, location on the slide (England Finder coordinates should be used — see p. 55) and magnification of all objects photographed. Negatives should be stored in standard protective sheets in a special file. Clear negative sheets are an advantage as they allow direct contact printing but they do have electrostatic properties which attract dust.

Problem solving

Slides

It will be evident from the plates that, although uniform photographic techniques were adopted, results are not of uniform quality. In particular, clarity and contrast between grain and background varies. This relates to two factors: (i) the quality of the type slide, and (ii) the nature of the grain. There are several steps which can be taken to help overcome these problems, although it is doubtful whether any slide collection could be produced which is uniformly ideal from the photographer's viewpoint.

In the initial phase of work the pollen analyst may doubt the need for photography. However, any publication or display of material, be it in a magazine, poster, project report or scientific journal, will benefit from photographic illustrations. The slides are at least semipermanent and in our experience even the earliest preparations are often re-examined. Therefore it is worth considering these points from the very start of the work.

Generally, the simpler the equipment the better the slides need to be in order to obtain usable photomicrographs. A slide which is perfectly adequate for routine study, identification and understanding of grains may be useless for photography.

The following problems can be circumvented:

1 *Variable response to acetolysis*. Inevitably if a standard acetolysis procedure is used (see p. 44) some grains will be under-acetolysed and others over-treated. With very little experience it is possible to recognize which these are likely to be; for example, wind-pollinated tree and grass pollen are easily over-acetolysed. Over-acetolysed grains are often crumpled and distorted; structural information may be lost and the grains become 'ghosted'. These cannot be used for photography. With type material a second preparation should be made using shorter acetolysis time. The small amount of extra effort

necessary will be of enormous future benefit. With fossil material, containing a range of grain types, it may also be necessary to make two preparations if grains susceptible to acetolysis are of particular interest. When reducing acetolysis times, care should be taken to avoid under-acetolysis which, for example, may leave cellulosic residues that can affect the appearance of pore structure.

2 *Under- or over-staining*. This problem is interlinked with acetolysis problems as thin-walled grains tend to under-stain whereas thick-walled material easily becomes over-stained. As the stain can be modified it is worth checking a temporary mount for grains which may be problematic before preparing the permanent slide. With fossil material, containing a range of grains, two preparations may be necessary. Most of the photographs in this volume where grey grains occur within grey, rather than white backgrounds (apart from those using phase contrast, e.g. Plate 16), are due to under-stained grains.

3 *Extraneous debris in the slides*. In fossil preparations slides are likely to include a variety of material other than pollen and spores (see Chapter 5). Much of this contains valuable evidence and should be retained, though the methods mentioned in this chapter can be employed to remove some unwanted material. However, with modern type slides it is always possible to achieve a good clean result. Problems arise if whole flowers or large pieces of plants are processed complete with dust, soil and other particles as well as additional plant parts. It is worthwhile dissecting only the grain-bearing anthers or sporangia and processing these alone. Coarse debris and fine debris can then be sieve-separated (see earlier in this chapter) and a debris-free slide produced. Debris remaining in slides renders photographs useless as it not only affects their aesthetic value but also frequently obscures crucial features of shape or exine structure. Even worse, debris can easily create false appearances of sculpture or apparent apertures. The observer at the microscope should be able to distinguish these from real features with

careful focusing, but they can be very misleading on photographs.

4 *Dirty and fragile slides.* Slides used for photography must be perfectly clean (coverslip and slide itself underneath) otherwise blurred and 'dirty' negatives result. Immersion oil and fingermarks are the main problems. Slides should always be cleaned after study. In particular, immersion oil should be wiped with a tissue then the slide polished with a clean tissue. Intractable dirt can be tackled with tissue or cotton buds soaked in alcohol. It is preferable that the slide edges are sealed (see earlier in this chapter) otherwise the edge of the mounting medium catches on the polishing tissue and spreads a fresh smear across the slide. Also sealed slides are more robust and withstand frequent polishing. Solid mounting media such as glycerine jelly are, in this respect, far preferable to those like silicone oil where grains may be mobile and where slide cleaning may affect the mount. Care should be taken not to break the thin slides (see point **5** below) by placing them under undue pressure during cleaning.

5 *Slide thickness.* Some laboratories may have thick slides and coverslips which are stronger for routine use. These must be avoided by pollen analysts because it is often impossible to focus the oil immersion objective (with very short working distance) on grains mounted on such slides. Slides should be no thicker than 1.0–1.2 mm and coverslips must be number 1 or thinner.

6 *Location markings on slides.* If it is necessary to mark the position of an object on the slide one method is to draw a circle around it on the underside of the slide with indian ink. Unfortunately this ink, like that in most 'indelible' marker pens, is affected on glass by solvents such as alcohol and by immersion oil. When cleaning a slide for photography it is almost impossible to avoid removing such markers. The England Finder should be used as an alternative method of locating objects on slides (see earlier in this chapter).

Photographic techniques

Initially, it is advisable to run one or several test films photographing the same grains with various conditions. A range of grains should be chosen, e.g. thick well stained, thin poorly stained and with flat or rounded surfaces. The performance of any automatic exposure system should be checked and compared with manual settings. A range of exposures on either side of that indicated by the camera can be used to check if the setting implied by the equipment is really the best for the material. Some systems calculate the exposure only on a small area of the field of view, which needs to be over the item of interest, not the background. Alternatively exposure calculated over the entire field of view may perform badly if there is a small grain with a large clear background. It is worth experimenting with the condenser iris diaphragm, varying the depth of focus, as the results on film can differ from that expected (perhaps due to accommodation in focus by the eyes of the observer).

It is useless to attempt photography with the microscope on an ordinary bench with people walking past, using other parts of the bench or slamming nearby doors. An ideal solution is to have a strong, heavy, stable worksurface isolated from external movements. Using FP4 film (which is fairly fast but not very fine-grained), we were able to obtain exposure times between half and one second for most grains in plain light. This minimizes the risk of blurring due to movement. Some finer grained films (which are slower speed but should provide better resolution) produce results with less contrast and this must also be considered when selecting film.

With phase contrast systems the image is much darker and much longer exposure times are needed (even with brighter illumination) than for plain light, so these need to be considered when selecting film speed. Generally we have found it impossible to produce phase contrast photographs which are aesthetically pleasing because the method of imaging results in grey backgrounds and ring effects around the grains.

Fluorescence microscopy can be useful for recognizing reworked grains, for study of very thin-walled material and for discriminating between particles of differing botanical origin (Traverse 1988). But fluorescence photography requires long exposure times and an exposure correction factor needs to be applied which varies according to the photomicroscopy system used.

It is essential that the photomicroscope is kept clean and dust free. Immersion oil should be removed from objectives after each session and dust covers should be replaced. During use all areas in the light path should be checked periodically for possible dust accumulation and if necessary cleaned with brush, air or lens tissue. If a microscope has been in general use and is then to be used for photomicroscopy it is well worth having it serviced, paying particular attention to cleanliness of the internal optical parts. Inexperienced users should avoid cleaning internal optics as they may have been specially coated.

SCANNING ELECTRON MICROSCOPY

Electron microscopes utilize a beam of electrons rather than light for illumination. The electron beam has a much shorter wavelength, greatly increasing the resolution of the microscope. The transmission electron microscope (TEM) utilizes transmitted electrons mainly for the study of thin sections and has provided detailed information on pollen wall structure (see Chapter 5). It is very unlikely that this detail of structural information will ever be routinely used in pollen analysis.

In contrast to the TEM the scanning electron microscope (SEM) utilizes secondary electrons, emitted when the electron beam strikes the specimen, to study surface topography. As will be clear from the keys and text, the surface sculpture of pollen is frequently used in routine pollen analysis. Thus, whilst SEM is unlikely to become a routine tool for the pollen analyst, a series of SEM illustrations should be of some

value, particularly as an aid to understanding the various types of surface sculpture (see Chapter 5). The SEM also has a considerably increased depth of focus and thus provides an effect of three-dimensional images. These are valuable for understanding grain shapes and distribution of surface features which need to be reconstructed from a series of focal planes seen under the light microscope (LM). The improved resolution of the SEM may reveal features unseen with the LM and it is important not to be confused by these. A good example is *Chenopodium* (Plate 63i, j) where minute surface perforations are seen in the wall (Plate 63j). These, measuring around 50 nm in diameter cannot be seen with the LM, which has an ultimate resolution of 100 nm (Watt 1985). In contrast the larger perforations (>150 nm) in the Caryophyllaceae (Plates 63g; 64e, f) are distinguishable with the LM (Plates 21 & 22). Further details of EM techniques and principles may be found in Watt (1985) and Postek *et al.* (1980).

Preparation techniques

The SEM illustrations in this book (Plates 59−71) were taken using a Philips 50B scanning electron microscope. In most cases grains were treated in the standard manner for type slide preparation (see earlier in this chapter) with KOH and acetolysis. Debris, which may be acceptable, though undesirable, for light microscopy is a complete disaster for SEM where even small particles trapped on or near the grains make the preparation unusable for photography. Samples were therefore sieved through coarse (125 μm) and fine (8−15 μm) mesh, retaining material between these sizes. The final pellet of pollen was resuspended in a small amount of distilled water (the amount depending on the apparent concentration of grains and experience, but ranging from 2 to 20 ml).

Cambridge specimen stubs were prepared by attaching a small square of unexposed, developed, black and white negative film, emulsion upwards, onto the stub using araldite. If only a small amount of SEM work is planned the ends of LM films (which remained in the cassette or which were wound on at the beginning after the camera was closed) are sufficient for the purpose. Care should be taken when cutting and mounting the film not to scratch the emulsion surface. After the araldite had cured, a small drop of pollen suspension was pipetted onto the film and the stubs left covered, undisturbed in a warm place for the water to evaporate. The moistened gelatine in the film is sufficient to cause slight adhesion of the grains but the stubs should always be handled carefully. Stubs were then coated with gold in a Polaron sputter coating unit and examined under the SEM. After study, stubs should be stored in a dust free environment preferably in constant (low) humidity.

The 'shelf-life' of stubs is largely unknown (Chapman 1985; Collinson 1987). During preparation of this volume we re-examined stubs of *Tilia* prepared in the above manner 3 years before. Although requiring recoating they were perfectly usable (Plate 66d). Fine cracks are evident in the grain surfaces at high magnification (Plate 66e−g). These are probably due to slight expansion and contraction of the grains, and hence the coating during storage and transfer in and out of the vacuum. This slight cracking has not affected the surface features important for distinction of *Tilia* species.

Problem solving

The major problem encountered when studying pollen with the SEM is that many grains become distorted, presumably due to a combination of drying in air and the subsequent vacuum in the coater and the EM. Some of these shape changes may reflect natural harmomegathy (Chapter 6) but some grains become totally flattened or deeply distorted. This is particularly prevalent in thin-walled grains such as those already mentioned above as vulnerable to over-acetolysis. Some improvement can be obtained by reducing the acetolysis time (e.g. to 2 min) or the KOH

time (e.g. to 5 min) or by not boiling the KOH. Even so it is often only a small percentage of grains which retain their three-dimensional shape. Reducing the KOH/acetolysis treatment can also result in residues remaining on the grains which obscure the very surface features which are of interest. We did not evaluate the 'sonication' method proposed as an alternative to acetolysis by Bredenkamp and Hamilton-Attwell (1988).

In some grains the problem of collapse occurs even in the absence of any pre-treatment. In these cases special techniques are necessary to retain the hydrated grain shape into the dry, vacuum conditions of the SEM. The most widely used are freeze drying (FD) and critical point drying (CPD) (for details see Postek *et al.* 1980; Watt 1985). An alternative method using Peldri II, a proprietary fluorocarbon, has also produced good results (Chissoe *et al.* 1990).

All these techniques rely on the rapid removal of water from the specimen by sublimation or phase change. In FD specimens are quickly frozen and the ice is sublimed under low temperature and high vacuum. Critical point drying employs the property of liquids (e.g. CO_2) to change from a liquid phase to a gaseous phase at a 'critical point' of temperature and pressure. The Peldri II method employs the property of this fluorocarbon to sublime from the solid state (the substance is solid below 23°C) under vacuum desiccation.

We used CPD for two grains which we felt were essential to illustrate here, but for which we failed (several times with varying treatments) to obtain acceptable results with air drying. These were the bisaccate grain (we chose *Pinus* as easily available) and the grasses which persistently showed distortion even without KOH or acetolysis.

The major problem with all methods is manipulating such small items as pollen grains during treatment and retrieval, handling and mounting (dry and isolated). We used modified BEEM capsules to solve this problem (see also Postek *et al.* 1980, pp. 143−144). The BEEM capsule, designed for embedding material for TEM study, is a plastic capsule with a clip-on cap and a conical base. If the conical base is cut away a second clip-on cap (removed from another capsule) can be attached resulting in a small cylinder capped at either end. This cylinder will fit into the CPD unit. In order to allow free passage of fluids during processing, the lids are pierced by large holes and then lined by a fine mesh. We used nylon mesh aperture size 8–15 μm.

Anthers and pollen sacs were collected from plants actively shedding pollen and fixed in 5% glutaraldehyde. The BEEM capsule was prepared with one end capped and the mesh wetted in advance (to reduce surface tension). A suspension of pollen was transferred to the capsule (which must be filled as full as possible with fluid) and the other end was capped. The capsules were placed in small jars and given three changes of distilled water followed by a series of 50–100% acetone. Experiments are needed to determine the minimum period required for each change to filter through the fine mesh. Initially the capsules tend to float, so the jars must be completely filled, capped and inverted periodically. We erred on the safe side, leaving each solution for 8–12 h. The capsule with 100% acetone was placed into the critical point drier, the acetone replaced by liquid CO_2, which after several changes is passed through the critical point. The capsule when removed contains loose, dry grains. If one end of the capsule is un-capped an SEM stub with a very thin film of adhesive (e.g. araldite) or a square of double-sided sellotape can be upended onto the open end of the capsule, which is slightly narrower in diameter than the stub. If the whole is then inverted, grains drop onto the stub. Reinversion, with a gentle tap on the bench displaces those grains that have not adhered, which can then be stored in the recapped capsule. The crucial factor in this operation is having a suitably thin film of adhesive, partially set hence sufficiently viscous to prevent 'creep' up the sides of the grains, and to prevent the grains sinking into the adhesive. The quickest method for non-critical work which we employed, is to use double-sided sellotape. Grains do sink into this in the long term (Chapman 1985) and the background is uneven, but it can produce acceptable results (Plates 59j; 60e, f) if numerous grains are present (some will land on a smooth area) and if grains are studied immediately.

In hindsight one refinement would have been valuable, that is to rinse the grains more thoroughly through the fine mesh after acetone treatment and before CPD. This should have removed the fine particles which are evident on Plate 60e and 60f. The *Pinus* and *Phleum* (Plates 59j; 60e, f) are the only unacetolysed grains we figure in this volume, but they do have clean surfaces. The acetone treatment was probably responsible for removing soluble material from the grain surface.

5

POLLEN GRAINS AND SPORES

In the seed plants, angiosperms and gymnosperms, the wall of the pollen grain has the important function of protecting the male gametophyte during its journey between anther and stigma. In lower plants, such as pteridophytes and bryophytes, the spore wall has a similar function in protecting gametophyte tissue, but this time during its dispersal from the sporophyte to a moist location suitable for germination. In view of these functions, one might expect the pollen and spore walls to be tightly sealed to prevent desiccation, yet this is seldom the case. They are usually equipped with apertures, and often their surfaces are sculptured in an elaborate way, sometimes with reticulately arranged perforations. It is very difficult to explain the functional significance of the various sculpture types of pollen and spores, but before considering views on this subject it is first necessary to examine the detailed structure of the wall, and the range of ornamentation found upon it.

COMPOSITION AND STRUCTURE

The living pollen grain of an angiosperm has a wall that is made up of two layers. The outer layer is called the *exine* and is composed of a very unusual substance, *sporopollenin* (Zetzsche 1932), together with smaller quantities of polysaccharides (Rowley *et al.* 1981), sometimes referred to as the *glycocalyx*. The inner layer, or *intine*, of the pollen wall is made of cellulose and is very similar in construction to an ordinary plant cell wall. During fossilization only the very resistant sporopollenin-containing exine

remains, and it is this that carries the characteristic form and sculpture which permits the identification of pollen grains. For a long time the chemical structure of sporopollenin remained a mystery, largely because of its relatively inert nature. All that was known about the substance related to its resistance to a range of chemical reactions. It can withstand treatment with hydrogen fluoride, and acetolysis (boiling in actetic anhydride and concentrated sulphuric acid). But it is affected by severe oxidation reactions and by some bases such as fused potassium hydroxide or 2-aminoethanol. Treatment with a strong oxidant, such as sulphuric acid mixed with hydrogen peroxide, or with 40% chromic acid or ozone, degrades pollen grains. Brooks and Shaw (1968) suggested that sporopollenin is a complex polymer of carotenoids and carotenoid esters with oxygen. They came to this conclusion by analysing the substances present in the early stages of anther growth in the lily *Lilium henryi*. Carotenoids extracted from lily anthers were then artificially polymerized, and the infrared spectrum of the substance produced was found to be remarkably similar to the spectra of sporopollenins from a wide variety of sources (see Fig. 5.1). This work has, however, been questioned and the involvement of carotenoid derivatives is now being further investigated (Guildford *et al.* 1988; De Leeuw & Largeau, in press). Brooks and Shaw (1968) claimed that true sporopollenin is much more widespread in the plant kingdom than was originally thought, but acetolysis may be confusing the analyses. Sporopollenin and sporopollenin-like materials are

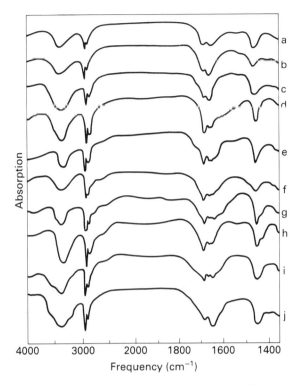

FIG. 5.1. The infrared spectra of some sporopollenins (after Brooks & Shaw 1971): (a) *Chara corallina* algal spore; (b) *Pediastrum duplex* algal spore; (c) *Mucor mucedo* fungal spore; (d) *Lycopodium clavatum* spore exine; (e) *Lilium henryi* pollen exine; (f), oxidative polymer from *L. henryi* carotenoids and carotenoid esters; (g) oxidative polymer of β-carotene; (h) *Selaginella kraussiana*, a modern megaspore; (i) *Valvisporites auritus* a fossil megaspore (250×10^6 years old); (j) *Tasmanites punctatus* fossil spore exine (350×10^6 years old).

found in the spores of such widely separated groups as algae, fungi, pteridophytes and angiosperms. Their chemical stability means that they survive even in ancient fossils, and a substance which appears similar to modern sporopollenin has even been recovered from some of the oldest sedimentary rocks in the world, the Precambrian rocks of the Onverwacht (3.7 billion years old) and the Fig Tree cherts of South Africa (3.2 billion years) (Brooks & Shaw 1968; 1971). The presence of this unique plant polymer in such very early sediments has important implications for investigations into the origin of life, and it

has been found subsequently in many key stages in plant evolution, such as the origin of land plants (Delwiche *et al.* 1989). Brooks and Shaw have also found small amounts of a sporopollenin-like substance in the organic matter of the Orgueil and Murray meteorites and they believe that this is powerful evidence for the existence of extraterrestrial life.

Although sporopollenin is so widespread, it is only in the higher plants (angiosperms and gymnosperms) that it is built into the complex wall structures evident in the photographs in this book. Much work has been carried out on angiosperm pollen grains and their structure and genesis is known in some detail. The terminology used differs among various authors, hence some definitions are necessary. The exine of pollen grains is divided into an outer sculptured *sexine*, and an inner unsculptured *nexine* (Erdtman 1966). The sexine commonly takes the form of a set of radially directed rods, supporting a roof. This roof may be complete, partially dissolved or completely absent. Following the suggestions of Reitsma (1970) we shall call the roof a *tectum*, a rod which supports a tectum or any part of it, a *columella*, and a rod which is not supporting anything, a *baculum* (see Fig. 5.2, system b). The term 'baculum' was first coined by Erdtman and was used to mean any rod in the sexine, whether it was supporting something or not; it is used in this sense in many of the papers referred to later on.

System (a) in the diagram differs from (b) in that it derives from the various staining characteristics of the exine layers, and not the sculpturing differences of the layers. If the wall is stained with basic fuchsin the 'ektexine', which is actually the sexine plus nexine 1, becomes heavily stained before the 'endexine' or nexine 2 (Faegri 1956). Southworth (1974) has shown that the ektexine of some pollen grains is soluble in hot 2-aminoethanol and several related substances while the endexine is left unaffected. The staining and solubility differences indicate that the sporopollenin of these two layers may differ chemically. In addition, Afzelius (1956) in

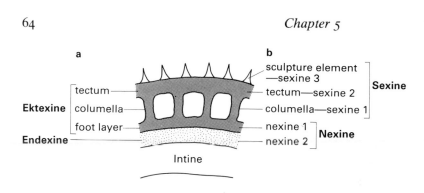

FIG. 5.2. Terminology of the pollen grain exine. (a) As defined by Faegri (1956), Faegri and Iversen (1964). (b) Modified from Reitsma (1970). Faegri has stressed the difference in the staining characteristics of the ektexine and the endexine. Reitsma's division (following that of Erdtman 1966) is a formal morphological one, dividing the exine into convenient descriptive units.

an early electron microscopical study found a difference in the way sporopollenin had been deposited in the layers of the exine. The endexine had a laminated appearance while the ektexine was amorphous – granular. Some recent studies show that this may indicate a difference in the consolidation of the sporopollenin in the two layers. The sporopollenin of the ektexine appears laminated in a radial direction when it is first laid down, but this disappears later as it consolidates. The endexine sporopollenin is deposited in tangential layers and retains this lamination when the grain is mature; the endexine is thus less consolidated than the ektexine.

Despite the fact that chemical and developmental studies of the pollen grain wall indicate that a division into ektexine and endexine is the most basic one, distinguishing these layers with a light microscope presents problems. A foot layer (nexine 1) is not always present and even when it is found, it is often very difficult to distinguish from the endexine (nexine 2). With the light microscope it is easy to distinguish between sculptured sexine and unsculptured nexine, thus these are better terms from the point of view of the pollen analyst, whose main concern is describing the characteristics of a pollen grain exine in order to identify the grain or to compare its characteristics with those of other grains. For this reason we shall adhere to the sexine – nexine division for the rest of this book.

Although gymnosperms appear to have a wall stratification similar to that of angiosperms (layers have been tentatively assigned to sexine and nexine), spores of pteridophytes and bryophytes do not resemble angiosperms in their wall structure. The walls of their spores often appear to be laminated throughout their thickness and bear no layer containing columellae. In pteridophytes a division on a different basis has been attempted, for there is often a loose outer layer surrounding an inner layer. The outer is called the *exosporium* (or *perisporium*) and the inner layer the *endosporium*.

THE DEVELOPMENT OF THE POLLEN GRAIN WALL

The method whereby the pollen cell can produce the complex wall which surrounds it is a subject that has fascinated people for many years. Just after the turn of the century, pollen grain development attracted the attention of the cytologist Rudolf Beer and, although a few other studies had been made previously, his was the most significant of the early works. In 1911 he presented an account of the development of the pollen grain wall of the morning glory (*Ipomoea purpurea*), a species possessing large pollen grains with elaborately sculptured walls. In the anther, each *pollen mother cell* undergoes meiotic divisions to produce a tetrad of four haploid *microspores*. Beer saw that the main features of the sculpturing pattern appeared on each microspore while it was still enclosed in the callose wall of the pollen mother cell, so the patterning was not produced as a result of contact with any of the parent plant tissue in the anther. Beer also observed fibrils running from the nuclear membrane towards the cell wall during this pattern determining period. They disappeared after the release of the spores from the tetrad, but the pattern remained. As far as he could see this meant that it was the haploid microspore nucleus which determined the template of the mature pattern. This was as far as the story could be taken with the light microscope; the details of the process had to await the invention of the transmission and the scanning electron microscopes. Since the diversity of pollen types is wide, one would expect there to be differences in the wall construction process, and this has been found to be the case. Excellent reviews of studies on the pollen wall development of various species are given by Echlin (1968), Godwin (1968) and Heslop-Harrison (1971a). These form the basis of the generalized account given here.

The process of pattern formation begins early in the tetrad period, as soon as each cell is independent and completely isolated from the other cells by a thick callose wall. A cellulose layer is then formed between the plasmalemma of the microspore and the investing callose wall. This cellulose layer forms a continuous sheath to the grain, except over those areas which are destined to be apertures (pores or furrows) in the mature grain. In these areas a plate of endoplasmic reticulum is closely apposed to the plasmalemma, which in turn lies directly against the callose tetrad wall. Thus the endoplasmic reticulum acts as a stencil, allowing cellulose to be deposited everywhere except those areas it has 'blocked'. The initial positioning of these endoplasmic reticulum plates is often correlated with the position of the microspores within the tetrad (Wodehouse 1935). The common tetrahedral disposition of the spores (see Fig. 5.5) means that each spore has three faces of contact with its sibs. It is on these three faces that apertures are often formed, e.g. the pores in

three-pored (triporate) grains. If the 'contact geometry' is destroyed, aperture formation is disrupted. In the lily, Heslop-Harrison has shown that when colchicine is used to block the two meiotic divisions of the mother cell, the whole mother cell then behaves as a single spherical spore in that either apertures are not formed or one irregular, randomly placed aperture is formed.

The cellulose layer in the non-apertural areas is soon seen to be traversed by radially directed rods named *probacula* by Heslop-Harrison (in our terminology they would be called procolumellae). These probacula at first appear to be lamellated in a radial direction and are composed of lipoprotein. Thus before any sporopollenin has come on the scene the three major characters of the future exine have appeared, the number and position of the apertures and the distribution of the bacula (columellae). From some investigations, e.g. Skvarla and Larson (1966), the sites of the probacula have been thought to be determined by some cytoplasmic membranes in a manner reminiscent of Rudolf Beer's 'fibrils'.

Rowley *et al.* (1981) have used chemical methods for the sequential degradation of the exine and they came to the conclusion that the original polysaccharides of the exine form a system of branching tubules on to which the sporopollenin is assembled. This system remains as the glycocalyx, enveloped in sporopollenin, in the mature exine.

As to the question of whether the haploid nucleus controls the patterning of the probacula, the basic 'programme' for exine pattern has been shown to be present in the cytoplasm even before cleavage of the mother cell (Heslop-Harrison 1971b). Thus any nuclear genes coding for the pattern must be transcribed in the diploid mother cell, not in the haploid microspore. This has been shown by disrupting normal cleavage and producing fragments of cytoplasm without a nucleus. These fragments go on to produce a normally stratified exine with the correct pattern of columellae.

After the appearance of the probacula, con-

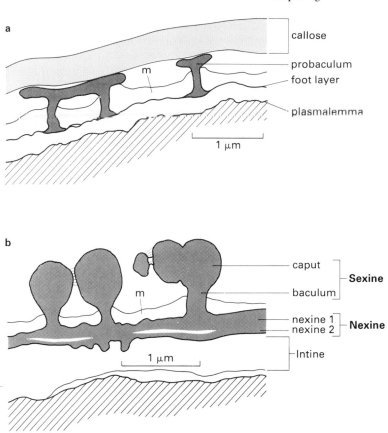

FIG. 5.3. Drawings from electron micrographs of *Lilium longiflorum* pollen grain exines (from Heslop-Harrison 1968). (a) *Late primexine*. Here the probacula are visible as electron-opaque rods with linking muri above them and matrix material (m) between them. The foot layer (nexine 1) is forming from the bases of adjacent probacula. The gap between this and the plasmalemma may be an artifact due to fixation. Note callose wall which envelopes microspore. (b) *Mature exine*. Here each baculum has a swollen head and the foot layer has fully developed, becoming divisible into nexine 1 and nexine 2. The indefinite layer below nexine 2 is the intine, which is composed of cellulose only and is the last of the walls to appear.

nections develop between their feet to form the future nexine 1 (foot layer), and between their heads to form the future tectum. The probacula become much more electron dense and their lamellated structure disappears. This is the time when the probacula become resistant to acetolysis so it is assumed that sporopollenin deposition has occurred. This is called the 'primexine' stage. Figure 5.3a shows approximately this stage in the lily pollen grain. Shortly after this the thick callose wall breaks down, and the pollen grain is released from the tetrad. On release the grain undergoes a rapid size increase, causing the primexine to stretch and become thinner. Consequently the cellulose layer (in which the probacula are embedded) is shredded and dispersed, leaving only a remnant lying on the foot layer between the probacula. The thinning of the primexine is not as noticeable as it might be

because sporopollenin accretion is occurring all the while, thickening the probacula, tectum and foot layer (see Fig. 5.3b).

The formation of nexine 2 (endexine) and the cellulose intine begins slightly before the dissolution of the callose wall and the break-up of the tetrad. Nexine 2 is built up by the apposition of tangentially orientated lamellae, produced at the plasma membrane and upon which sporopollenin has been deposited. These lamellae resemble those seen in the early probacula. In some species the lamellated appearance of this layer persists in the mature grain. Rowley (1963) showed that this laminated material forms a large part of the thickenings around the pore of grass pollen grains. Thus nexine 2 can contribute to the architecture of the grain by variations in its thickness.

When considering the part that parent plant

tissue plays in the formation of the exine it is interesting to look at the origins of the sporopollenin in the exine layers. The sporopollenin of primexine and nexine 2 are thought to come from within the haploid spore. At this time the callose wall cuts off all exchange of materials with the anther cavity and the other spores. However, after dissolution of the callose wall the spore is set free in the anther space, which often means close contact with the tissue known as the tapetum. Within the tapetal cells are small granules or droplets of what appears to be lipid material. These have been called *orbicules* or *Übisch bodies*. Brooks and Shaw (1971) believe that this lipid material consists of carotenoids and carotenoid esters, the precursors of sporopollenin. Many fine-structural studies, especially those of Rowley (1963) and Echlin and Godwin (1968) show that these orbicules are released into the anther fluid as the walls of the tapetal cells break down, and on release they seem to gain a sporopollenin skin. The role of these orbicules is not yet clear; one view is that they actually come into contact with the spore exine and transfer the sporopollenin skin to the structures of the developing exine. Banerjee and Barghoorn (1971) have found that the orbicules appear to form the outermost spinules on the tectum of some grasses. Another view is that they have no role in sporopollenin transfer and just happen to become coated with sporopollenin because, like the pollen grains, they are free in the anther fluid and sporopollenin deposition is going on over all available surfaces.

One thing that is certain is that many other substances are coated on to the mature pollen grain from the disintegrating tapetum. These substances have been given the names tryphine or 'pollenkitt' and they include the characteristic pigments and sticky or odorous materials that are important in the pollination process. The tapetum imparts yet another class of compounds to the outer part of the exine (and to the spaces between the columellae) of some pollen grains, namely the 'recognition' proteins which ensure that the pollen grain germinates only on a com-

patible stigma. If the stigma is incompatible the recognition proteins instigate a reaction which stops pollen tube growth.

APPLICATION OF MORPHOLOGY TO THE IDENTIFICATION OF FOSSIL POLLEN GRAINS AND SPORES

In the study of pollen grains and spores, the complexity of their structure and patterning has necessitated a formidable terminology. In addition, structures have been given different names by different investigators. We have already come up against this sort of problem in the naming of the layers of the exine (see pp. 63−64). Unfortunately there is now a situation where the two major works for pollen grain identification (Erdtman *et al.* 1961, 1963; Faegri & Iversen 1989) use different terminologies. We shall here adopt the suggestions of Reitsma (1970), who has attempted to unify these terminologies. All the terms we think it necessary to use are in the glossary on pp. 163−166.

Apertures

For the identification of any fossil grain or spore the first features to note are those of the apertures. An *aperture* is any thin or missing part of the exine which is independent of the exine pattern e.g. in *Buxus* (Plate 19) the apertures could not be confused with the lumina of the reticulate pattern because the apertures are much bigger and cut across the pattern. There are two sorts of aperture named *pori* (pores) and *colpi* (furrows). Colpi are thought to be more primitive than pori and are distinguished from the latter by being long and boat shaped with pointed ends. Pori are usually isodiametric apertures, but can be slightly elongated with rounded ends. Grains with pori are called *porate*; with colpi, *colpate*; and with both colpus and porus combined in the same aperture, *colporate*. In the living grain apertures are not actually open, but are covered by a thin and delicate layer of exine

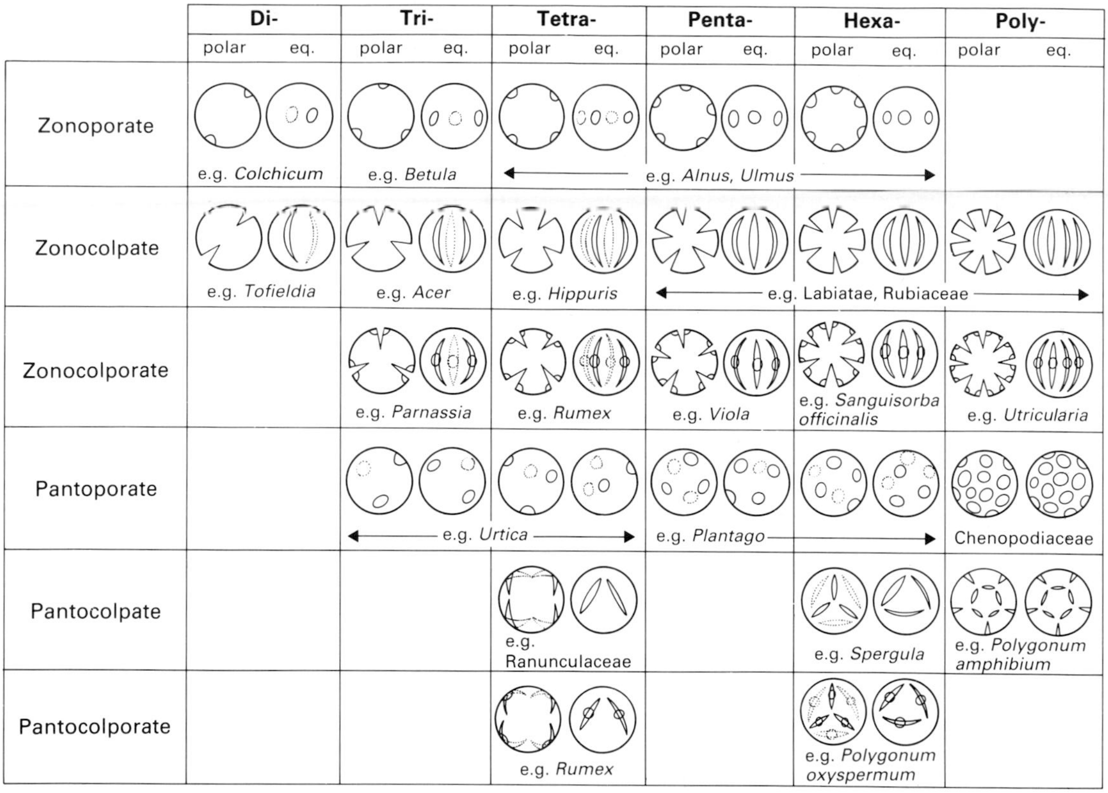

	Di-		Tri-		Tetra-		Penta-		Hexa-		Poly-	
	polar	eq.	polar	eq.	polar	eq.	polar	eq.	polar	eq.	polar	eq.
Zonoporate	e.g. *Colchicum*		e.g. *Betula*		◄——————— e.g. *Alnus, Ulmus* ———————►							
Zonocolpate	e.g. *Tofieldia*		e.g. *Acer*		e.g. *Hippuris*		◄——— e.g. *Labiatae, Rubiaceae* ———►					
Zonocolporate			e.g. *Parnassia*		e.g. *Rumex*		e.g. *Viola*		e.g. *Sanguisorba officinalis*		e.g. *Utricularia*	
Pantoporate			◄——— e.g. *Urtica* ———►				e.g. *Plantago* ————►				Chenopodiaceae	
Pantocolpate					e.g. Ranunculaceae				e.g. *Spergula*		e.g. *Polygonum amphibium*	
Pantocolporate					e.g. *Rumex*				e.g. *Polygonum oxyspermum*			

FIG. 5.4. Diagram showing the range of aperture number, position and character. Some of the possible combinations have no example within the British flora. Classification of pollen types based upon the number and arrangement of apertures. Examples are shown in polar and equatorial views. Dotted lines indicate a different focal plane. Empty positions denote the lack of a North West European example.

material. In contrast, the intine found under apertures is usually thicker than that elsewhere on the grain.

An aperture (colpus or porus) is demarcated by a line most often caused by changes in the thickness of the sexine or nexine or of both at once. Thus it can be said to be situated in the sexine or in the nexine or in both. In the detailed analysis of pollen grain structure an aperture which is a feature of the sexine is called an *ectoaperture* (*ectocolpus, ectoporus*) and an aperture which is a feature of the nexine is called an *endoaperture* (*endocolpus, endoporus*). In some cases ecto- and endoapertures are of the same type (i.e. pori or colpi) and occur in the same place, in other cases they may be of differ-

ent types occurring in slightly different positions. For example in *Centaurea cyanus* (Fig. 5.6e(i) and e(ii)) there are three ectocolpi running meridionally and one continuous endocolpus running equatorially and forming a complete girdle to the grain. Other examples are *Polygonum aviculare* where there are three meridional ectocolpi each crossed by an equatorial endocolpus, and *Oxyria* or *Fagus* where there are three meridional ectocolpi each underlain by an equatorial endoporus.

Pollen grains and spores can be divided into groups on the basis of the number, position and character of their apertures. The classification is basically simple and consistent. The number of apertures is indicated by attaching the prefixes

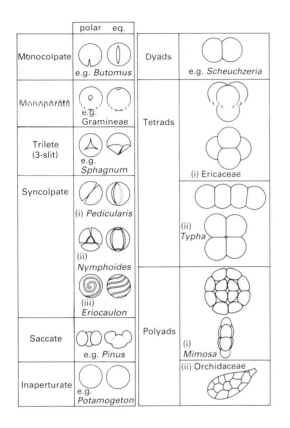

mono-, *di-*, *tri-*, *tetra-*, *penta-*, and *hexa-* before the terms *colpate*, *porate* and *colporate*. More than six apertures is indicated by the use of the prefix *poly-*. In most cases the pori and/or colpi are arranged equidistantly around the equator of the grain. This situation is indicated by the prefix *zono-*. If the apertures are scattered all over the surface of the grain the prefix *panto-* is used.

For example:

Polyzonoporate — with more than six pores, these pores being situated in equatorial zone.

Polypantoporate — with more than six pores, these pores being scattered all over the surface of the grain.

Pentazonoporate — with five pores arranged in an equatorial zone.

Pentapantoporate — with five pores scattered all over the surface of the grain.

Figure 5.4 shows the whole range of pollen types possible. There are some types which do not fit neatly into the system outlined above. One of these is the *syncolpate* grain. Here two or more colpi may fuse, usually at the poles of the grains (e.g. *Nymphoides*, *Pedicularis*, Plate 28) but occasionally elsewhere. An example of the latter case is given by *Eriocaulon* (Plate 28), where a set of colpi are fused giving the appearance of a set of spirals surrounding the whole grain. Another odd grain is that belonging to the Lactuceae tribe of the Compositae family. Here there is a simple aperture system (trizonocolporate) but it is obscured by the very unusual sexine pattern. There are large apparent gaps in the sexine which are separated by high spiny ridges or crests (e.g. *Taraxacum*, Plates 6a,b; 61). This group was originally termed 'fenestrate' by Faegri and Iversen (1975), but is more correctly termed *echinolophate* (Blackmore 1984). In the present work such grains are keyed out under '*crested*'. In some genera or families the pollen grains are characteristically present as aggregates, e.g. *tetrads* in the Ericaceae or Typhaceae and *polyads* in the Orchidaceae.

Spores of pteridophytes and bryophytes possess very different wall structure from angiosperms and gymnosperms (see p. 64). As there is no sexine−nexine division any aperture that they possess cannot be assigned strictly to colpi and pori, as these are sexine features. Some spores have one long slit-shaped aperture, e.g. Polypodiaceae, while others have a three-branched slit forming a Y shape, e.g. *Sphagnum*. These various slit marks on spores are actually tetrad break-up scars but the slits are in no way related to the aperture system of angiosperms.

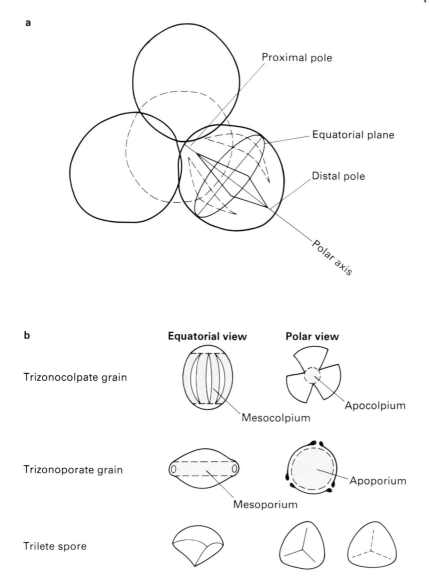

FIG. 5.5. (a) Tetrahedral tetrad showing how the parts of a grain are named. The proximal pole of a grain is that which was nearest the centre of the tetrad, the distal pole that which was farthest away from the centre of the tetrad. The polar axis is a line which passes from the centre of the proximal pole to the centre of the distal pole. The tetrad also shows how aperture position (in this case three colpi) can be influenced by the orientation of the grain within the tetrad. (b) Examples of how grains with differing aperture types would appear in polar and equatorial view. The diagrams also show how the apertures partition the surface of the grain into areas which are given names for descriptive purposes.

The slits always occur at the proximal pole of the spore whereas angiosperm pori and colpi occur on the distal pole, or variously over the surface but rarely on the proximal pole. Figure 5.5 shows a tetrahedral tetrad and the way the parts of a grain or spore are named according to the orientation of that grain or spore in the tetrad. Once separated from the tetrad it is often impossible to tell the orientation of a grain and thus it is very difficult to determine whether an aperture is on the distal or proximal face, and consequently whether it is, for example, a true colpus or a slit in a spore. For this reason all one-slit spores have been included under the heading *monocolpate* in the key. For three-slit spores the problem does not arise because their aperture type is almost unique, they therefore have a separate class on their own (*trilete*). Very rarely, angiosperm pollen grains may be seen with a three-slit aperture resembling a trilete scar, e.g. *Simethis* in the Liliaceae (Andrew 1984).

Those areas on a grain which are not occupied by apertures are given names depending on whether they are adjacent to pori or colpi. The area bordered by two colpi is called the *mesocolpium*, and that bordered by two pori is called the *mesoporium*. If the pori or colpi are in the zono-arrangement, at each pole there is an area where no apertures occur. This polar area is called the *apocolpium* if the zonally arranged apertures are colpi, and the *apoporium* if the zonally arranged apertures are pori (see Fig. 5.5b).

Having dealt with the number and position of the apertures, the next feature to note is their character. The exine often shows a slightly altered structure in the vicinity of apertures. When this happens, the aperture is said to be *bordered*. Borders can be a feature either of the sexine or of the nexine, and a few examples of them are shown in Fig. 5.6. A sudden thickening or thinning of the sexine around an ectoporus is called an *annulus* (Fig. 5.6c & d), and around an ectocolpus is called a *margo* (e.g. *Salix*, Fig. 5.6g). Thickenings of the nexine around an endoaperture (*Centaurea cyanus*, Fig. 5.6e(i) &

e(ii)) or below the edge of an ectoaperture (Fig. 5.6a) are called *costae*. Thinnings of the nexine are also known (e.g. *Myrica*, Fig. 5.6c). In some grains the two layers of the exine become separated from one another in the vicinity of the apertures. The cavity so formed is commonly found around pori (e.g. *Betula* Fig 5.6b) and is called a *vestibulum*. If such a cavity forms around each porus of a zonocolporate grain it is termed a *fastigium*. Other grains have the central part of the aperture membrane with a sexine layer as thick as that occurring on the main body of the grain. This thickened centre is called an *operculum* and can be seen on the pori of *Plantago lanceolata* (Fig. 5.6d), and on the colpi of *Potentilla* type. As the membrane is still very thin around the edges of the aperture, opercula are frequently lost during fossilization (e.g. as in grasses).

Shape

Sometimes the shape of a grain or spore is useful in identification. However it is wise not to lay too great an emphasis on the shape because it can vary considerably within one grain type or even within one species. Variation in shape is also caused by the choice of extraction methods, and embedding media. Although pollen grains and spores are three-dimensional objects, (see SEMs) this fact is often difficult to observe using the light microscope, where only one plane is in focus at any time. Thus pollen grains and spores are described by the shape of their outline in polar and equatorial views. Figure 5.5b shows some examples of polar and equatorial views. Figure 5.7 shows the naming of shape classes that we shall use.

Sculpturing

After separating pollen grains and spores into classes based on the features of their apertures, these classes can be divided by consideration of the fine structure and pattern of the exine. The sexine has been described as being composed of

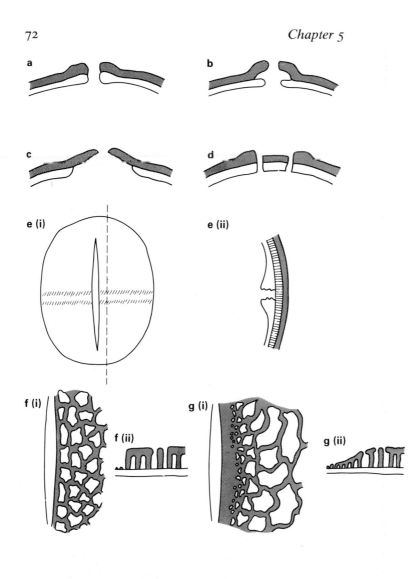

FIG. 5.6. Examples of exine features associated with pori and colpi (sexine stippled, nexine plain): (**a**) porus with a *costa* (thickening of the nexine), e.g. Gramineae; (**b**) porus where sexine separates from nexine to form a *vestibulum*, e.g. *Betula*; (**c**) porus with an *annulus* formed by a slight thickening of the sexine, e.g. *Myrica*. The nexine is here absent in the vicinity of the porus; (**d**) porus with an *operculum* (thickening of the middle of the aperture membrane) and an annulus (sexine thickening), e.g. *Plantago lanceolata*; (**ei**) equatorial surface view of *Centaurea cyanus* to show *equatorial endocolpus*; (**eii**) section of the exine along the dotted line in (**ei**). This shows the *endocolpus* to be bordered by heavy thickenings (costae) of the nexine; (**fi**) colpus without a border or *margo*, i.e. the lumina of the reticulum remain the same size right up to the colpus edge; (**fii**) the sexine also remains the same thickness right up to the colpus edge, e.g. *Fraxinus*; (**gi**) colpus with a *margo*, i.e. the lumina become smaller towards the colpus edge and disappear at the edge, giving a tectate margin. (**gii**) The sexine becomes gradually thinner towards the colpus edge as is shown in the section, e.g. *Salix*.

small radially directed rods which sit on the nexine and are called columellae if they support something (e.g. the tectum, a plate or a small knob) and bacula if they do not support anything and are cylindrical in shape. In some cases the rods are obviously free at their heads, but are non-cylindrical in shape. In such cases they are called *clavae* if they are club-shaped, *echinae* if they are sharply pointed, *pila* if they have swollen heads and *gemmae* if they are short and globular. Sometimes the sexine elements which sit on the nexine do not resemble rods at all. They may appear to be small hemispherical warts (*verrucae*) or tiny irregular lumps (*scabrae*) or other smaller elements (*granules*). In Table 5.1 there is a more detailed description of all these sexine elements. When the heads of the columellae are joined by a complete tectum, the grain is described as *tectate*. The columellae are usually simple, but in some genera, e.g. *Convolvulus* (Plate 31a) and *Geranium* (Plate 15e,f), they branch in a quite complex manner. A grain with

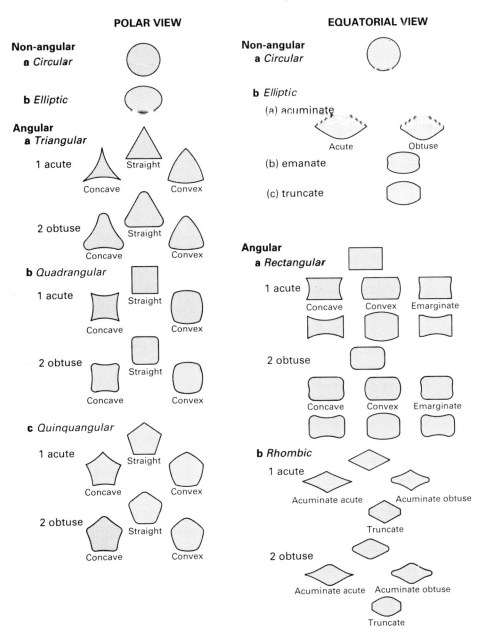

FIG. 5.7. Shapes of symmetrical grains in polar and equatorial views (from Reitsma 1970).

free rods, i.e. bacula, clavae, echinae, etc. is described as *intectate*. Tectate and intectate are two extreme conditions and grains exist which have only a partial tectum, the *semitectate* condition. Examples of semitectate grains are *Salix* or members of the Cruciferae, where the heads of the columellae are joined in two directions only to form winding walls or *muri*. The muri in these cases form a network or *reticulum*. The gaps between the reticulum walls are called *lumina*. Another semitectate pattern is *striate* (e.g. *Acer*) where the muri and lumina run parallel to each

TABLE 5.1 Description of sculpturing types and examples of each. Occasionally, appearances under light microscopy can be deceptive. For example, although *Mercurialis perennis* appears pilate under light microscopy in an exine optical section, SEM photographs reveal that the capita of the pilae are actually joined by a reticulum. With care, this can be detected using phase contrast microscopy.

Type	Description	Example	Plate
Psilate	With surface completely smooth	*Aconitum* under LM 217	29
Perforate	With surface having small holes >1 μm in diameter	*Calystegia* *Cerastium* *Taraxacum*	63 63 61
Foveolate	With holes or depressions <1 μm in diameter; distance between them always greater than their breadth	*Tilia* SEM exine (part)	66
Fossulate	With sideways elongate holes, i.e. grooves which can be straight or sinuous	*Huperzia selago*	10
Scabrate (syn. Granulate)	General term for all sculpturing elements less than 1 μm diameter; shape may vary	*Populus* *Phleum* *Urtica*	8 & 60 60 16
Verrucate	With wart-like sculpturing elements, usually broader than high and never constricted at the base; size >1 μm	*Plantago* *Fumaria*	20 20
Papillate	With hollow, finger-like projections, longer than broad and always >1 μm high	*Tsuga* *Polystichum*	26
Baculate	With rod-shaped sculpturing elements, longer than wide and >1 μm high	*Viscum* *Nymphaea* (some elements)	38 & 69 69
Gemmate	With sculpturing elements higher than 1 μm, the same width as height and constricted at their bases	*Nymphaea* (some elements) *Linum* *Juniperus*	69 38 & 69 9 & 59
Clavate	With club-shaped sculpturing elements, height >1 μm; like a baculum which is thicker at the apex than at the base	*Ilex* (some elements)	9 & 69
Pilate	With element's rods that have swollen or knob-like heads (*capita*); taller than 1 μm	*Nymphaea*, *Ilex* (some elements) *Mercurialis* (with LM only — SEM shows, capita joined laterally)	38 & 69 48 & 65
Echinate	With acutely pointed sculpturing elements, >1 μm high	*Lonicera* *Cirsium* *Malva*	34 53 19
Rugulate	With sculpturing elements elongated sideways >1 μm long and arranged in an irregular pattern	*Nymphoides* *Apium* SEM *Erodium* SEM	28 71 71

TABLE 5.1 (contd)

Type	Description	Example	Plate
Striate	With sculpturing elements elongated sideways (length at least 2 × breadth) and running more or less parallel; ridges = muri, gaps between = grooves	*Menyanthes* *Saxifraga* *Potentilla*	70 39 & 70 56
Reticulate	With sculpturing elements ridges arranged in a network which has gaps (lumina) >1 μm; breadth of ridges (muri) equal to, or narrower than, the width of the lumina	Cruciferae *Potamogeton* *Hedera* SEM	32 8 65

other. Patterns that fall between reticulate and striate are called *rugulate*. Illustrations of these types are given in Fig. 5.8.

The number of fine structural variations is theoretically infinite, thus any attempt to segregate pollen grains into neat classes on the basis of fine structure is bound to run into problems. This point can be illustrated easily. A tectum may have perforations of any shape or size in it. If these perforations are large it is often difficult to say whether the grain is then tectate-perforate with large perforations, or semitectate-reticulate with small lumina. If the lumina are equal to or wider than the muri then the grain is described as reticulate, and if the muri are wider than the lumina then the grain is described as tectate-perforate or tectate-foveolate. The situation is complicated by the fact that tectate grains may have distinct structures within and upon the tectum. Any structures upon the tectum are described in the same way as free sexinous rods, i.e. according to their shape (e.g. echinae, clavae, bacula, pila, gemmae, verrucae and scabrae). It is also possible for there to be low walls on top of the tectum forming reticulate, striate or rugulate patterns. In tectate grains there is again the problem of types running into one another, for example, *Quercus* falls between scabrate and verrucate and thus has to be designated scabrate-verrucate.

It can be seen from Fig. 5.8 that two grains with the same surface sculpture — say, for example, reticulate, can have very different fine structure. One grain could be tectate with the reticulum walls on top of the tectum and the other could be intectate where the reticulum walls are formed by columellae connected at their heads. An optical section of the grain would determine which was the case. In the key the two types might be distinguished by calling one tectate, reticulate (or *suprareticulate*) and the other intectate, reticulate (or *eureticulate*).

An optical section does not always make the fine structure of the exine as clear as one might wish, so investigators often deduce a good deal from careful focusing through the sculpturing and patterning presented in a surface view of the grain. The surface types depicted in Fig. 5.8 show any holes or lower areas to be dark and any raised areas or projecting elements to be light. This is how they would appear at highest focus. On focusing carefully down through the exine their appearance would change due to a change in the diffraction images produced. For instance any raised areas or projecting elements might appear dark and any holes, light. Thus a perforation and a columella may both appear as dark spots, but *at different focal planes*. By concentrating attention on a small area of the exine and observing the apparent changes in it as you

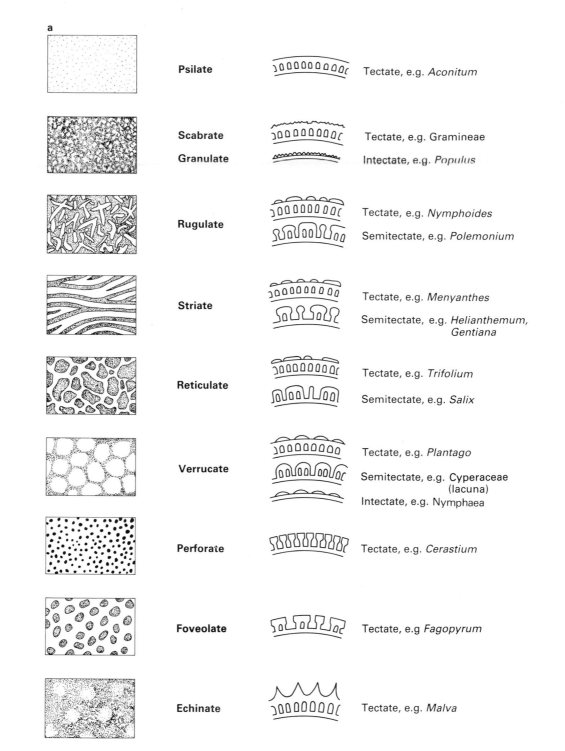

FIG. 5.8. (a) Diagrams of sculpturing types visible in surface view and optical section showing possible underlying exine types. In the sculpturing types all raised areas are shown light, all lower areas or holes are shown dark. It is possible that one sculpturing type, e.g. verrucate, may be produced by three different exine structures. Other sculpturing types, e.g. perforate, can be produced by only one exine structure. (b) Here the same surface pattern is produced by four different sculpturing types. Gemmate, baculate, clavate and pilate all refer to the shape of the projecting processes (see Table 5.1). Theoretically all these types of process could occur on top of the tectum of a tectate grain.

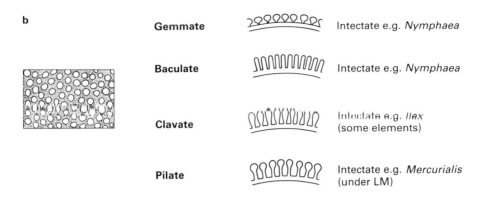

b

Gemmate		Intectate e.g. *Nymphaea*
Baculate		Intectate e.g. *Nymphaea*
Clavate		Intectate e.g. *Ilex* (some elements)
Pilate		Intectate e.g. *Mercurialis* (under LM)

focus down, the structure of the sexine can be deduced. Figure 5.9 shows some of the light and dark patterns that result from focusing down through different exine types. This type of investigation was called 'LO'-analysis (from the Latin *lux*, light and *obscuritas*, darkness) by Erdtman (1956). 'LO' was the term he gave to the sequence: light islands and dark channels (high focus) followed by dark islands and light channels (low focus). If the reverse sequence occurred, i.e. dark islands and light channels followed by light islands and dark channels, it was given the term 'OL'. This system works very well as long as pollen grains and spores are embedded in a medium with a lower refractive index than that of the sporopollenin in the exine.

The SEM illustrations (Plates 59–71) may assist in the assembling of three-dimensional

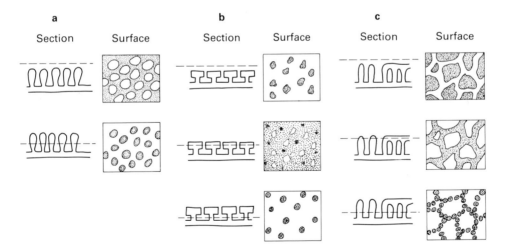

FIG. 5.9. Sculpturing types at different focal planes. The focal plane is indicated by the dashed line on the exine sections. Focusing down through the exine produces diffraction images which vary according to the position of the focal plane. Exine structure may be inferred from the sequence of these images (see 'LO' pattern, this page). (a) Tectate, clavate — the clavae appear as white islands when plane is above them, dark islands when plane sections them. (b) Tectate, perforate — perforations appear as dark islands when plane is above the tectum, diffuse white islands when plane is at tectum level. Columellae become distinct dark islands when plane is below the tectum. (c) Semitectate, reticulate — the reticulum appears as a white network with dark islands (lumina), when plane is above it, dark network with white islands when plane is at the reticulum level. Further downward focusing shows the columellae as distinct dark islands sitting in a reticulate pattern.

structures for certain pollen types. The use of phase contrast can make the picture of exine fine structure even clearer. Under phase contrast, differences in refractive index are converted to differences in light intensity. Thus the greater the difference in refractive index there is between a solid structure and its surrounding medium, the darker the solid structure will appear to be. This means that very small echinae, scabrae, striae and perforations may become visible where, without phase contrast, they would be indiscernible. Phase contrast also makes easier the detection of a tectum and the investigation of the pattern of the columellae underneath a tectum.

By the use of these features of pollen grain morphology and sculpturing it is possible to identify recent and fossil grains with a fairly high degree of resolution. To this end, Chapter 6 consists of a key based upon these structural features.

THE FUNCTIONAL SIGNIFICANCE OF POLLEN MORPHOLOGY

The exine of pollen grains is frequently sculptured in an elaborate manner and many attempts have been made to relate these morphological features to specific functions. It is most unlikely that such structures as echinae, pori, colpi, striae and the lumina of a reticulum are of neutral value and have no selective advantage to their possessors. Many possible functions can be postulated, but the picture is not a simple one and it is entirely likely that each feature may have several functions. A valuable collection of papers discussing these issues in some detail has been edited by Blackmore and Ferguson (1986).

The pollen wall itself is responsible for the protection of living reproductive cells during their transport from anther to stigma. This transport entails a certain degree of mechanical stress, so the coat must be tough enough to withstand battering. It must also be resistant to desiccation, thus preventing undue loss of water from the

sensitive cells. It is possible that some dehydration is inevitable, and this may induce a dormant state in the pollen grain (Heslop-Harrison 1979), which is subsequently broken by rehydration on the receptive stigma. When water is lost, the wall must be sufficiently flexible to cope with contraction, and also with the expansion experienced on rehydration. These changes in volume and shape in response to alterations in hydration are termed harmomegathy (Wodehouse 1935), and the demands these changes place upon the mechanical structure of the exine have been investigated by Blackmore and Barnes (1986). Colpi may alter their width and degree of folding in response to water relations, and changes may occur in the configuration of elastic pore membranes, allowing changes in cytoplasmic volume. The entire shape of pollen grains may change with hydration, often involving an extension of the polar axis.

One obvious function of an aperture is to allow the passage of the germinating pollen tube, and one could argue that a number of apertures provides a better opportunity for the pollen tube to emerge close to the stigma surface than does a single one. But the apertures also provide exits for recognition proteins, as described earlier in this chapter. The function of a reticulum's lumina may also be explicable in terms of recognition signals. Horner and Pearson (1978) have shown that substances held within the exine are released from the pollen grain via these lumina in the pollen grain of the sunflower (*Helianthus annuus*). Yet these same lumina perform a totally different function during pollen development, when they permit the movement of sporopollenin from the external tapetum to enter the exine and accumulate in the foot layer. The possible significance of exine structural features during pollen development has been reviewed by Barnes and Blackmore (1986).

One might expect the conspicuous processes found on some pollen grains to be associated with adhesion to pollination vectors but, though there are some cases where such relationships seem to exist, they are seldom simple. The assumption that psilate grains are anemophilous, while highly sculptured grains are entomophilous is often applicable, but is certainly not invariably the case (Chaloner 1976). In a study of pollen grains from the monocot family Araceae, for example, Grayum (1986) found that species with verrucate, striate and reticulate pollen grains were pollinated by a wide range of insects including flies, beetles and bees, but species with psilate or with echinate grains (both of which are believed on taxonomic grounds to be the most evolutionarily advanced members of the family) were pollinated largely by beetles (psilate grains) or by flies (echinate grains). Grayum considers that the spiny grains are better adapted for attachment to the hairy bodies of fast-flying insects like flies, whereas the psilate grains become stuck to the smooth surfaces of slow-moving beetles (such as scarabs) with the aid of sticky secretions from the stigmas. Many entomophilous pollen grains (though few in the Araceae) are equipped with their own adhesive materials, termed pollenkitt. This is removed by acetolysis, so is not evident in treated grains.

Corbet *et al.* (1982) examined the possibility that electrostatic forces were involved in the transfer of pollen grains. They were able to show that when a charged body (such as a bee) approaches a flower of oilseed rape (*Brassica napus*) it induces an opposite charge in the floral parts. The highest charge is found on the stigma and the lowest on the anther, which means that pollen on the bee's surface will be attracted to the stigma. Chaloner (1986) has discussed this effect in relation to pollen sculpturing and suggests that a smooth grain would rapidly lose its charge when landing on a new surface, whereas a sculptured grain would retain it longer, and hence be more adhesive. A highly ornamented grain might thus be at an advantage both in its cleavage to an insect vector, and in its adhesion to its ultimate goal, the stigma.

From this brief account it should be evident that the relationship between pollen form and function is still an area where much research remains to be done. The complexity of pollen

grain structure may well be matched by the multiple functions of apertures and ornamentation.

NUMERICAL APPROACHES TO POLLEN IDENTIFICATION

There are occasions when the identification of individual pollen grains is impossible but when the identification of a population can be achieved with some confidence. Success depends upon isolating an attribute of the pollen grains that can be expressed numerically and that provides a minimum of overlap between the taxa in question. It may be that one dimension of the grain can simply be measured, or some aspect of the sculpturing assessed, such as echina or lumen density.

Birks (1973) has used this approach in the identification of *Isoetes* spores in sediments from the Isle of Skye, Scotland. The two species involved, *Isoetes echinospora* and *I. lacustris*, differ in the length of their spores, *I. echinospora* having a mode at 25 μm and *I. lacustris* at 34 μm, but there is some overlap between the two species. This means that an individual spore of size, say, 30 μm cannot be consigned with confidence to either species. But if all spores in a fossil count are measured, then it can be determined whether the population represents one or other of the species, or (if bimodal) a mixture of both. In his study, Birks was able to trace a change through time from a pure *I. echinospora* population to one consisting solely of *I. lacustris*, passing through a mixture of the two.

In this kind of study it is important to ensure that all samples have received the same treatment, as different reagents or mounting media can cause swelling of grains (see Chapter 4).

Criteria other than size can be used. Figure 5.10 shows the number of lumina in a transect across the mesocolpium of pollen grains of *Olea europaea* and *Phillyrea angustifolia* (Stevenson 1981). These two pollen types are often found in the same sediment samples in Mediterranean Europe and can be difficult to distinguish when

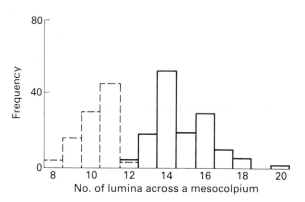

FIG. 5.10. Frequency histogram of the number of lumina counted across the mesocolpium of a sample of *Olea europea* (----) and *Phillyrea angustifolia* and *P. latifolia* (——) (redrawn from Stevenson 1981).

they are in poor condition. *Olea* can be seen to have fewer lumina, but again there is some overlap and individual grains in this size range may be difficult to place. A study of whole populations of the pollen grains can be of assistance.

Where the overlap between populations of grains is very considerable, then simple inspection of histograms may not permit their identification. Statistical methods for effecting such separations have been described by Gordon and Prentice (1977) and have been applied with considerable success by Prentice (1981) to the determination of fossil mixtures of *Betula* species (*B. nana* and *B. pubescens*). Numerical techniques have also been applied to the differentiation of *Cannabis* from *Humulus* pollen (Whittington & Gordon 1987).

The use of pore number in pollen identification may be of value even in cases where the number of pores is relatively small and the overlap considerable. Stockmarr (1970), for example, has found that different species of *Ulmus* differ in their pore numbers. He expresses this in the form of a triangular projection in which each point of the triangle represents a pore number (4, 5 or 6) and the location of a population mean can be plotted on the triangular graph. The technique has provided sufficiently precise data

to allow Tolonen (1980b) to separate fossil elm pollen populations in Finnish sediments.

Langford *et al.* (1990) have investigated the feasibility of pollen type separation by textural analysis. They have experimented with six pollen types of different texture (*Quercus robur, Corylus avellana, Plantago lanceolata, Lolium perenne, Ulmus glabra* and *Pinus sylvestris* — see Plate 59; Plate 60; Plate 62 & Plate 64) and have used a computerized system for analysing the grain texture as shown in SEM photographs. After some data manipulation they claim a 94% success rate in identification.

The use of automated techniques in the determination of different types of organic matter is being applied to petroleum exploration studies. (R. Tyson, pers. comm.) and it may prove useful for routine pollen analysis in the future.

OTHER MICROFOSSILS

One problem that will be encountered in the course of pollen analysis is the separation of pollen grains from spores and other types of microfossil remains. The key and the illustrations in this book concentrate on pollen grains and pteridophyte spores, together with a limited number of the more characteristic and common spores of bryophytes (e.g. *Sphagnum*) and fungi (e.g. *Tilletia sphagni*). But many other microfossils will be found in the analysis of lake sediments, peats and other fossil sources that may cause confusion, but which also must be regarded as valuable residues of palaeoecological information. It is beyond the scope of this book to provide identification keys and illustrations of all the varied types of microfossil that may be encountered, but the analyst should be aware of the following groups of microfossils and seek further information in the references provided.

Bryophyte spores

Apart from the sphagna, the spores of mosses are very difficult to identify with any degree of precision (see Dickson 1973; 1986), but some genera, such as *Polytrichum* and *Encalypta* have been tentatively identified on the basis of their spore sculpturing. Some of the hornworts (*Phaeoceros* and *Anthoceros* — see Rybnickova 1973 for illustrations, also Plate 11d), together with some liverworts (e.g. *Riccia* — see van Geel *et al.* 1983 and van Geel 1986 for illustrations, also Plate 11e) have been recorded fossil. These bryophytes have trilete spores with a range of reticulate and echinate sculpturing which can lead to confusion with the spores of *Lycopodium* and *Selaginella* species.

Algal spores and other remains

Some algal cells are often encountered in microfossil preparations, such as *Pediastrum*, *Botryococcus* and various other phytoplankton types, especially desmids (see Cronberg 1986). In preparations where HF has not been employed, diatoms are often present (Battarbee 1986).

Spores of algae are commonly found and their identification and interpretation have been greatly assisted by the work of van Geel (1972, 1986). Perhaps the most important group in this respect is the Zygnemataceae (a range of illustrations and proposals concerning ecological implications is given by van Geel 1976, 1979, 1986; Ellis & van Geel 1978; van Geel & van der Hammen 1978).

Fungal spores

Some fungal remains, i.e. the spores of *Tilletia sphagni* (van Geel 1972) and *Microthyrium* (Godwin & Andrew 1951) have long been recognized in peat samples. Van Geel and his co-workers have developed a catalogue of microfossil types, mainly of fungal origin, which they have published and illustrated in a series of papers (van Geel 1972; Pals *et al.* 1980; van Geel *et al.* 1981, 1983; Bakker & Smeerdijk 1982). A number of these microfossil types have now been identified, though many more remain simply with a type number. Even the latter, however, can often be interpreted in palaeoecological terms

as a result of their stratification in particular assemblages with known indicators of local conditions. Thus, such factors as wetness, pH and fire occurrence can be inferred from fossils even when their precise identity remains unknown.

Rhizopod tests

The shelled, or testate rhizopods are frequent components of peat deposits, are generally well preserved, have very distinctive structures (see for example the illustrations of Corbet (1973)), and can often provide information about the local mire environment, particularly its wetness. A useful introduction to the analysis, identification and interpretation of this group is given by Tolonen (1986b).

Phytoliths

These are mineralized microfossils and may be lost if the sample is treated with HF or HCl. They form inside plant cells (especially the epidermal cells) and often exhibit diagnostic shapes. The most common are plant opal phytoliths, formed from silica, well known from the epidermal cells of grasses. These are of particular value in the study of grassland communities or grass cultivation when grass pollen may not be specifically diagnostic, or in studies of soil where pollen is poorly preserved (Kurmann 1985; Pearsall 1989). Phytoliths also occur in a wide variety of other plants and plant organs. Phytoliths can be used, for example, to recognize former cultivation of arrowroot, chufa and squash in tropical sites (Piperno 1989). Further details of phytoliths may be found in Pearsall (1989) and Rovner (1986).

Cuticles

Many cuticle fragments can be found in palynological preparations, though the larger pieces may be removed by coarse sieving. Palmer (1976) has demonstrated the value of grass cuticles in reconstructing grassland communities (also in association with phytoliths — see above). In older rocks, dispersed cuticles have been used to aid interpretation of climatic changes at major boundary events (Wolfe & Upchurch 1987). Cuticles offer an additional source of evidence, potentially of value for recognizing taxa for which pollen are not diagnostic and for identifying local rather than long distance input. Further details may be found in Kerp (1991).

6

ILLUSTRATED POLLEN AND SPORE KEY
WITH GLOSSARY

USE OF THE KEY

To identify any pollen grain or spore it must first be classified according to its aperture type, number and position by the use of the key to pollen classes on p. 86. Once assigned to a class, the grains or spores within that class can be separated by reference to sculpturing type, details of aperture structure, etc. Chapter 5 discusses the range of sculpturing and aperture types which may be encountered, and a glossary on pp. 163–166 explains the terminology used.

The key is dichotomous, thus at every point of division there are alternatives marked (a) and (b). As far as possible only positive characters have been used, rather than the absence of certain features. Having keyed a grain out to a particular species or type, the relevant photograph will provide confirmation or indicate when a wrong turning has been taken in the key. The photographs should never be used alone for identification as they provide only one (or two) views of a three-dimensional object, and cannot always show the critical features mentioned in the key. This is true particularly where delicate 'LO' analysis (see p. 77) with the microscope is necessary to determine exine construction. In some cases, such as when a grain is in a bad condition (crumpled, obscured or eroded) or if it is a member of a difficult morphological group (e.g. Rosaceae, Compositae) then identification may prove impossible without type grains for comparison. Ideally all grains identified by means of the key and subsequently compared with the photographs should be checked against type

material, since no verbal description or photograph can provide a complete alternative to a type specimen.

When dealing with grains mounted in glycerine jelly a special problem occurs when a grain is presented in an awkward position. With zonocolporate grains an orientation such that one of the poles is uppermost will make the detection of the porus at the equator of each colpus difficult if not impossible. It is possible to change the orientation of individual grains on a slide by pressing gently on the coverslip above the grain with a hot needle. If this proves impossible, taking the grain through both the -colpate and -colporate classes may achieve identification, particularly if the grain has an especially distinctive sculpturing type and is confirmed by the photograph. The detection of a porus in the middle of a colpus may be difficult even if a grain is presented in a good equatorial view. Thickened, circular pori present no problems but unthickened irregular pori are easily missed, and it has been thought that these latter should not be regarded as true pori. In this key any slight constriction, bridge or rupture in the equatorial region of the colpus is taken to represent a porus. Thus, most of the Rosaceae are included in trizonocolporate while other authors, e.g. Faegri and Iversen (1964), feel that they should be placed in trizonocolpate.

Spores of some pteridophytes and bryophytes are placed here in the aperture classes which were originally devised to separate angiosperm pollen grains. For instance, pteridophyte spores with one groove are included under monocolpate instead of being placed in a class on their own (*monolete* in other texts such as Erdtman *et al.* 1963). Although a groove on a pteridophyte spore is not homologous with an angiosperm colpus, when one is faced with a fossil grain it is often difficult to distinguish the difference. It is hoped that this classification will make some identifications easier. For similar reasons *Ephedra* has been placed in hexazonocolpate and polyzonocolpate rather than in polyplicate, which is the term used by Faegri and Iversen (1989).

The use of size measurements has been avoided as much as possible because size varies greatly between grains of the same species, and between grains given different treatments in the extraction process. It should be emphasized that the key is constructed for grains which have been pre-treated with KOH and then acetolysed using an acetic anhydride and sulphuric acid mixture, followed by mounting in glycerine jelly. Acetolysis is known to swell grains to varying extents depending on the duration of the treatment (Reitsma 1970) and they may be up to 25% larger than grains which have been simply KOH treated. This may not be a problem unless a size-based alternative, like 'grain greater than 30 μm or less than 30 μm', occurs in the key. A purely KOH treated grain might fall below the measurement while an acetolysed grain would fall above it. This would place the grain treated simply with KOH in the wrong part of the key.

Grains mounted in glycerine jelly are always slightly larger than grains mounted in silicone oil, and workers using the latter medium should employ the size measurements in the key with extreme caution. Glycerine jelly mounted grains may be up to 1.5 times the size of silicone oil mounted grains. The conversion factor varies, however, with the species involved and it is not possible to give a simple figure here which would enable those using silicone oil to convert all the measurements in the key.

The key which follows here, whilst based on that in Moore and Webb (1978) has been completely rewritten in order to extend its coverage and improve its accuracy and efficiency. It still, of course, owes a debt to other previously published keys, e.g. Erdtman *et al.* (1963); McAndrews *et al.* (1973); Faegri and Iversen (1964; 1975; 1989). It includes many taxa of North America, northwestern Europe and some of the Mediterranean areas. We must emphasize, however, that it is not meant to have a complete coverage of these regions; only some of the more important and commonly found taxa are included (especially the ones that are now introductions to the British flora). The key certainly contains a great many more words than in its earlier version (Moore & Webb 1978), but this should contribute considerably to its precision. We feel that it should still prove easy for the inexperienced worker to use. Difficult or variable taxa are keyed out in several places and photographs are provided for confirming the identity of almost every taxon. However, even with the help of photographs, it is worth reiterating that all critical identifications should be confirmed by consulting type slides. Serious research in palynology demands an extensive type collection and workers must on no account rely on keys and photographs alone.

Abbreviations used in the key

agg.	Aggregate, including 2 or more species which resemble each other closely
=	Equal to
>	Greater than
<	Less than
LM	Light microscope
μm	Micrometres
P/E ratio	Ratio of polar axis length to equatorial axis length
SEM	Scanning electron microscope
ssp.	Subspecies
*	Taxon not native to northwestern Europe today. This applies mostly to North American species, but also possibly Japanese, Australian or

North African species which may be modern introductions

Nomenclature follows Clapham *et al.* (1987).

NOTES ON THE ILLUSTRATIONS

The number of illustrations in this edition is considerably expanded, trebling that in the earlier work (Moore & Webb 1978). We have increased the systematic coverage, now illustrating over 450 species including examples for over 95% of taxa keyed out. In most cases we include more than one view of the grain and we have used several views where appropriate.

A standard magnification of ×1000 has been used for most of the light micrographs. Exceptions occur in cases where grains are very large yet needed to be grouped on one plate of illustrations (e.g. Plate 5 of the saccate grains). In all except two of these cases (*Malva* and *Abies* at ×400), ×500 has been used, i.e. half the size of the standard examples. Magnifications are clearly indicated in the plate explanations.

The standard ×1000 magnification permits direct comparison between grains and also allows an easy concept of 'large' and 'small' grains which we hope will benefit the user. Furthermore, it is very easy to measure the accurate size of an illustrated feature such as grain size, width of reticulum, length of echinae, etc. as 1000 μm = 1 mm, so a measurement on the grain illustration in mm is the actual true size in microns (μm). For example, in the separation of *Limonium vulgare* type A and *Armeria maritima* type A grains (trizonocolpate reticulate key couplet 15) Plate 33 photograph a (*Limonium*) shows grain diameter 72 mm; exine thickness 9 mm and sculptural elements less than 1 mm long. The grains are therefore 72 μm in diameter with exine thickness of 9 μm and microechinae (less than 1 μm long). Comparable features of *Armeria* (Plate 33 photograph c) are 85 μm, 12 μm and with echinae (2 μm).

Illustrations using scanning electron microscopy (SEM) have also been increased. No attempt has been made at comprehensive coverage but grains have been selected to illustrate the variety of surface sculpture. We hope this will assist the user in interpreting the features described and used in the keys, and illustrated in the light micrographs.

Magnifications for the SEMs have also been standardized but necessarily with a greater range, i.e. ×1000, ×2000, ×4000 and ×8000. This should allow easy measurement and comparison. For example, on the Caryophyllaceae (Plate 63) the echinae seen in profile measure less than 2 mm in length on photographs at ×2000. They are therefore shorter than 1 μm and hence are microechinae. Grain sizes on SEMs may differ from those on LMs as the former grains are dry, and hence contracted compared with the latter, which are hydrated and expanded.

Master key to pollen and spore classes

(see Fig. 5.4, pp. 68–69 for illustrations
of these classes)

1a Grains united into groups of:
    two grains: DYADS (p. 87)
    four grains: TETRADS (p. 87)
    more than four grains: POLYADS (p. 90)

1b Grains single: 2

2a Grains without apertures (pori or colpi):   3

2b Grains with apertures (pori, colpi or a combination of the two):   5

3a With bladders (sacci) projecting from the body of the grain: SACCATE (p. 90)

3b Without such bladders: 4

4a With grain shape obscured by a coarse, high, network of crests or ridges separated by depressed areas: CRESTED (p. 91)

4b Without such coarse, high ridges separating lower exine areas: INAPERTURATE (p. 92)

5a With one three-branched, slit-like aperture in the shape of a Y: TRILETE (p. 97)

5b With more or less circular apertures (pori) or elongate apertures (colpi). An aperture can be a combination of colpus + porus: 6

6a Apertures pori only: 7

6b Apertures either colpi only, or a mixture of pori and colpi: 9

7a With one porus: MONOPORATE (p. 99)

7b With more than one porus: 8

8a Pori arranged in an equatorial zone with:
    two pori in zone: DIZONOPORATE (p. 101)
    three pori in zone: TRIZONOPORATE (p. 101)

    four pori in zone: TETRAZONOPORATE (p. 103)
    five pori in zone: PENTAZONOPORATE (p. 104)
    six pori in zone: HEXAZONOPORATE (p. 104)

8b Pori scattered all over the grain surface, usually equidistant from each other:
    four pori: TETRAPANTOPORATE (p. 105)
    five pori: PENTAPANTOPORATE (p. 105)
    six pori: HEXAPANTOPORATE (p. 105)
    more than six pori: POLYPANTOPORATE (p. 107)

9a With colpi only: 10

9b With either a porus and a colpus combined in each aperture or some apertures colpi, some colpi + pori (note that any bridge, constriction or rupture along a colpus is taken to indicate the presence of a porus):   13

10a With one free colpus: MONOCOLPATE (p. 113)

10b With more than one colpus. Colpi may or may not be fused: 11

11a Colpi fused, either straight and intersecting, or curving and joined into spirals or circles: SYNCOLPATE (p. 112)

11b Colpi with free ends: 12

12a Colpi arranged in an equatorial zone with:
    two colpi in zone: DIZONOCOLPATE (p. 118)
    three colpi in zone: TRIZONOCOLPATE (p. 118)
    four colpi in zone: TETRAZONOCOLPATE (p. 130)
    five colpi in zone: PENTAZONOCOLPATE (p. 130)
    six colpi in zone: HEXAZONOCOLPATE (p. 131)
    more than six colpi in zone:
    POLYZONOCOLPATE (p. 131)

12b With colpi scattered over the grain surface:
    four colpi: TETRAPANTOCOLPATE (p. 132)
    five colpi: PENTAPANTOCOLPATE (p. 132)
    six colpi: HEXAPANTOCOLPATE (p. 132)

more than six colpi: POLYPANTOCOLPATE (p. 132)

13a With some apertures colpi, some colpi com-
    bined with pori: HETEROCOLPATE (p. 161)

13b With all apertures colpi combined with pori: 14

14a Colpi + pori arranged in an equatorial zone:
    three colpi + pori: TRIZONOCOLPORATE (p. 133)
    four colpi + pori: TETRAZONOCOLPORATE (p. 158)

five colpi + pori: PENTAZONOCOLPORATE (p. 159)
six colpi + pori: HEXAZONOCOLPORATE (p. 160)
more than six colpi + pori:
    POLYZONOCOLPORATE (p. 160)

14b Colpi + pori scattered over the grain surface:
    four colpi + pori: TETRAPANTOCOLPORATE (p. 162)
    five colpi + pori: PENTAPANTOCOLPORATE (p. 161)
    six colpi + pori: HEXAPANTOCOLPORATE (p. 162)

The key

DYADS (Plate 1)

Grains inaperturate, finely reticulate with reticulum
continuous from grain to grain, not influenced by
the join between them. Wall separating the two cells
made of nexine only: *Scheuchzeria*
[NB many fungal spores occur in Dyads and these
may frequently be encountered on slides made from
peats. They generally are a darker brown or take up
stain differently to pollen grains — see van Geel,
1978]

TETRADS (Plates 1–4; 59)

Grains of some of the Juncaginaceae e.g. *Juncus*
and *Luzula*, occur in tetrads but they are very thin-
walled and tend to disintegrate during acetolysis of
type slides, so they are unlikely to survive as fossils
and are not included in the key below.

1a Tetrad with grains psilate, granulate, scabrate or
    verrucate: 2
1b Tetrad with grains reticulate or echinate: 3
2a Each grain trizonoporate, with large vestibulate
    pori. Viscin threads may be attached to the
    exine: *Epilobium* type
    [includes most species in this genus e.g.
    E. hirsutum, E. montanum, E. palustre and *E.
    parviflorum*. The closely related *Chamaenerion
    angustifolium* has single grains]
2b Each grain either trizonocolpate or trizonocol-
    porate: Ericaceae, Empetraceae and Pyrolaceae
    [see the special key on this page]
3a Grains echinate, clavate or baculate: 4
3b Grains reticulate: 6
4a With long, sparse echinae, >6 μm long. No
    folding where the grains touch each other:
        Selaginella selaginoides
4a With densely packed echinae, clavae or bacula;

or a mixture of these. Processes <5 μm long.
Exine where grains touch thrown into folds which
converge towards the innermost point of the
tetrad: 5
5a Processes all of similar length: *Drosera intermedia*
5b Processes of several lengths — the larger ones
    usually echinae, between which are smaller and
    more numerous echinae, baculae or clavae:
        Drosera rotundifolia type
    [includes *D. rotundifolia* and *D. anglica*]
6a Reticulum very coarse, muri always simplico-
    lumellate with thick columellae. Individual
    grains without distinct apertures, but the area of
    exine on each which has no reticulum may
    represent the porus: *Listera* type
    [includes *Listera, Neottia, Epipactis, Goodyera*
    and possibly others. See Andrew, 1984. The
    spores of the rare hepatic *Sphaerocarpos* exist
    as coarsely reticulate tetrads, but may be dis-
    tinguished from *Listera* type by their large size
    of 80–110 μm, combined with the absence of
    columellae and poral areas. See Macvicar, 1971]
6b Reticulum less coarse (some grains may be
    microreticulate). The reticulum may tend
    towards a more rugulate sculpture. Muri often
    duplicolumellate, especially at the corners of the
    reticulum. Individual grains with distinct pori
    which are roughly circular, with margins not
    sharply delimited: *Typha latifolia* type
    [includes *T. latifolia* and *T. minima*. See Punt,
    1975, for further descriptions and for distinction
    of *T. minima*]

SPECIAL KEY TO THE ERICACEAE, EMPETRACEAE AND PYROLACEAE

This key should be taken merely as a guide to what
to look for on your type slides. Several collections of
each species should be examined as there is variation

within some species as regards surface sculpturing, shape of tetrad, width of colpi and abundance of endocracks — see the comments in the studies by Oldfield (1959), Beug (1961), Foss (1988), Faegri and Iversen (1989). This key tries to encompass the variation, but it still may be the case that it does not work for all fossil tetrads. Perfect preservation and expansion of fossil tetrads is rare, and it is worth noting that tetrads that are slightly degraded nearly always seem to assume the surface sculpture of *Calluna vulgaris* and *Erica tetralix* as the outer layers of the exine are gradually eaten away. Incompletely expanded grains will always have narrow, slit-like colpi and a more globular or triangular-obtuse shape.

Note that *Erica terminalis* in the Ericaceae and *Orthilia secunda* in the Pyrolaceae have single grains and are thus not covered in the following key.

1a Tetrad with all grains in one plane, surface coarsely scabrate-verrucate-gemmate:
Calluna vulgaris
[occasional tetrads only. Some tetrads from *Pyrola* type may also key out here]

1b Tetrad with grains tetrahedrally arranged: 2

2a Colpi **short**, well defined or crack-like, diffuse, sinuous and may or may not be obscured by granules: 3
[the decision on colpus length must, to a certain degree, be subjective. No measures are given because 'long' or 'short' colpi are descriptions relative to the whole tetrad size. As a rough guide 'short' colpi are ones where the length of a colpus up the side of a grain from the grain boundary with an adjacent grain is = or < the distance between adjacent colpus ends at the pole]

2b Colpi **longer**, narrow and slit-like or widening towards the equator, but not obscured by granules: 10
[longer colpi are, as a rough guide, where the length of a colpus up the side of a grain from the boundary with an adjacent grain is >the distance between adjacent colpus ends at the pole]

3a Tetrad markedly lobed and grains so loosely held together that a small gap is visible in the centre between them. Colpi not coincident from grain to grain at the surface junctions. Each colpus commonly crossed by a long, narrow, transverse endocolpus: *Chimaphila umbellata*

3b Tetrad lobed, globular or triangular-obtuse, but never with a gap between the grains. Apertures always coincident from grain to grain: 4

4a Sculpture very coarsely scabrate-verrucate-gemmate. Tetrad lobed and often **irregularly shaped**. Colpi often unevenly spaced (sometimes 4 to a grain) with costae undetectable. Pori undetectable or a mere rupture:
Calluna vulgaris

4b Sculpture psilate, scabrate or verrucate but not so coarse and without gemmae. Tetrad lobed, globular (with flattened apocolpia) or triangular-obtuse (with pointed apocolpia); usually neatly symmetrical. Colpi evenly spaced, with or without costae. Pori may or may not be clear: 5

5a Colpi granulate, outline and positions therefore obscure: 6

5b Colpi not granulate, outlines clear: 7

6a Costae detectable along the colpi in a surface view. Sculpture varies from psilate to scabrate-verrucate: *Phyllodoce coerulea*
[some collections only. The grains in this species seem to be very variable in sculpture, clarity of colpi and size on type slides]

6b No (or insignificant) costae to the colpi: *Pyrola* type
[includes *Pyrola* and *Moneses*]

7a Tetrads large (>40 µm). Colpi with prominent costae outlined by a series of endocracks. Colpi wide or narrow: 8

7b Tetrads smaller (<40 µm). Costae may or may not be clear, may or may not be outlined by endocracks. Colpi usually narrow: 9

8a Tetrads very large (>50 µm, commonly much more) and thick walled, lobed to globular, colpi always clearly widening towards the equator. Sculpture psilate to scabrate, uniform over each grain surface. A network of endocracks usually visible under each apocolpium:
Rhododendron ponticum*
[Beug, 1961, quotes several other *Rhododendron* species as smaller than 50 µm; for example *R. lapponicum* goes down to 38.5 µm]

8b Tetrads usually <60 µm, globular to triangular-obtuse, mesocolpia sometimes concave. Apocolpia with coarser sculpture than the mesocolpia. Colpi usually narrow and slit-like. No (or few) endocracks under the apocolpia:
Andromeda polifolia
[*Empetrum nigrum* ssp. *hermaphroditum* may key out here due to its larger size of up to about 46 µm, but it does not have the characteristic endocracks around the costae and coarser sculpture at the poles. Some tetrads of *Erica cinerea* with rather short colpi may key out here, but here the apocolpia are flatter and the sculpture uniform over the tetrad surface or coarser in the mesocolpia]

9a Tetrad lobed, surface sculpture scabrate. Coarse endocracks abundant across the apocolpia and sometimes across the mesocolpia as well. Size usually <37 μm:

Loiseleuria procumbens

[some collections of this variable species only — see also 16b. *Erica lusitanica* has short enough colpi to key out here as do occasional tetrads of *Erica tetralix* type and *E. ciliaris* type. *E. tetralix* type may be distinguished from *Loiseleuria* by the colpi commonly widening towards the equator and the clear pori. See also note under 8a about small *Rhododendron* species]

9b Tetrad usually triangular-obtuse, never lobed. Inner walls in the centre of the tetrad quite strong, being darker staining and as thick, or thicker than, the outer walls of the tetrad. Surface sculpture psilate to granulate. No, or few endocracks in the apocolpia:

Empetrum nigrum ssp. *nigrum* type

[includes *Empetrum nigrum* ssp. *nigrum*, *E. nigrum* ssp. *hermaphroditum* and *Corema album*. *Corema* can have rather more abundant endocracks and coarser sculpture than *Empetrum*. *E. nigrum* ssp. *nigrum* and *Corema* have smaller tetrads than *E. nigrum* ssp. *hermaphroditum*, but there is considerable size overlap. *E. nigrum* ssp. *nigrum* is quoted by Beug, 1961 as <38 μm, and by Oldfield, 1959, as <34 μm; both authors state that *E. nigrum* ssp. *hermaphroditum* can reach up to 46 μm. *Ledum palustre* has a similar shape to this type and also has strong, thick inner walls, but its colpi are usually longer. Some of the smoother tetrads of *Phyllodoce* may key out here, but can be separated by their thinner inner walls]

10a Tetrad inner walls very thin, faint and appearing **abundantly perforate** or **foveolate**. Tetrad globular, colpi narrow and slit-like. Endocracks absent or very rare: *Arctostaphylos alpinus*

10b Tetrad inner walls may or may not be thin and faint, but with no perforations or only a few perforations. Tetrad lobed, globular or triangular-obtuse, colpi wide or narrow: 11

11a In polar view each individual grain of the tetrad appears triangular-obtuse or triangular-obtuse-concave, with the colpi between the angles. Tetrads commonly globular, always >37 μm: 12

11b In polar view, each individual grain of the tetrad appears more or less circular, never markedly triangular-obtuse. Tetrads lobed, globular or triangular-obtuse: 13

12a Inner walls of the tetrad with a few perforations, colpi usually long and narrow with indistinct or absent endopori. Endocracks mostly outlining the costae or a few present at the poles. Size usually <50 μm: *Arctostaphylos uva-ursi*

12b Inner walls of the tetrad thin but not perforate. Colpi narrow or widening towards the equator. If widening then each crossed by a large clear elliptic-acute endoporus which sometimes branches. Endocracks most prominent at the poles, where there is sometimes one, prominent Y-shaped one. Size >48 μm: *Arbutus unedo*

13a Tetrads globular, >45 μm and with apocolpia flat. Colpi sometimes narrow but often widening towards the equator. Costae rather clear and endocracks prominent. Porus represented by a narrow elliptic-acute endoporus or transverse endocolpus: *Erica cinerea*

[plus large *Vaccinium* tetrads. The latter are possibly distinct because of their insignificant costae]

13b Tetrads lobed to globular or triangular-obtuse: 14

14a Tetrads with colpi commonly widening towards the equator where a clear porus is seen: 15

14b Tetrads with long, narrow, slit-like colpi, pori not usually clear: 16

15a Costae to the colpi apparent in surface view (use phase contrast) usually not tapering towards the polar ends of each colpus. Sculpture coarsely scabrate-verrucate. Tetrad commonly lobed with flattened apocolpia: *Erica tetralix* type

[includes *E. tetralix*, *E. umbellata* and occasional grains of *Erica* species with wider colpi within *Erica ciliaris* type]

15b Costae to the colpi insignificant or undetectable in surface view (use phase contrast). If present then tapering towards the polar ends of each colpus. Sculpture psilate, scabrate or verrucate. Tetrad lobed to globular: *Vaccinium*

[*V. myrtillus* can show anything from an almost psilate to a scabrate-verrucate surface but it is quite thin-walled compared to its size and commonly lobed. The other *Vaccinium* species are more commonly globular. *V. vitis-idaea* seems to be scabrate or verrucate. *V. uliginosum* has apparently the largest grains — up to 53 μm according to Beug, 1961]

16a Costae to the colpi apparent in surface view (use phase contrast) not tapering towards the polar ends of each colpus. Sculpture scabrate to rather coarsely verrucate. Tetrads lobed, globular or triangular-obtuse: *Erica ciliaris* type

[includes *Erica ciliaris*, *E. mackaiana*, *E. vagans*, *E. erigena*, *E. australis*, *Chamaedaphne calyculata*, *Daboecia cantabrica*, *Cassiope hypnoides* and *Ledum palustre* and others. The longest colpi are seen in *E. ciliaris* and *E. vagans*. The smallest tetrads were observed in *E. vagans* and *Cassiope*, the largest in *E. australis*]

16b Costae to the colpi insignificant or undetectable in surface view (use phase contrast). If present then tapering towards the polar ends of each colpus. Sculpture psilate, scabrate or verrucate. Tetrad lobed to globular: *Vaccinium* [see comments under 15b. In addition some collections of *Loiseleuria procumbens* may key out here. Grains of this species tend to be clearly lobed and with endocracks across the apocolpia]

POLYADS (Plate 1)

1a Mass of grains not any particular shape. Individual grains of the group not regularly arranged, being so pressed together that the outline of individual cells becomes irregularly angular (see Fig. 5.4):
Orchidaceae (excluding those species that have tetrads. See note after 1b below)

1b Mass of grains with a definite shape of a convex disc. Individual grains regularly arranged within the disc and if pressed together, then quadrangular in outline (see Fig. 5.4): Mimosaceae [occasionally, grains of many species are seen still united in tetrads or larger masses on type slides, but they appear so as fossils only if the parent plants may have been part of the sediment-producing vegetation, and whole anthers have become encorporated. Some examples of taxa where this has been seen are: *Potamogeton*, *Myriophyllum*, *Salix*, Cruciferae, *Filipendula*. Fungal spores occur in what appear to be polyads. They are usually a dark brown colour and without columellate wall structure. See van Geel, 1978, for photographs and descriptions]

SACCATE (Plates 5, 59)

1a Grain with two separate sacci, more or less equal in size and usually with a coarse or fine, irregular reticulum on their internal surfaces: 2

1b Grain with a continuous or interrupted ring-shaped rudimentary saccus like a 'frill' around it. Body of grain with irregular hollow verrucae or papillae: *Tsuga** [most species]

2a Sacci not constricted at their point of attachment to the body of the grain, being roughly hemispherical (widest at the join): 3

2b Sacci constricted at their point of attachment to the body of the grain, so that they 'balloon out' (widest above the join): 5

3a Distal portion of the grain body between sacci finely rugulate. Proximal body wall away from the sacci with markedly undulating outermost layer, giving uneven exine thickness: *Cedrus*

3b Distal portion of the grain body between the sacci psilate, verrucate or microverrucate. Proximal body wall away from the sacci not, or only slightly, undulating: 4

4a Conspicuous verrucae on the inner side of the distal body wall between the sacci (phase contrast helps): *Pinus* subgenus *Strobus* = *Haploxylon* [see notes under 6b]

4b Distal body wall between the sacci psilate, granulate or microverrucate (phase contrast): *Picea* [see notes under 6b; also Birks, 1978, on morphological variation within *Picea abies*]

5a Length of grain body (see Fig. 6.1 for measurement position) in size range approximately 70–100 μm. Proximal body wall varying gradually in thickness; being thinnest next to the sacci and in the centre opposite the gap between the sacci. Never any marginal crests: *Abies*

5b Length of grain body (see Fig. 6.1) in size range approximately 40–60 μm. Proximal body wall may show a slightly undulating outer layer, sometimes developed into a frill-like marginal crest under the sacci: 6

6a Conspicuous verrucae on the inner side of the distal body wall between the sacci (phase contrast helps to detect these):
Pinus subgenus *Strobus* = *Haploxylon*

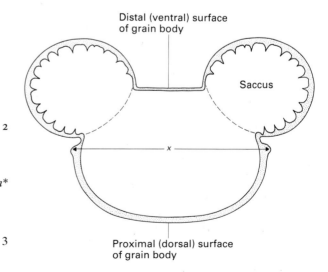
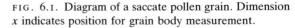

Distal (ventral) surface of grain body

Saccus

Proximal (dorsal) surface of grain body

FIG. 6.1. Diagram of a saccate pollen grain. Dimension *x* indicates position for grain body measurement.

[includes such species as *P. strobus**, *P. cembra*, *P. peuce*, *P. aristata** and others. Note that microverrucae may occur between the sacci in *Picea*. See references under 6b]

6b No conspicuous verrucae on the inner side of the distal body wall, this area is either completely psilate or faintly granulate:

Pinus subgenus *Pinus* = *Diploxylon*

[includes most species in this subgenus e.g. *P. sylvestris*, *P. nigra*, *P. rigida**, *P. pinaster*, *P. ponderosa**, *P. mugo*, *P. resinosa** and others. A few members of this subgenus possess verrucae — see Klaus, 1978. Readers interested in identifications within the Pinaceae should see Ueno, 1958; Ting, 1965, 1966; Hansen and Cushing, 1973]

CRESTED (Plates 6, 61)

Although not strictly an aperture system, it is useful to group grains under the heading 'Crested' here because the most obvious features are folds, crests or high echinate ridges of thick or folded exine; which divide the grain into large depressed areas. The true aperture system of these grains is more or less obscured and has been detected to be either trizonoporate or trizonocolporate. This section replaces the one headed 'Fenestrate' in the first edition of the present work. Blackmore and Heath (1984) point out that 'Fenestrate' (see Faegri & Iversen, 1989) is not strictly an accurate description of the sculpturing in the Lactuceae tribe of the Compositae.

1a With 3 non-echinate ridges or wrinkled folds which run meridionally and fuse to a high triradiate ridge at each pole: *Trapa natans*
[This grain is actually trizonocolporate. The colpi are situated under the ridges and the pori under the thinner equatorial part of each ridge, so the aperture system is obscured. It is, however, so distinctive that this hardly matters]

1b With more than 3 echinate ridges. Together these form a coarse network over the grain surface separating sunken areas or lacunae which are arranged in a geometrical pattern (i.e. the grain is echinolophate). Grain triangular-obtuse to hexangular (or rarely quadrangular-obtuse to octangular) in polar view. Columellae usually detectable under the ridges: most members of the Lactuceae tribe of the Compositae except *Scorzonera humilis* — see Blackmore and Heath (1984). This taxon is roughly equivalent to the 'Compositae, Liguliflorae' mentioned in the first edition of the present work.

NOTES ON DISTINCTION WITHIN THE LACTUCEAE

According to Blackmore and Heath the Lactuceae which key out here can be separated into 6 pollen morphological types. Four of these types might be expected to be identified frequently from sediments as they contain the commonest genera in plant communities today. A brief outline of the distinguishing features of these types is given below (although to confirm identification, readers should refer to the original paper and its excellent photographs. See this paper also for the *Arnoseris minima* type and the *Scorzonera laciniata* type). The terminology used is explained in Fig. 6.2. Figure 6.3 depicts the four types. It may be necessary to be able to roll a grain over to permit identification.

1 *Cichorium intybus* type — *Taraxacum* type of Wodehouse (1935)

— equatorial ridges present and narrow, with a single central row of echinae
— polar lacunae absent
— polar areas variable in size but always with some central echinae in addition to ones bordering the lacunae
— total of 15 lacunae on grain surface
Blackmore and Heath include:
all *Hieracium* species examined,
all *Hypochoeris* species examined,
all *Crepis* species examined,
all *Leontodon* species examined,
all *Picris* species examined,
all *Taraxacum* species examined,
3 out of the 4 *Lactuca* species examined,
both *Cicerbita* species examined,
Cichorium intybus,
Andryala integrifolia.
Further subdivision of this type is apparently possible. It is also noted that *Hieracium* and *Taraxacum* produce a proportion of variant (e.g. tetrazonocolporate) grains associated with their apomixy. It is expected that by far the majority of fossil grains will be referable to group A.

2 *Sonchus oleraceus* type

— 3 small lacunae at each pole (polar lacunae) or 4 in tetrazonocolporate grains
— equatorial ridges present, with one or two rows of echinae on each
— total of 21 lacunae over the grain surface
Blackmore and Heath include:
all *Sonchus* species examined,
Aethorhiza bulbosa.

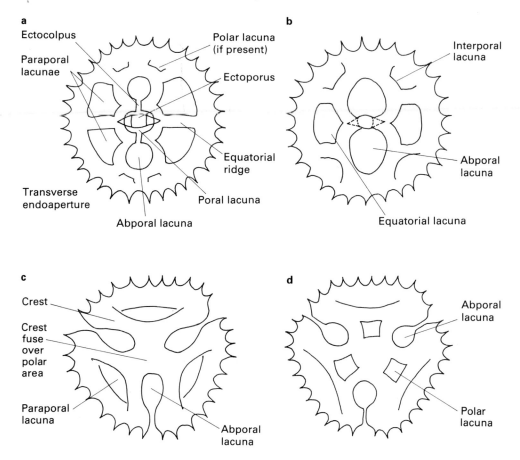

FIG. 6.2. Generalized equatorial views of Lactuceae pollen grains (a and b equatorial views with one colpus facing observer; c and d polar views): (a) with equatorial ridges; (b) without equatorial ridges; (c) without polar lacunae; (d) with polar lacunae.

3 *Lactuca sativa* type

— no polar lacunae
— polar area reduced to a narrow triradiate ridge, usually with one row of echinae on it, sometimes two rows
— equatorial ridges present
— total of 15 lacunae over the grain surface
Blackmore and Heath include:
Lactuca sativa,
L. serriola,
Mycelis muralis.

4 *Tragopogon pratensis* type

— equatorial ridges absent, a single equatorial lacuna in each mesocolpium
— poral lacunae absent
— polar area large and hexagonal, with numerous echinae
— total of 15 lacunae over the grain surface
Blackmore and Heath include:
all *Tragopogon* species examined.

INAPERTURATE (Plates 7–9, 59, 60, 69)

1a With echinae; or psilate; or possessing two coats. If with two coats, the outer one variously wrinkled or folded and more or less separate from the inner: 2
1b Without echinae or wrinkled outer coat. Grains may be granulate, gemmate, clavate or reticulate but not psilate: 13
2a Echinate: 3
2b Psilate or possessing two coats: 7

Polar view   **Equatorial view**

a

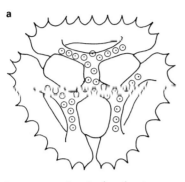

b

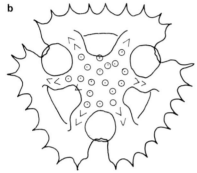

c

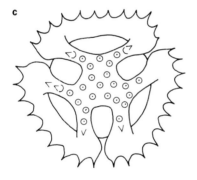

d

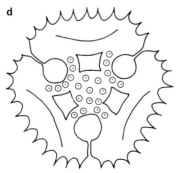

FIG. 6.3. Four types of Lactuceae pollen grains: (a) *Lactuca sativa* type; (b) *Tragopogon pratensis* type; (c) *Cichorium intybus* type; (d) *Sonchus oleraceus* type.

3a With largest echinae 5−8 μm long, cylindrical-conical and solid, often with one or two constrictions along their length. No columellae visible in the area between the echinae. The bases of the echinae project downwards to produce verrucae on the inner surface of the nexine. Close examination may reveal the presence of a single, wide, operculate colpus that is sometimes difficult to see as its operculum bears echinae that are nearly as large as those on the rest of the grain surface. Also the grains may often be partially collapsed so that the colpus is deeply inrolled and the junction of the edge with the operculum mistaken for a fold. Longest dimension of grain >45 μm: *Nuphar*
[see Jones and Clark, 1981]

3b Longest echinae <5 μm long, columellae may or may not be visible between the echinae (×1000):   4

4a Echinae small and situated on the peaks of tectum undulations that give the grain the appearance of being coarsely, irregularly verrucate. Columellae dimorphic: very coarse, long ones support the peaks of the undulations; finer, shorter ones occupy the trough areas (see exine optical section ×1000). Grain large, >50 μm:
                             Ranunculus arvensis
[some species of *Tsuga** may key out here if the echinae on top of the papillae are prominent; however no columellae of any sort are visible in this genus]

4b Echinae not on the tops of verrucae, no such dimorphic columellae: 5

5a Grain >35 μm long, elliptic and usually with a single split. Echinae mostly cylindrical-conical (sharpened pencil-shaped) occasionally conical or cylindrical-obtuse (like bacula). Columellae visible in the area between the echinae. Possibly a faint suggestion of lines joining the bases of the echinae: *Stratiotes aloides*
[other rare grains which might possibly key out here are *Elodea*, *Crocus* and *Laurus nobilis*]

5b Grain <35 μm long, more or less circular. Columellae not or only faintly visible in the area between echinae (×1000, phase contrast):   6

6a Echinae sharply conical. No suggestion of columellae between the echinae. Grain never cracked or irregularly ruptured although it often crumples. Careful examination should reveal one faint, unthickened porus: *Lemna*
[see note under 6b]

6b Echinae cylindrical-conical (pencil-shaped) conical or cylindrical-obtuse. A very faint suggestion of columellae between the echinae (×1000, phase contrast). Echinae sometimes appear to be arranged in a coarse irregular reticulate pattern. Exine often cracked or irregularly ruptured: *Hydrocharis morsus-ranae*
[other rare grains which may key out here are *Zannichellia palustris* and *Asarum europeaum*. Careful consultation of type slides should always accompany identification of *Hydrocharis* as there may be many bryophyte, fungus or alga spores that agree with the above general description, but are distinct to the eye. For example Dickson, 1973 illustrates spores of the moss *Physcomitrium pyriforme* (a likely subfossil on ecological grounds) which has a surface approximating to the description of *Hydrocharis* given above]

7a Grain psilate, lacking a wrinkled additional outer coat: 8

7b Grain with a variously wrinkled additional outer coat: 10

8a Grains usually circular but very often crumpled or split in the manner of *Juniperus*, >50 μm and commonly over 100 μm in diameter:
                                        Larix type
[includes *Larix* and *Pseudotsuga**. Type slides should always be consulted. *Equisetum* spores that have lost their outer coat may key out here, but they measure <50 μm. Many other objects like this can be found on slides made from sediments. They may be of fungal, animal or algal origin. They are generally a darker brown colour or stain differently to pollen grains, see van Geel, 1978]

8b Grains bean-shaped, D-shaped or elliptic; usually <40 μm long: 9

9a Grains <25 μm long, exine tectate with thick tectum and very fine columellae (see wall section ×1000, and phase contrast of surface). Small, rather sparse irregular perforations may be visible on the tectum: *Acorus calamus*
[there is nothing much to distinguish this grain, comparison with type slides is essential]

9b Grains usually >25 μm long, completely psilate, without any discernible wall structure (no columellae, tectum or perforations). Careful analysis may reveal a single furrow (monolete tetrad scar):   Polypodiaceae
[spores minus the sculptured perine which has been lost during fossilization or during acetolysis]

10a Grain approximately circular, as regards the inner coat: 11

10b Grain bean-shaped, particularly as regards the inner coat: 12

11a Grain covered with irregular, hollow verrucae or rugulae or papillae produced by foldings of the outer coat. Projections often sinuous in surface view, with or without small echinae on

the tops. One side of the grain concave, making it rather saucer-shaped. The concave exine has finer verrucae, gemmae or granules and the rim of the concavity is with or without a frill of longer, larger, papillae. Grain >50 μm in diameter and in some species over 90 μm: *Tsuga**
[some distinction between the species can be achieved by comparison with type slides. The thickness of the outer coat, the shape and prominence of the verrucae and the presence and shape of the echinae are all important. *T. canadensis* is without the echinae but there are several species that have them e.g. *T. diversifolia*. See Erdtman 1943, 1957]

11b Grain with outer coat extremely thin-walled and faint, being completely loose around the inner coat, never thrown into verrucae, but variously wrinkled and folded: *Equisetum*

12a Outer coat with few folds or wrinkles, separated by a gap from the inner coat and following the contour of the inner wall except where it projects as a ridge, keel or torn area from the straight side of the grain. Careful analysis should reveal the presence of a monolete scar: *Isoetes*
[see Birks, 1973 for differentiation of *I. lacustris* from *I. echinospora* on size criteria]

12b Outer coat distinctly folded, wrinkled or ridged. Careful analysis should reveal the presence of a monolete scar: Polypodiaceae
[see 'Monocolpate' on p. 113 to identify types further]

13a Grain reticulate, or at least with sculpturing elements arranged in a reticulate pattern: 14

13b Grain without reticulate pattern. Sculpturing elements either arranged randomly or in a dense, more or less uniform, carpet. Sculpturing elements may be gemmae, bacula or clavae; or there may be columellae under a tectum: 20

14a Grain very small (approximately 14 μm in diameter) reticulum with very large meshes for the size of grain. Hexangular lumina, no columellate structure under the muri:
    Tilletia (*Bryophytomycetes*) *sphagni*
[this is the spore of a fungus. According to Dickson, 1973; in the living state the spores are found inside sporophyte capsules of *Sphagnum* replacing the moss spores. It is often common on pollen slides made from *Sphagnum* peat and its abundance parallels that of *Sphagnum* spores]

14b Grains usually larger and with reticulum meshes smaller in comparison to the size of the grain, always with some indication of columellate structure under the muri: 15

15a Tops of columellae joined to form clear muri over at least some of the grain (use LO analysis or phase contrast — a black network appears in phase):   16

15b No clear muri in LO analysis or phase contrast. Tops of columellae appear free (i.e. they are bacula which are arranged in a reticulate pattern) or possibly they are connected by very faint, broken muri: 19

16a Grain large, cylindrical and curved like a boomerang (although the shape may be more or less obscured by the delicate exine being crumpled). Columellae very widely spaced at the corners of a fine reticulum which becomes even fainter and shallower over three areas representing pori (one at each end and one in the middle of the grain). Long axis over 70 μm:
    Ruppia spiralis

16b Grain circular or elliptic, <40 μm in diameter: 17

17a Exine thick, >3 μm, mostly made up of sexine. Long, coarse columellae united to thick muri which bear small echinae (×1000). Careful analysis may reveal the presence of small indistinct pori which are only slightly larger than the lumina: *Daphne*
[*D. laureola* has a much bigger grain with thicker exine than *D. mezereum*. Distinction is possible by reference to type slides. *Matthiola sinuata* and *M. incana* are reported by Andrew, 1984, to be inaperturate, and they may thus key out here]

17b Exine thin, <2 μm. Reticulum less coarse and columellae finer. Never any echinae on muri: 18

18a Reticulum broken and incomplete, muri wider than lumina, parts of grain granulate: *Populus*
[e.g. *P. alba*, although not all grains of this species show a reticulum]

18b Reticulum complete, no free granules visible. Sculpture a very neat pattern of angular, mostly regular sized, lumina:
    Potamogeton subgenus *Potamogeton* type
[includes the members of this subgenus which includes e.g. *Potamogeton natans*. It also includes *Groenlandia densa* and the genus *Triglochin* in the Juncaginaceae. Compare also *Zannichellia palustris* which may key out here]

19a Grain with bacula or slim clavae sitting in a reticulate pattern, exine >1 μm thick but with nexine very thin and delicate. Grain usually broken or ruptured: *Callitriche*

19b Exine thinner, approximately 1 μm thick, with bacula. Exine not usually broken:
    Potamogeton subgenus *Coleogeton*
[*P. filiformis* and *P. pectinatus*]

20a Most of surface tectate-perforate (×1000 phase contrast helps) but with circular or elliptic areas where the surface is semitectate, verrucate

(with isolated islands of sexine). Grains may be pear-shaped or isodiametric, often collapsed or crumpled: Cyperaceae
[further identification may be possible if sufficient type slides are available. The most distinctive grains are the thick-walled, almost spherical, ones of *Rhynchospora alba*; and the large grains of *Cladium mariscus* in which one end of the grain is narrowed to a finger-like projection. See Faegri and Iversen, 1989]

20b Grain apparently intectate; no such lacunae. Shape circular or more or less isodiametric: 21

21a With at least some sculptural elements long enough to be defined as clavae or bacula: 22

21b Sculptural elements gemmae, microgemmae or minute granules or a mixture of all three; but none long enough to be clavae or bacula: 23

22a Grain small, <35 µm, with elements all clavae or bacula and all of the same height. Nexine very thin and delicate, grain often broken or ruptured: *Callitriche*
[note that some bryophyte spores approximately agree with this description — see SEMs of *Polytrichum* spores e.g. *P. formosum*, in Dickson, 1973; and the drawings of many species in Erdtman, 1957. In some bryophytes the exine processes are more or less lost during fossilization or acetolysis]

22b Grain may be larger or smaller than 35 µm, sculptural elements varying in thickness and height, i.e. mixed large and small clavae, bacula, gemmae and verrucae. Elements crowded or scattered but with a large area of much finer granulate-gemmate sculpture on one side (careful examination may reveal that this is actually the operculum to a very large, single, porus). An optical section of the grain wall may reveal large irregular verrucae or masses on the *inner* surface of the nexine: *Nymphaea alba* type
[includes *N. alba* and *N. candida*. Distinction may be made between these species by consulting type slides. The latter is larger, with exine processes finer, more uniform and crowded. Bacula are reported to be more rarely present by Jones and Clarke, 1981]

23a Grains large, >50 µm: 24
23b Grains <50 µm: 25

24a Surface with very small granules, irregularly spaced. Grains commonly split and crumpled in the manner of *Juniperus*. Often over 100 µm in diameter: *Larix* type
[includes *Larix* and *Pseudotsuga**]

24b Surface with elements resembling gemmae or microgemmae, regularly spaced in a more or

less even carpet. Size not over 80 µm:
                                    Meesia triquetra
[data from Dickson, 1973; van Geel *et al.*, 1989]

25a Elements all microgemmae, appearing singly or in very irregularly scattered groups (×1000, use phase contrast if possible). The microgemmae may be more or less absent, some grains thus well covered, others almost bare. Grain often split and then folded in a characteristic manner: 26

25b Elements gemmae, microgemmae or minute granules; present in a more or less even carpet, with no bare areas: 27

26a Grain with a single prominent papilla which is sometimes slightly curved or hooked:
                                    Taxodiaceae
[*Taxodium**,   *Sequoia**,   *Sequoiadendron**, *Cryptomeria**, *Cunninghamia**, *Metasequoia**. Not all grains will key out here, because in many the papilla is difficult to detect or absent]

26b No such papilla present: *Juniperus* type
[includes all members of the Cupressaceae i.e. *Juniperus*, *Cupressus*, *Chamaecyparis**, *Thuja** and others; also split members of the Taxodiaceae in which the papilla is obscured or absent]

27a Grains approximately circular with dense, but rather irregularly disposed granules or microgemmae (see SEM) which are only clear in phase contrast. Granules varying in width, height and arrangement. In phase contrast, small perforations may be visible between the granules: *Populus*
[consultation with type slides is always necessary. The surface of *Populus* is distinctive to the eye, but difficult to describe. This key exit covers very thin-walled, fine-sculptured grains of species such as *P. tremula* and the larger, more robust grains of *P. nigra* and *P. canescens* which have coarser sculpture. See also note under 27b]

27b Grain obtusely angular (sometimes rectangular-obtuse or triangular-obtuse). Sculptural elements larger, more like gemmae and noticeably varying in size. Grain frequently splitting, and then resembling *Juniperus* type, except for the fact that it does not have such sparse elements: *Taxus*
[consult a type slide and compare under phase contrast. Some *Orobanche* species and some bryophyte spores conform to the descriptions given for *Populus* and *Taxus*. The bryophyte spores are usually smaller and distinct to the

eye e.g. *Polytrichum, Funaria*. See also Vitt and Hamilton, 1974, who describe gemmate or papillate spores in *Encalypta* species. It would be worth also consulting the drawings of moss spores in Erdtman, 1957]

TRILETE SPORES (Plates 10–12, 50)

Simethis planifolia is reported by Andrew (1984) to have a three-branched colpus which resembles the three-slit aperture of some spores

1a Spores reticulate, foveolate, fossulate or with irregular creases: 2

1b Spores psilate, granulate, echinate, micro-gemmate, gemmate, baculate, verrucate or verrucate-rugulate: 11

2a Foveolae or fossulae present, at least on the proximal surface: 3

2b Foveolae or fossulae absent, spores reticulate or with irregular creases: 5

3a With echinae between the foveolae:
Hymenophyllum

3b No echinae between the foveolae: 4

4a Spore triangular with the angles truncated. Foveolae conspicuous over proximal and distal surfaces. Trilete scar with long 'arms', reaching almost to spore outline in face view:
Huperzia selago

4b Spore circular, foveolae may be clear only on proximal surface. Distal surface may have foveolae or a mixture of foveolae and irregular fissures or channels. Trilete scar with short 'arms' (<or = $\frac{2}{3}$ of the spore radius):
Ophioglossum vulgatum type A
[includes some specimens of *O. vulgatum* and *O. lusitanicum*. The former seems to be rather variable. See Knox, 1951, and Moe, 1974. Some *Botrychium* species e.g. *B. multifidum* may also key out here but may be distinguished by their more triangular-obtuse shape. See Moe, 1974]

5a With large, irregular, zig-zag creases on the distal surface and small rounded verrucae or gemmae on the proximal surface. Spores >35 µm in diameter: *Lycopodiella inundata*

5b With reticulum on the distal surface and proximal surface with variable structure: 6

6a With large, bifurcated echinae situated over the intersections of the muri. Proximal surface often foveolate: *Anthoceros punctatus* type
[includes *A. punctatus* and *A. agrestis*. See Macvicar, 1971, Dickson, 1973 and Watson, 1981]

6b No such large echinae on the muri: 7

7a Spores very large (>50 µm and maybe up to 100 µm in diameter) usually with very coarse reticulum on distal face (commonly <12 lumina across the distal face): *Riccia* type
[includes *Riccia* and *Ricciocarpus*. See descriptions and drawings in Macvicar, 1971. There is quite a lot of variation in the spore morphology of these genera. It would appear also that spores of *Targionia, Reboulia* and *Preissia* fall into this size class. Their sculpturing is distinct — see Macvicar, 1971]

7b Spores <50 µm in diameter: 8
[Lycopodiaceae, Ophioglossaceae; and possibly also some spores of Hepaticae such as *Pallavicinia, Moerkia, Petallophyllum, Fossombronia* and *Cryptothallus*. For the Hepaticae see Macvicar, 1971 and for the Pteridophyta see Moe, 1974. Dickson, 1973, illustrates spores of the moss *Oedipodium griffithianum* which might key out here. It is possible to distinguish all the above mentioned bryophytes by their spores, but there is as yet no evidence that any are likely to be found as fossils]

8a Muri varying greatly in height and thickness, in cross section often wide at the base and narrowing towards the top. Proximal surface foveolate, trilete scar with short 'arms' (arms <or = $\frac{2}{3}$ of a radius) spore always circular:
Ophioglossum vulgatum type B
[some forms of this species. See note under 4b above]

8b Muri all the same height and approximately the same thickness, narrowly rectangular in cross section. Proximal surface reticulate, psilate, or a combination of these. Spores often triangular-obtuse: 9

9a Most of proximal surface psilate, reticulum visible only at the edges of the face:
Lycopodium annotinum type
[includes *L. annotinum* and, according to McAndrews *et al.*, 1973, *L. obscurum**. The latter may apparently be distinguished by its smaller trilete scar and lumina. Jones and Blackmore, 1988, include *L. dubium* in this type]

9b Reticulum extends well onto the proximal surface (it is over $\frac{1}{2}$ reticulate): 10

10a Proximal face almost completely reticulate. Reticulum with thickened corners at the intersections of the muri: *Lycopodium clavatum*
[most grains. See note below 10b]

10b Proximal face less reticulate — a small psilate area appears between the 'arms' of the trilete scar. Reticulum without such thickened corners: *Diphasiastrum* type [includes *D. alpinum* and *D. complanatum*. According to Jones and Blackmore, 1988, *D. issleri* and *D. tristachyum* belong to this type. These authors also state that some grains of *L. clavatum* are not distinguishable from *Diphasiastrum* type]

11a Spores psilate, granulate or microgemmate: 12

11b Spores echinate, gemmate, verrucate or verrucate-rugulate: 19

12a Grains psilate or almost totally so: 13

12b Grains granulate to microgemmate: 16

13a Spore wall relatively thin compared to grain size, (approximately 2 μm thick) and of uniform thickness all over spore surface. Spore shape triangular-obtuse: 14

13b Spore wall thicker, >2 μm; may or may not be uniform in thickness (maybe thickest at corners of triangular shape seen in face view). Shape circular or triangular-obtuse: 15

14a Spore shape usually triangular-obtuse with concave sides, trilete scar with relatively short 'arms'. Spore size averages below 40 μm:
Pteridium aquilinum
[spores minus their sculptured outer coat, the most likely condition for a fossil *Pteridium*]

14b Spore shape triangular-obtuse with convex sides, trilete scar conspicuously ridged and with longer 'arms' that reach almost to spore outline in proximal view. Spore size averages well over 40 μm: *Adiantum capillus-veneris*
[spores minus their sculptured outer coat. Always consult a type slide, as the vast majority of spores that reach couplet 14 are going to turn out to be *Pteridium*. Spores of *Salvinia* may be more or less similar to *Adiantum* and *Pteridium* but significantly smaller — see Andrew, 1984]

15a Spore shape usually triangular-obtuse with convex sides. Exine thickest at the corners of the triangular shape. Actual trilete aperture with rather short 'arms', although the proximal face can be strongly angled and thus ridges may extend to the spore outline: *Sphagnum*
[spores minus their outermost sculptured coat. The species vary in size, 25−40 μm encompassing most of them]

15b Spores usually circular in shape with a thick wall that stains darkly. 'Arms' of trilete scar long, prominent and raised. Size usually >40 μm: *Pilularia* microspores
[spores minus their outer coat. *Azolla* microspores are circular and with thick, darkly staining walls, but they are much smaller, averaging <20 μm — Andrew, 1984]

16a Spore shape usually triangular-obtuse with concave sides. 'Arms' of trilete scar reach approximately halfway to spore outline in proximal view. Sculpture varies from densely granulate to sparsely granulate depending on the degree of loss of the outer coat:
Pteridium aquilinum

16b Spore shape triangular-obtuse with convex sides. Trilete scar 'arms' long or short: 17

17a Exine thickest on the angles of the triangular shape when seen in face view. Actual trilete aperture with rather short 'arms', <or = to halfway to the spore outline. However, the proximal face can be strongly angled and thus ridges may extend to the spore outline. Sculpture varies from densely granulate-microgemmate to sparsely so, depending on the degree of loss of the outer coat: *Sphagnum*
[Tallis, 1962, illustrates the variation between the spores of different *Sphagnum* species. Size variations 25−40 μm. It is felt to be unlikely that the sculptural differences survive the effects of fossilization and acetolysis)

17b Exine of similar thickness all over the spore. Trilete scar 'arms' long, reaching nearly to the spore outline: 18

18a Close examination (×1000) shows most exine processes sharply pointed. Exine uniformly thin all over. Average size 32 μm (Knox, 1951; Andrew, 1984). Consult a type slide:
Trichomanes speciosum

18b No sharply pointed processes on close examination; spore closely covered in small, irregular granules. Average size >40 μm (Knox, 1951; Andrew, 1984). Consult a type slide:
Adiantum capillus-veneris

19a Spores echinate: 20

19b Spores gemmate, verrucate or verrucate-rugulate: 24

20a At least some of the echinae on the distal surface forked and bent. 'Arms' of trilete scar long and appearing as prominent ridges: 21

20b Echinae not forked. Proximal surface psilate or with same sculpture as distal surface: 22

21a Echinae very coarse, large foveolae on proximal surface. Bases of echinae often joined by a coarse reticulum: *Anthoceros punctatus* type
[includes *A. punctatus* and *A. agrestis*. See Macvicar, 1971, Dickson, 1973 and Watson, 1981]

21b Echinae smaller, no foveolae but gemmae and granules on proximal surface. Bases of echinae not joined by reticulum: *Phaeoceros laevis*
[see Macvicar, 1971 and Watson, 1981]

22a Echinae long (>5 μm on distal face) and sparse. No echinae on proximal face:
Selaginella selaginoides

sculpture similar on distal and proximal face: 23

23a Echinae small (up to 2 μm long). Spore <40 μm:
Trichomanes speciosum

23b Echinae larger (2 μm or longer) foveolae and
perforations may be visible between their bases.
Spores commonly larger, >40 μm:
Hymenophyllum

24a Sculpture papillate, or irregularly gemmate-
baculate. Spore usually circular in face view
and >50 μm in diameter: 25

24b Sculpture verrucate or verrucate-rugulate.
Spore triangular-obtuse to circular, less or more
than 50 μm: 26

25a With thick (>5 μm) outer coat thrown into
closely packed papillae or folds which are the
same on both faces. Inner coat also thick and
densely staining. Trilete scar 'arms' reach <$\frac{2}{3}$ of
the distance to the spore outline:
Pilularia globulifera microspores

25b With thinner outer coat, sculpture appearing
to be an irregular mixture of solid verrucae,
gemmae and bacula with elements sometimes joining sideways to form a rugulate or
reticulum-like pattern. Trilete scar 'arms'
usually reach >$\frac{2}{3}$ of the spore radius:
Osmunda regalis
[this description may also cover certain moss
spores e.g. *Oedipodium*, see Dickson, 1973]

26a Proximal surface has verrucae markedly smaller
than those on the distal surface. Spore shape
usually triangular-obtuse with **concave** sides,
proximal surface often sunken-in towards the
grain centre: *Cryptogramma crispa*

26b Verrucae usually of similar size on proximal
and distal surfaces or proximal surface with
large ridges. Spore shape triangular-obtuse with
straight or **convex** sides: 27

27a Proximal surface with two large smooth ridges
joining the ends of trilete scar 'arms'. Long
narrow furrows between the ridges. Trilete scar
raised and coarsely papillate:
Anogramma leptophylla
[data from Knox, 1951]

27b No large smooth ridges, instead, proximal surface has similar sized verrucae to those on the
distal surface: 28

28a Spore wall thickest at the corners of the triangular shape when seen in face view, verrucae
low and indistinct, not elongated. Actual trilete
aperture with rather short 'arms', <or = to
half the spore radius. However, the proximal
face can be strongly angled, and thus ridges
may extend to the spore outline. Size 25–40 μm:
Sphagnum
[see note after 17a above]

28b Spore wall of approximately the same thickness

all over. Verrucae usually prominent, sometimes slightly elongated to form irregular
rugulae. Trilete scar 'arms' reach over halfway
to spore outline. Spore diameter commonly
>40 μm: *Botrychium lunaria* type
[according to Moe, 1974, this includes *B. lunaria*, *B. lanceolatum*, *B. matricariifolium*, *B.
boreale*, *B. simplex*, *B. virginianum*. There is
considerable variation between these species
and differentiation appears possible. According
to the photographs in Dickson, 1973, spores of
the mosses *Encalypta vulgaris*, *E. rhabdocarpa*,
Breutelia chrysocoma and *Oedipodium griffithianum* approximately resemble the description given above of *Botrychium* species — at
least as far as the distal surfaces go —
the proximal surfaces differ, and provide distinguishing features when combined with the
smaller moss spore sizes. *Selaginella rupestris**
is quoted by McAndrews *et al.*, 1973, as being
densely verrucate and thus may key out here.
It is also noted that some convex-sided *Cryptogramma* spores may key out here]

MONOPORATE (Plates 7, 9, 13, 14, 60, 69)

1a With very large, operculate porus; diameter
more than half that of grain diameter. Surface
of grain with various proportions of clavae,
gemmae, verrucae and bacula; all varying in size
and either crowded or rather scattered. Operculum with much finer granulate-gemmate sculpturing and often on a concave portion of exine.
Grain >25 μm: *Nymphaea alba* type
[includes *N. alba* and *N. candida*. Distinction
may be made between these species by consulting
type slides. The latter is larger (although size
ranges overlap) with exine processes finer, more
uniform and crowded. Bacula are reported rarely
present — see Jones and Clarke, 1981]

1b Porus may or may not be operculate. Grain
surface psilate, granulate, verrucate, reticulate
or echinate but never clavate, gemmate or
baculate: 2

2a Porus well-defined and circular. Nexine thickened around it to form a costa. Grains tectate
with verrucate, rugulate, scabrate, microverrucate or psilate sculpturing. Columellae may or
may not be visible (×1000, phase contrast). Columellae may be evenly distributed or aggregated
in groups: Gramineae
[some distinction may be made within this large
family, see page 100. Occasionally the papilla
may appear to be a porus in Taxodiaceae — see
Inaperturate key]

2b Porus not so well defined and without thickened

margin. The porus edge may be ragged and the
porus itself difficult to locate: 3

3a Grain tectate-perforate (×1000, phase contrast
helps) columellae visible and evenly distributed
under the tectum. Areas (lacunae) that are semitectate, verrucate (with isolated islands of sexine)
are visible on the sides of the grain. Grains pear-
shaped or roughly spherical, more or less
collapsed or crumpled: Cyperaceae
[further identification is possible if sufficient
type slides are available. The most distinctive
grains are the thick-walled, almost spherical
ones of *Rhynchospora alba*; and the large grains
of *Cladium mariscus* in which one end of the
grain is narrowed to a finger-like projection]

3b Grains intectate, either echinate or eureticulate: 4

4a With sharply conical echinae. Porus very obscure
and no columellae visible between the echinae
(×1000, phase contrast). Grain shape circular
although often crumpled: *Lemna*

4b With surface eureticulate, tectate-perforate or
microreticulate. Grain shape varies from circular
to obtusely angular: 5

5a Porus large, $>\frac{1}{4}$ grain diameter; represented only
by a large area of non-reticulate exine on one
side of the grain: *Cephalanthera damasonium*
[other species in the genus may be similar]

5b Porus smaller = or $<\frac{1}{4}$ grain diameter: 6

6a Microreticulate or tectate-perforate, width of
muri approaches or equals the size of the lumina:
                        Sparganium erectum
[data from Punt, 1975]

6b Reticulum with larger meshes, muri narrower
than lumina. Muri sometimes duplicolumellate,
especially at junctions. Lumina varying considerably in width and outline but with the largest
always in the area opposite the porus. Sometimes
the muri are rather winding and the sculpture
approaches the rugulate condition:
                        Typha angustifolia type
[includes *Typha angustifolia*, *Sparganium
emersum*, *S. angustifolium*, *S. friesii*, *S. fluitans*,
S. hyperboreum, *S. glomeratum*, and *S. minimum*. Data from Punt, 1976]

NOTES ON GRAMINEAE

Many authors have attempted distinctions within
the pollen grains of this family; for examples see the
works of: Beug (1961); Grohne (1957); Leroi-
Gourhan (1969); Andersen (1979); Faegri and
Iversen (1989). From the palaeoecological point of
view it is of great interest to know if any of the
grains encountered are from the major cultivated

cereal species in the genera *Triticum*, *Avena*,
Hordeum and *Secale*, rather than from wild grasses.
Very generally, it may be said that cereal grains are
always larger, with bigger annuli, in comparison to
these features in most wild grasses. Careful size
measurements need to be taken and the grains
therefore need to be uncrumpled, not suffering coverslip compression, and observed in freshly made-up
slides (if glycerine jelly is the mountant). If these
conditions are not met, attempts to identify cereal
grains based on size will generate inaccurate results.
Phase contrast is essential for detecting the surface
features and columellae arrangement.

Readers interested in Gramineae identifications
are referred to the above works, but as an example
the study by Andersen, (1979) may be mentioned.
He recognizes the groups which follow. It should be
noted that all the following measurements are in
silicone oil and not glycerine jelly. Faegri and Iversen
(1989) note that to convert silicone oil measurements
to glycerine jelly measurements, for the same species
conversion factors vary between ×1.1 and ×1.3 for
grain size and between ×1.1 and ×1.5 for porus
data.

1 A wild grass group
        Mean annulus diameter $<8\,\mu$m, mean
        grain size $<37\,\mu$m, surface scabrate or
                                    verrucate.

2 A *Hordeum* group
        Mean annulus diameter $8-10\,\mu$m, mean
        grain size $32-45\,\mu$m, surface scabrate.

3 An *Avena-Triticum* group
        Mean annulus diameter $>10\,\mu$m, mean
        pollen grain size $>40\,\mu$m, surface
                                    verrucate.

4 *Secale cereale*   Mean annulus diameter $8-10\,\mu$m,
        oblong grain outline (high pollen
                index), surface scabrate.

The above list should not be used as an identification aid without reference to the original paper
by Andersen; it is presented merely to encourage
readers to attempt useful distinctions within a
difficult group.

The actual species included in each group are
listed in Andersen's paper and the group headings
should be used in pollen diagram interpretation with
caution, for example, pollen falling within *Hordeum*
group could as well derive from *Glyceria* species as
from *Hordeum vulgare*.

Faegri and Iversen (1989) attempt further distinctions using characters such as whether or not the
columellae are aggregated into groups or are individually free; also whether the free columellae are
crowded or more sparse.

DIZONOPORATE (Plate 14)

Dizonoporate (and Dipantoporate) grains occur occasionally in taxa that usually produce three or more porate grains e.g. *Tilia, Betula, Corylus, Campanula* type, *Myriophyllum*. These grains rarely make up a significant proportion of those encountered during analysis of sediments but it is worth noting that Engel (1978) reports that up to 13% of *Myriophyllum alterniflorum* grains were observed to be Diporate on type slides. Such unusual grains can be recognized by the type of porus structure and exine sculpturing.

Colchicum normally produces Dizonoporate (as well as Trizonoporate) grains which are large, elliptic, with a delicate faint reticulum and two large ragged-edged pori. They are unlikely to survive fossilization and acetolysis. *Morus alba* produces Dizonoporate grains which appear irregularly scabrate or microechinate, with rather irregular-edged unthickened pori.

TRIZONOPORATE (Plates 14–17, 62, 64, 66)

In polar view, several taxa may appear trizonoporate to the inexperienced observer, when in fact, critical examination of the grain, or rolling it over, will reveal the fact that it is trizonocolpate or trizonocolporate with rather short colpi. Examples where such initial confusion may occur are: *Tilia, Lonicera, Linnaea, Symphoricarpos*, some *Elaeagnus* species and *Ludwigia*. The most easily confused ones are mentioned in the key below.

1a Grains slightly heteropolar, with pori in a zone slightly to one side of the equatorial zone (in a polar view, one or more pori will appear in face view, the others in optical section, occasionally all pori in face view). A circular, wide, faint area of dissolved nexine around each porus. Grain shape triangular-obtuse, size >40 µm:
    *Carya**

1b Grains never heteropolar, all pori clearly in an equatorial zone: 2

2a Grains large, >55 µm in greatest diameter: 3

2b Grains small to medium, <55 µm in greatest diameter: 6

3a With vestibulate pori (sexine and nexine separating at the porus) each vestibulum large and cylindrical or conical. Each vestibulum may or may not have thickened walls, but the actual opening of the porus to the outside is a small, thin, and possibly ruptured area of sexine at the vestibulum apex. No echinae or reticulate sculpture. Viscin threads may be attached to

the exine: *Chamaenerion angustifolium* type [includes *Chamaenerion, Oenothera** and occasional grains of *Epilobium*. Most *Epilobium* grains occur in tetrads. *Oenothera* may be separable by its larger size, characteristic porus-structure and coarse, irregularly-spaced columellae. Consult type slides. See also *Ludwigia palustris*]

3b Without vestibulate pori, grains either reticulate or echinate, pori often elongated and elliptic: 4

4a Grains semitectate, reticulate; with each porus elongated into an ellipse. Exine structure complex, with clavate or baculate projections on top of the reticulum muri. These projections are large and of similar thickness to the columellae (see the results of LO analysis and exine optical section ×1000):   *Geranium*

4b Grains echinate, with a dense carpet of microechinae between the sparsely spaced large echinae. Porus membrane (when present) with long, branched echinae which form the operculum: 5

5a Pori elliptic, each surrounded by a smooth annulus and a broad diffuse 'halo' or zone where both columellae and nexine are absent. Microechinae rather indistinct in optical section ×1000:   *Dipsacus* [see Clarke and Jones, 1981a, for full descriptions and further differentiation of Dipsacaceae pollen]

5b Pori circular to slightly elliptic with an indistinct annulus and no such 'halo' or zone lacking columellae and nexine. Microechinae clearly visible in optical section, ×1000:   *Knautia* [see ref. above under 5a. Compare *Lonicera*, p. 152]

6a Grains reticulate: 7

6b Grains psilate, scabrate or echinate: 8

7a Pori large, ill-defined, ragged ruptures:
    Colchicum

7b Pori very well-defined, with large nexinous thickenings (costae):   *Tilia* [*Tilia* grains are included here as the colpi are very short and may not be apparent in a polar view. See Trizonocolporate, p. 149 for separation of *Tilia* species]

8a Grains with protruding, vestibulate pori: 9

8b Grains with pori protruding or not protruding, non-vestibulate: 10

9a Each vestibulum more cylindrical than conical and very large compared to grain size. Walls of the vestibulum gradually thinning towards the apex, where the actual porus opening to the

outside is a small area of thin sexine which may or not be ruptured. Inner walls of vestibulum appearing ragged or covered in gemmae or irregular globules. Vestibulum floor complete, being formed by a thick, amorphous, granular layer: *Circaea*
[NB *Ludwigia* may key out at this exit if its short colpi are not detected in polar view. It agrees well with the description of *Circaea*, except that its vestibula are definitely conical, not cylindrical]

9b Each vestibulum more conical than cylindrical, smaller in relation to grain size. Sexine with tectum clearly thickest at the apex of the vestibulum (see porus in optical section ×1000) forming a well-defined annulus to the porus as seen in surface view. Vestibulum floor is a thin, ring-like shelf of nexine, rather than a complete, thick, amorphous, granular layer: *Betula*
[some separation of types within the genus is possible by consulting type slides and considering the relationship between grain size and porus size. Many authors consider *Betula nana* separable from the tree birches — see Birks, 1968; Prentice, 1981. In *B. nana* the vestibulum is rather low and inconspicuous, so it may not even key out here at all — see exit 20a. See also note on *Ludwigia* under 9a. Erdtman and Nordborg, 1961, report that grains of some races of *Sanguisorba minor* ssp. *minor* appear trizonoporate, vestibulate; and it is thus possible that such grains key out here]

10a Grain circular in polar view; pori sunken, flat on the surface or slightly protruding: 11
10b Grain between circular and triangular-obtuse in polar view, pori situated on the angles and may or may not be protruding: 20
11a Grains echinate, pori flush with the grain surface or sunken: 12
11b Grains psilate, scabrate or microechinate. Pori can be: slightly protruding, flush with the grain surface, or very slightly sunken: 14
12a No columellae, only infratectal bacula visible in optical section of wall ×1000. Echinae sparse, wide-based. Pori elliptic, indistinct, not heavily thickened: *Ambrosia** type
[includes *Ambrosia* and some species of *Iva* and *Xanthium* where the colpi are very short and may be mistaken for pori]
12b Columellae clear, no infra-tectal bacula. Echinae dense, narrow-based. Pori distinct, sunken, circular, large in relation to grain size and with thick costae: 13
13a Grain <30 μm in diameter: *Jasione* type
[includes *Jasione* and *Wahlenbergia*. These two

may be separated on relative echina density if type material is available]
13b Grain >30 μm in diameter: *Campanula* type
[includes *Campanula* and some species of *Phyteuma* and some grains of *Legousia hybrida*. The last two are also tetrazonoporate. Some separation is possible within this type on size of grain and size of echinae]

14a Pori not protruding, (sometimes a little sunken) non-annulate. Scabrae (or microechinae) may be present on the surface, if so, they are unequal in size and irregularly distributed (×1000, use LO analysis and phase contrast if possible):   15
14b Pori protruding slightly, an annulus present. Scabrae (microechinae) evenly or unevenly distributed (×1000, use LO analysis and phase contrast if possible):   17
15a Grain small, <15 μm; circular to elliptic with 3 large, granulate pori arranged around the equator of the ellipse (pori $\frac{1}{4}$ to $\frac{1}{3}$ of grain diameter). Exine thickest at poles of ellipse where the columellae are distinct and longest: *Corrigiola litoralis*
[includes *Corrigiola litoralis* and occasional grains of *Herniaria glabra* and *H. ciliolata* that do not have 4 pori]
15b Grains larger, 13 μm or more. Pori $<\frac{1}{4}$ of grain diameter, columellae indistinct: 16
16a Each porus large (diameter of actual hole 2.5 μm or more — can be up to 5 μm, or $>\frac{1}{7}$ of grain diameter) elliptic to circular, with irregular margins. Pori not always equidistantly spaced. Grain more or less than 20 μm in diameter: *Morus*
[includes some grains of *Morus alba*, which is mostly 2-porate and some grains of *M. nigra*, which is also 2, 4 or 5-porate. See Punt and Malotaux, 1984]
16b Pori smaller (diameter of actual hole <2.0 μm) always equidistantly spaced. Grain <20 μm in size (consult a type slide):   *Parietaria*
[see ref. above under 16a]
17a Pori only slightly protruding, annulus narrow, grain small, <21 μm: 18
17b Pori slightly to distinctly protruding in optical section, annulus a broad and clear ring (especially in phase contrast). This ring in surface view produced by the tectum near the porus rising up and then diving down into the porus cavity to, or below, the level of the nexine; so that the porus cavity is characteristically U- or C-shaped (see porus in optical section, ×1000. This is a difficult feature to see if one is inexperienced):   19
18a Scabrae (microechinae) of equal size, very reg-

ularly or evenly distributed (consult a type slide): *Urtica urens*
[see ref. above, under 16a]

18b Scabrae (microechinae) unequal in size, ir regularly distributed (consult a type slide): *Urtica dioica*
[see ref. above, under 16a]

19a Porus cavity C-shaped. Grain with scattered coarse and fine columellae giving a falsely scabrate impression. Careful examination of the grain wall in optical section reveals a thick, psilate tectum on top of the columellae. Pori sometimes with coarsely granulate opercula: Grain size usually >30 μm, exine >2.5 μm thick away from pori: *Celtis*

19b Porus cavity U-shaped. Grain with columellae fine to invisible, exine rather thin (<2 μm) surface psilate or very faintly scabrate in LO analysis and phase contrast. Grain <31 μm, pori without granulate opercula: *Cannabis* type
[includes *Cannabis sativa* and *Humulus lupulus*. Several authors report distinctions between these two important species e.g. Punt and Malotaux, 1984; French and Moore; 1986; Whittington and Gordon, 1987. Greater pore protrusion and larger grain size in *Cannabis* is probably the best character combination, but there is an overlap in the measurements. Punt and Malotaux report that the hollow area inside the annulus produced by the tectum rising up and diving down inside the porus (the 'internal void' described by Churchill, in Godwin, 1967); is not apparent in *Humulus* and is not consistently present in *Cannabis*, but the present authors have observed this character frequently on type slides of both species]

20a Nexine may be slightly reduced or split into a few lamellae immediately adjacent to the porus, but is never absent for a substantial area (see optical section ×1000). The sexine may be slightly thickened adjacent to the porus but there is no clear annulus. Surface sculpture (×1000, phase contrast helps) is faintly rugulate, with microechinae or scabrae on the ridges: *Corylus*
[NB It is possible that *Betula nana* and some specimens of *B. pubescens* may key out here if the vestibula are inconspicuous. Doubtful grains must be compared with type slides]

20b Nexine dissolved or absent in a ring around the porus such that in optical section, ×1000, the solid line of the nexine is seen to end well before the porus: 21

21a Grain surface rugulate, with microechinae on the ridges (×1000, phase contrast helps). Sexine

not appreciably thickened near the porus: *Ostrya* type
[includes *Ostrya* and very occasional grains of *Carpinus. Casuarina* may key out here — see Erdtman *et al.*, 1963, and consult type slides]

21b Grain surface regularly microechinate or scabrate (×1000, phase contrast helps). Sexine near the porus thickened in the area of nexine absence: *Myrica*
[*Engelhardtia* resembles *Myrica* but is <20 μm, *Platycarya* is similar but with curved streaks or grooves across the poles. *Comptonia** is very similar to *Myrica* but the sexine near the pori appears unthickened]

TETRAZONOPORATE (Plates 17, 62)

Tetrazonoporate grains are very occasionally found in taxa that are normally always trizonoporate e.g. *Betula, Myrica, Lonicera*.

1a Grains echinate or with broad, winding rugulae or reticulum (outline undulating in optical section): 2

1b Grains psilate, scabrate, verrucate, microechinate or microrugulate: 3

2a Grains echinate, with pori in slightly sunken areas and each with a distinct costa: *Phyteuma* type
[includes *Phyteuma tenerum* and *Legousia hybrida*, which are nearly always tetrazonoporate, and occasional grains of some *Campanula* species which are more commonly trizonoporate]

2b Grains with broad winding ridges or shallow, wide-walled reticulum. Pori not sunken, being slightly protruding: *Ulmus*
[grains of *Zelkova* are more or less similar and type slides should be consulted if this genus seems likely to occur. See the works of Stockmarr, 1970, Tolonen, 1980b, for differentiation within *Ulmus*]

3a Neighbouring pori connected by arcs or bands of nexinous thickening, pori vestibulate: *Alnus*

3b No arcs or bands connecting pori. Pori may or may not be vestibulate, with or without thickened margins: 4

4a Nexine apparently dissolved in a **wide** circular region around each porus (see optical section of the wall ×1000 and phase contrast of surface view): 5

4b Nexine not dissolved in a wide circular region around each porus, but continuing up to the porus edge: 6

5a Most grains <40 μm, pori all usually in an equatorial zone. Observation of the surface

under ×1000 may reveal faint rugulae with microechinae on the ridges: *Carpinus* type
[includes *Carpinus* and some occasional grains of *Ostrya*]

5b Most grains >40 µm, often with pori in a zone slightly to one side of the equatorial zone (i.e. grains are slightly heteropolar). Surface is regularly microechinate or scabrate, one apo porium with thicker exine than the other:
                                              Carya type
[includes some species of *Carya**, e.g. *C. glabra*; and occasional 4-pored *Pterocarya*]

6a With coarse, irregular, internal verrucate-granulate surface pattern. This pattern is produced by columellae or by columellae-like structures under a psilate tectum (use LO analysis and an optical section of the grain wall, ×1000, to discern these):   7

6b Columellae either fine, uniform and in an even carpet; or barely detectable. Tectum either completely psilate or scabrate (microechinate):   8

7a With pori circular in outline, projecting slightly and strongly ringed (especially in phase contrast). Ring in surface view produced by the tectum near the porus rising up and then diving down into the porus cavity below the level of the nexine (see optical section of wall, ×1000). No nexine projections; columellae rather sparse, variable in thickness and scattered in an irregular pattern. Porus membrane granulate: *Celtis*

7b Pori elliptic in outline; projecting slightly, but not so strongly ringed. Innermost layer of wall (? nexine) thrown into coarse irregular projections resembling rods and verrucae. Columellae extremely fine to undetectable. Pori may or may not be granulate: *Nerium oleander*

8a Pori asymmetrically distributed around the circumference of the grain (two close together at one side, two close together on the other). Pori with thick annuli: *Myriophyllum alterniflorum*

8b Pori symmetrically distributed around the grain circumference, with or without thick annuli: 9

9a With wide annuli around the pori (width of a porus + annulus >6 µm). Annuli either rounded or low and flat (see optical section). Pori elliptic or represented short colpi: 10

9a With either narrow annuli or no annuli to the pori (width of one annulus + porus <5 µm). Pori always circular: 11

10a With low, flattened, indistinct annuli and pori elongated to short colpi:
                                    Myriophyllum verticillatum
[consult type slides and see the work of Engel, 1978. This author quotes the short colpi to be 1 × 3.5 µm]

10b With thicker, more protruberant and distinct annuli; pori merely elliptic:
                                    Myriophyllum spicatum
[consult type slides and see the work of Engel, 1978. This author quotes the elliptic pori to be 1.5 × 3.0 µm]

11a Pori not protruding (sometimes a little sunken) with or without annuli. Scabrae (or microechinae) on the surface unequal in size and irregularly distributed (×1000, use LO analysis and phase contrast):   12

11b Pori protruding slightly, with a narrow annulus visible on each. Scabrae (or microechinae) either evenly or unevenly distributed (×1000, use LO analysis and phase contrast):   13

12a Pori rather large (diameter of actual hole 2.5 µm or larger — can be up to 5 µm) irregularly elliptic in shape and not always equidistantly spaced. Porus membrane with granules. Grains 18 µm or more: *Morus nigra*
[data from Punt and Malotaux, 1984]

12b Pori very much smaller (diameter of actual hole <2 µm) equidistantly spaced. Grains <20 µm:
                                              Parietaria
[see Punt and Malotaux, 1984]

13a Scabrae (microechinae) of equal size, evenly distributed: *Urtica urens*
[see Punt and Malotaux, 1984]

13b Scabrae (microechinae) unequal in size, irregularly distributed: *Urtica dioica*
[see Punt and Malotaux, 1984]

PENTAZONOPORATE, HEXAZONOPORATE (Plates 17, 62)

1a Pori vestibulate. Grain with arcs or bands of nexinous thickening joining adjacent pori: *Alnus*

1b Pori non-vestibulate, with margins either unthickened or solidly thickened. No arcs or bands of thickening joining adjacent pori: 2

2a Grain with broad rugulae or shallow wide-walled reticulum. Tectum undulating in an optical section of the wall: *Ulmus*
[the pollen grains in *Zelkova* are more or less similar, so consult type slides if this genus seems likely to occur. See the works of Stockmarr, 1970, Tolonen, 1980b, for differentiation within *Ulmus*]

2b Grain psilate, scabrate or faintly microrugulate: 3

3a Nexine present in a region immediately adjacent to each porus. Pori thickened, annuli may or may not be distinct: 4

3b Nexine dissolved in a circular region around each porus (to detect this, view porus in optical section, ×1000). Pori may protrude slightly, but are usually unthickened: 6

4a Pori asymmetrically distributed around the grain: two or three close together on one side; and two or three close together on the other side. Margins of pori very heavily thickened with large annuli. Pori circular: *Myriophyllum alterniflorum*

4b Pori symmetrically distributed around the grain. Margins may or may not be well thickened. Pori circular to elliptic, or elongated to short colpi: 5

5a With low, flattened, indistinctly thickened annuli. Pori elongated to short narrow colpi:
                        Myriophyllum verticillatum
[consult type slides and see the work of Engel, 1978. This author quotes the short colpi as being $1 \times 3.5\,\mu m$]

5b With thicker, more protruberant and distinct annuli. Pori circular to elliptic:
                        Myriophyllum spicatum
[consult type slides and see the work of Engel, 1978. This author quotes the elliptic pori as $1.5 \times 3\,\mu m$]

6a Surface regularly microechinate (scabrate) best seen in phase contrast: *Pterocarya*

6b Surface microrugulate with microechinae on the ridges, best seen in phase contrast: *Carpinus* type [includes occasional grains of *Carpinus* and *Ostrya*]

TETRAPANTOPORATE (Plate 18)

Very occasional grains of *Carpinus* type, *Urtica* or *Parietaria* may apparently fall into this class — see Trizonoporate

1a Grains small, <16 μm; circular to obtusely angular in shape and psilate. Pori slightly sunken, with granular membranes and outlines clear or rather obscure. Columellae distinct, exine markedly thickest and columellae longest in the middle of the mesoporium. Exine thinning towards porus margins. Consult type slides:
                        Herniaria type
[includes *Herniaria glabra*, *H. ciliolata* and rare grains of *Corrigiola litoralis*]

1b Grains larger, >16 μm; circular in shape and microechinate to verrucate. Pori with outlines clear or rather obscure, with or without annuli: grains of   *Plantago maritima*, *P. media*, *P. major* or *P. coronopus* — consult 'PENTAPANTO-PORATE, HEXAPANTOPORATE' key and type slides

PENTAPANTOPORATE, HEXAPANTOPORATE (Plates 18−20, 64, 68)

1a Each porus with either an annulus or a costa (i.e. there is a difference in the thickness of

sexine or nexine near the porus). Pori always circular with edges well-defined, with or without opercula: 2

1b No such differentiated area around each porus, edge of porus well defined or ragged and diffuse, porus membrane granulate: 8

2a Pori in flattened or concave areas of exine, each surrounded by a more or less irregular, well defined intectate-granulate zone where the tectum is clearly missing. Sometimes adjacent zones join so that two pori share the same one. The rest of the surface is psilate. Grain circular to obtusely angular: *Ribes rubrum* type [includes *R. rubrum*, *R. nigrum*, *R. petraeum*, *R. spicatum*, *R. aureum* and *R. sanguineum*. See Verbeek-Reuvers, 1977b]

2b Pori never surrounded by such irregular intectate-granulate areas, the areas of differentiated exine are always circular: 3

3a Pori not protruding in optical section of the grain, exine thinner at the porus edge than in the middle of the mesoporium. Surface tectate-perforate: Caryophyllaceae [this key exit includes a few species e.g. *Stellaria holostea*, which normally have a low number of pori; and occasional grains of species with normally a higher number of pori. See the specialist key on p. 110]

3b Pori protruding slightly in an optical section of the grain. Exine may or may not be thicker at the porus margins than in the mesoporium: 4

4a Nexine dissolved or absent in a broad band around each porus (see porus in optical section, ×1000) exine in this area thus slightly thinner than elsewhere. Surface sculpture scabrate or faintly rugulate with microechinae on the ridges (×1000, use LO analysis or phase contrast):   *Carpinus* type [includes grains of *Carpinus* where the zono-arrangement is not apparent; also rare grains of *Pterocarya* and *Juglans*]

4b Nexine not dissolved near the pori, exine always thicker here than in the mesoporium because of the annulus which appears as a solid or inter-rupted ring around each porus. Grain scabrate, verrucate, microechinate, or a combination of these: 5

5a With large pori surrounded by very thick annuli. Each porus covered by a thin psilate membrane which is often ruptured and protruding from the aperture. Distinct radial endocracks visible under the annuli. Usually six pori. One porus plus its annulus may measure $\frac{1}{4}$ of the grain diameter. Mesoporium verrucate: *Fumaria* [see Kalis, 1979]

5b Grains with smaller pori and annuli. Porus

membrane covered in granules, never psilate and never protruding. Annuli with or without radial endocracks: 6

6a With each annulus produced by the tectum near the porus rising up and then diving down into the porus cavity below the level of the nexine (see optical section of porus, ×1000). Columellae coarse, sparse, variable in thickness, and scattered in an irregular pattern under the psilate tectum. Porus membrane granulate:
Celtis
[occasional grains that have a high porus number]

6b Each annulus not produced by tectum rising up and diving down in such a fashion. Columellae finer, more uniform and regularly distributed. Sculpturing always verrucate (although sometimes the verrucae may be very low and faint — only detectable in phase contrast):   7

7a Each annulus interrupted, being formed by a ring of denser verrucae congregated around the porus. Microechinae or scabrae **distinct** and regularly spaced over the verrucae. Columellae visible as a dense and even carpet underneath the tectum: *Plantago maritima* type
[according to Clarke and Jones, 1977a, this includes *P. maritima*, *P. alpina* and *P. arenaria*. There is considerable variation within this type in the prominence of the annuli and verrucae. Grains of *P. alpina* and *P. maritima* are most likely to be annulate and thus key out here]

7b Annulus not interrupted, being a very prominent, solid, broad ring not formed by separate verrucae. Verrucae in the mesoporium large and well defined. Microechinae or scabrae **indistinct** even in phase contrast, but columellae clear as a dense and even carpet:
Plantago coronopus
[see ref. under 7a]

8a Surface regularly baculate-gemmate, processes all of uniform height and quite densely packed so that some of the capita (heads) appear polygonal in surface view. Pori large, often irregular or elongate in outline. Not all endoapertures correspond with ectoapertures so some pori are nexine features only i.e. the baculate-gemmate sexine continues over them. Grain diameter >50 μm: *Linum perenne* ssp. *anglicum*
[long-styled individuals have dimorphic processes — sparse thick ones and abundant thin ones; short-styled individuals have monomorphic processes — all rather thick. See Punt and den Breejen, 1981. *L. perenne* ssp. *perenne* has similar sculpture — see *L. austraicum* type, pp. 128−129]

8b Surface not baculate-gemmate, being verrucate, microechinate, scabrate or some combination of these. Pori may or may not be large and irregular: 9

9a Grain verrucate (verrucae may sometimes be so low and faint as to be detectable only with phase contrast) with or without microechinae: 10

9b Grain microechinate to scabrate: 12

10a Microechinae **indistinct** in bright field, ×1000; barely detectable under phase contrast. Verrucae prominent. Pori not sharply delimited, their presence indicated merely by the cessation of verrucate sculpturing. <28 μm. Consult a type slide: *Plantago major*
[see ref. under 7a according to the data in Bassett and Crompton, 1968, the American species *P. rugelii** may key out here]

10b Microechinae **distinct** in bright field and phase contrast — easily distinguishable from the more closely spaced columellae underneath. Verrucae prominent or indistinct, pori may or may not be well delimited: 11

11a Pori usually circular and with margins clearly delimited. Verrucae relatively distinct to rather faint. Consult type slides: *Plantago maritima* type
[according to Clarke and Jones, 1977a the type includes *P. maritima*, *P. alpina* and *P. arenaria*. There is considerable variation within this type in the prominence of annuli and verrucae. Grains of *P. arenaria* and *P. maritima* are most likely to be non-annulate and thus key out here. Note that *P. maritima* and *P. alpina* are closely related in one subgenus whilst *P. arenaria* is in a different subgenus]

11b Pori with irregular and not sharply delimited margins — presence indicated merely by the cessation of verrucate sculpturing. Verrucae always very large and dense. Consult a type slide: *Plantago media*
[see ref. under 7a]

12a Grain 30 μm or more with very large pori (>6 μm in diameter). Pori may or may not be circular, having granulate membranes and irregular margins. Usually 6 pori, grain shape that of an obtuse cube with one porus per face. Surface covered in regularly arranged microechinae which are rather more sparse than the columellae: *Papaver argemone* type
[includes *P. argemone* and *Roemeria hybrida*. See Kalis, 1979]

12b Grain <30 μm, pori <6 μm in diameter: 13

13a Columellae very fine and overlain at higher focus by regularly arranged microechinae. Granules on the porus membranes coarser than the columellae or microechinae. Exine not markedly thinning from mesoporium to the margins of the pori. Pori may be situated in slightly sunken areas of the exine giving the grain a 'dimpled' appearance: *Thalictrum* [rare grains with a lower than average porus number — most likely to have been produced by *T. alpinum* or *T. minus*. See also p. 108. Grains of *P. maritima* where the verrucae are undetectable may key out here. Consult type slides]

13b Columellae coarser, no microechinae visible at high focus. Exine thickest, and columellae longest, in the mesoporium; exine thus thinning towards porus margins. Porus membrane granules relatively fine compared to columellae. Pori may or may not be slightly sunken. Grain size <22 μm and often as small as 14 μm. Consult a type slide: *Illecebrum verticillatum* [occasional grains with a low porus number. Some *Herniaria* species have a similar appearance, but the North West European ones examined have <5 pori]

POLYPANTOPORATE (Plates 18−22, 63, 64)

1a Grains with surfaces psilate-scabrate, tectate-perforate or microechinate **without** any verrucae: 2

1b Grains with surfaces: echinate, baculate, verrucate, foveolate, eureticulate, rugulate-striate or microechinate **with** verrucae: 15

2a Each porus with a differentiated area around it e.g. an annulus. This area may have either sexine or nexine thicker or thinner than the rest of the mesoporium and e.g. columellae more densely or less densely spaced than in the rest of the exine. Pori usually circular with edges well-defined, perhaps with a ring or band: 3

2b No such clearly differentiated area around each porus, exine adjacent to each porus the same as that in the mesoporium. Edge of porus either well-defined or faint, possibly irregular and diffuse: 9

3a Exine tectate, tectum **without perforations or microechinae** (phase contrast may be necessary to detect this). Usually with >40 pori:
Chenopodiaceae and Amaranthaceae
[fewer than 40 pori are quoted by Andrew, 1984, in some *Chenopodium* species e.g. *Chenopodium polyspermum*, *C. vulvaria*, also in *Beta vulgaris*, *Salsola kali* and possibly others. In such grains with low porus numbers, very careful surface examination is necessary in order to verify that there are no perforations. On the other hand, Fredskild, 1967, quotes well over

40 pori in some *Sagina* species. Perforations in such grains are usually obvious under phase contrast]

3b Exine **tectate-perforate**, with or without microechinae (phase contrast may be necessary to detect these features). Usually <40 pori: 4

4a Grain surface psilate. Pori (actually endopori) on flattened or concave areas of exine, each surrounded by a more or less irregular, well-defined, intectate-granulate zone. Sometimes adjacent zones join so that two endopori share the same one. Grain circular to obtusely angular: *Ribes rubrum* type [includes *R. rubrum*, *R. nigrum*, *R. petraeum*, *R. spicatum*, *R. aureum* and *R. sanguineum*. See Verbeek-Reuvers, 1977b]

4b Pori never surrounded by such well-defined, **irregular** intectate-granulate zones: 5

5a Margins of pori slightly protruding in optical section of the grain; tectum with regularly spaced microechinae or scabrae: 6

5b Pori not protruding in optical section. Exine markedly thinner at the porus edge compared to the mesoporium (pori may appear sunken). At least some of the columellae are coarse. Grains tectate-perforate, with or without microechinae. Porus membranes commonly present and granulate: 7

6a Nexine fragmented or dissolved in a broad zone around each porus. Columellae distinct, uniform and fine. Grain with 6−20 pori, sometimes aggregated towards one part of the grain: *Juglans* type [includes *Juglans* and *Pterocarya*. *Pterocarya* has the smaller number of pori and the common arrangement here is to have most pori in a ring or circular zone. As the pori are on the angles this gives the grain an angular appearance]

6b Nexine not dissolved in a zone around each porus. Narrow annuli are clearly present. Columellae indistinct; exine thin (approximately 1.5 μm) not having a clear distinction between sexine and nexine. Pori 6−12, never aggregated to one part of the grain: *Urtica pilulifera* [see Punt and Malotaux, 1984]

7a Grain very large (>70 μm in diameter) exine thick (>4 μm). Minute perforations are clearly detectable on the tectum, but no microechinae are present. Tectum supported by very coarse columellae which **clearly branch** at $\frac{2}{3}$ of their length: *Calystegia*

7b Grains smaller (<70 μm in diameter) columellae may or may not be branched. Microechinae **and** perforations should be detectable (view under ×1000, phase contrast may be necessary to detect these two features):   8

8a A large zone around each sunken porus with fine, short, regularly arranged columellae forming a diffuse annulus. Columellae coarser and longer in the middle of the mesoporium and sometimes arranged in a reticulate pattern under the tectum here (an infrareticulum). Tectum with perforations mostly <1 μm. Exine thickened into ridges between the adjacent pori, this giving the grain a polyhedral appearance. With 15–20 pori: *Alisma* type [according to Punt and Reumer, 1981 this includes *Alisma*, *Baldellia* and *Luronium*. It is noted that some members of the Caryophyllaceae such as *Scleranthus perennis* and *Stellaria holostea* may give the impression of *Alisma* type, but may easily be distinguished from it by reference to type slides]

8b Grains with clear-cut annuli, usually seen as a thinning of the exine, accompanied by a neat ring or band around the edge of each porus. The ring is due to the thinning and consolidation of the sexine. Up to about 40 pori: most genera in the Caryophyllaceae [many species are included here and the special key on p. 110 attempts to separate out some useful types]

9a Grain clearly tectate-perforate, with no microechinae or scabrae (view ×1000, use phase contrast if possible). Columellae distinct and more densely spaced than perforations. Pori large, often with irregular margins and with coarsely granulate membranes: *Liquidambar styraciflua** [note that *Sagina* may key out here as it has little or no annulus, but it is distinct because of the small pori and the presence of microechinae]

9b Grain not perforate, but microechinate or scabrate (×1000, phase contrast):   10

10a Pori large in relation to grain size, porus diameter in face view being $>\frac{1}{8}$ of grain diameter. An infrareticulum may be visible in the mesoporium: 11

10b Pori smaller in relation to grain size: 12

11a >9 pori, each one roughly circular with a coarsely granulate membrane. Pori regularly spaced and in slightly sunken areas of the exine so that grain is slightly to markedly polyhedric. Less densely microechinate: *Damasonium* type [according to Punt and Reumer, 1981; this type includes *Damasonium* and *Caldesia*. Consult type slides]

11b Up to 9 pori, irregularly spaced out over the grain. Each >6 μm in diameter and with a very ragged outline. Porus membranes often missing

but if still present, then granulate. Tectum densely microechinate. Grain 30 μm or more: *Papaver argemone* type [includes *P. argemone* and *Roemeria hybrida*. See Kalis, 1979]

12a Grain small, <22 μm and often as small as 14 μm. Columellae relatively coarse, being easily detectable and appearing longest and most spaced out in the mesoporium. No microechinae detectable at high focus. Pori may or may not be slightly sunken, may or may not be clearly delimited and with membrane granules fine compared to the columellae. Consult a type slide: *Illecebrum verticillatum* [some *Herniaria* species have a similar appearance, but the North West European ones examined (*H. glabra* and *H. ciliolata*) have <5 pori, whereas *Illecebrum verticillatum* has >6 pori. *Sagina* may key out here, and it would be worth checking other small Caryophyllaceae, see key on p. 110]

12b Grains may or may not be >22 μm. Columellae finer — less coarse than the membrane granules. Regularly spaced microechinae distinct at high focus: 13

13a Grain with 9 or more pori. Pori may be in slightly sunken areas giving the grain a 'dimpled' appearance: *Thalictrum* [most grains of species with higher average porus numbers e.g. *T. flavum*. Note that *Plantago tenuiflora* and *Littorella uniflora* may key out here if the verrucae are undetected. See type slides and Clarke and Jones, 1977a. Also check *Sagina* and see Caryophyllaceae key on p. 110]

13b Grain with 8 or less pori: 14

14a Pori with margins clearly defined, more or less circular in shape. Careful analysis in phase contrast may reveal low faint verrucae under the microechinae. Pori not usually in sunken areas. Consult a type slide: *Plantago maritima* type [the type includes *P. maritima*, *P. alpina* and *P. arenaria* but only some grains of *P. maritima* are likely to key out here as their verrucae are too low to be detected with bright field illumination — see discussion under exit 23a and consult type slides]

14b Porus margin more faint. Pori may be in slightly sunken areas, giving the grain a dimpled appearance. Consult a type slide: *Thalictrum* [most grains of the species like *T. alpinum* and *T. minus* which have lower average porus numbers]

15a Grains echinate, baculate, clavate, microechinate or verrucate (or with verrucae covered in microechinae or scabrae):   16

15b Grains foveolate, eureticulate or rugulate-
striate: 28

16a Grain echinate, baculate, clavate or microechinate, sometimes with a single verruca under
each echina: 17

16b Grain verrucate, sometimes with microechinae
(or scabrae) visible on top of the verrucae: 20

17a Grains large, being >60 μm in diameter with
many >100 μm; covered with sharply conical
echinae or occasionally with thick bacula or
clavae (grain sizes include projections). The
longest echinae are >6 μm: *Malva* type
[includes *Malva*, *Lavatera* and *Althaea*. Some
distinction between these genera may be possible, see Culhane and Blackmore, 1988. *Althaea
officinalis* is distinct because of the subglobose
bases of the echinae which give the pollen
grains an undulating outline. *Malva pusilla* is
distinct because some grains of this species are
echinate and some baculate-clavate or verrucate. This was reported by Erdtman *et al.*,
1961. Our own observations indicate that such
dimorphic grains may occur rarely in other
Malva species]

17b Grains usually smaller, most <60 μm and all
<80 μm in diameter including projections.
Echinae shorter (<6 μm). Surface often microechinate: 18

18a Pori circular and well-defined, with clear-
cut annuli, each usually seen as a thinning of
the sexine, accompanied by a neat ring or
band around each porus. Echinae small, usually
0.5 μm or less (microechinae). Grains tectate-
perforate: most genera in the Caryophyllaceae
[many species within this family key out here
and the special key on p. 110 separates out
types which may be useful]

18b No well-defined circular edge to the porus and
no annulus present. Pori often with irregular
margins and thus difficult to distinguish from
the mesoporium. Echinae more prominent,
>1 μm tall: 19

19a Pori not operculate, tending to have very irregular margins and to be elongated like small
colpi. Echinae conical but not broad at the
base: *Koenigia islandica*

19b Pori operculate, each operculum consisting of
several large islands bearing echinae in the
same way as the rest of the exine. Echinae
conical and broad-based: *Sagittaria*
[*Ranunculus arvensis* may key out here if the
pori are detectable. It is recognizable by a
characteristic exine structure with dimorphic
columellae. Also *Sagittaria* has up to 15 pori
according to Punt and Reumer, 1981; and
R. arvensis has 24 or more pori according to

Punt, personal communication of unpublished
data]

20a Each porus with an area of thickened sexine
around it (an annulus). This may be either a
solid ring or be composed of a ring of denser
verrucae congregated around the porus. Pori
with solid opercula or with granulate porus
membranes: 21

20b No such annulus to each porus, presence of
pori indicated sometimes merely by the cessation of verrucate sculpturing. Pori never
with solid opercula, but having granulate
membranes: 24

21a Grain with very large pori surrounded by very
thick solid annuli. Each porus covered by a
thin, psilate porus membrane which is often
ruptured and protruding from the aperture.
Distinct radial endocracks visible under the
annuli. Commonly with 6 pori, some species
having up to 12. In the case of the lower number, one porus + annulus may measure up to
$\frac{1}{4}$ of the grain diameter: *Fumaria*

21b Grains with smaller pori and with or without
solid annuli. Each porus membrane either with
thick operculum or granules; but never psilate,
ruptured and protruding from the aperture.
Annuli with or without radial endocracks: 22

22a Pori with distinct, solid opercula. Each operculum with similar structure to the rest of the
exine and covering most of the porus membrane, but leaving a ring of thin membrane
next to the annulus. Thin radial endocracks
under the annuli. Verrucae large, irregular,
and rather diffuse. Verrucae and spaces
between covered with distinct microechinae or
scabrae (view ×1000 and under phase contrast):
Plantago lanceolata type
[according to Clarke and Jones, 1977a, this
includes *P. lanceolata* and the North American
P. altissima; although Bassett and Crompton,
1968, quote the latter as distinctly larger. In the
rather unlikely event of all the opercula being
missing in a fossil grain, *P. lanceolata* is most
likely to be confused with *P. coronopus*. It can
be distinguished from this by the observation
that *P. coronopus* has a lower porus number
and broader, thicker annuli]

22b Pori with granulate membranes, microechinae
on the verrucae may be clear or indistinct: 23

23a Annulus interrupted, being formed by a ring of
verrucae around the porus. Microechinae or
scabrae distinct and regularly distributed over
the verrucae. Columellae distinct and in an
even carpet (×1000, all these features best seen
under phase contrast): *Plantago maritima* type
[includes *P. maritima*, *P. alpina* and *P. arenaria*.

See Clarke and Jones, 1977a. There is considerable variation between these species in the prominence of the annuli and verrucae. *P. alpina* is the most likely to show a clear annulus and *P. maritima* often has very indistinct verrucae. Note that *P. maritima* and *P. alpina* are closely related within one subgenus whilst *P. arenaria* is in a different subgenus]

23b Annulus not interrupted; being a very prominent, thick, broad ring not formed by separate verrucae. Verrucae in the mesoporium large and well defined. Microechinae or scabrae **indistinct** but columellae clear and regularly arranged (×1000, these features best seen under phase contrast):   *Plantago coronopus* [see ref. under 23a]

24a Grains with >or = 8 pori. Verrucae may or may not be well-defined, appearing small in relation to grain size.   25

24b Grains with <or = 8 pori (in rare cases there may be more) verrucae distinct. Grain size commonly <30 μm: 26

25a Verrucae distinct. Grain diameter >30 μm and commonly >35 μm. 8−13 pori present. Consult a type slide: *Littorella uniflora* [a small proportion of the smaller grains may overlap in size with the *Plantago maritima* type. See ref. under 23a. *Ranunculus parviflorus* may key out here according to data in Santisuk, 1979 and unpublished data supplied by Punt, personal communication. It may be distinguished by its dimorphic columellae and larger porus number of 20+]

25b Verrucae indistinct. Grain diameter <30 μm. 9−17 pori present. Consult a type slide: *Plantago tenuiflora* [see ref. under 23a]

26a Microechinae (or scabrae) **indistinct** and scarcely visible in phase contrast. Verrucae rather small compared to grain size. Consult a type slide: *Plantago major* [see ref. under 23a]

26b Microechinae (or scabrae) **distinct**, usually in bright field as well as in phase contrast: 27

27a Verrucae large and coarse compared to grain size. Margins of pori poorly defined. Consult a type slide: *Plantago media* [see ref. under 23a]

27b Verrucae rather smaller compared to grain size. Margins of pori more clearly delimited. Consult type slides: *Plantago maritima* type [the type includes *P. maritima*, *P. alpina* and *P. arenaria*, but grains that key out at this point are most likely to have been produced by

P. arenaria or *P. maritima*, as these most often have non-annulate grains. See discussion and ref. under 23a]

28a Grains with rugulae or winding striae, muri wide. Columellae sitting in a reticulate pattern underneath the ridges (use careful LO analysis, ×1000). Pori >50 in number: *Polemonium*

28b Grains tectate-perforate to eureticulate. Pori <50 in number: 29

29a Each porus small, just filling the bottom of a lumen in the reticulum. Pattern thus not interrupted by the pori: 30

29b Pori not in the lumina of the reticulum. Porus diameter much greater than that of the lumina, pori thus interrupt the reticulate pattern: 31

30a Non-porate lumina with baculate or granulate floors: *Polygonum persicaria* type [includes *P. persicaria*, *P. lapathifolium*, *P. mite*, *P. minus* and *P. hydropiper*. McAndrews *et al.*, 1973, report this type in several North American *Polygonum* species]

30b Non-porate lumina without bacula or granules. Small echinae visible on top of the muri: *Daphne* [*D. laureola* has a much bigger grain with thicker exine than *D. mezereum*. Distinction is possible by reference to type slides]

31a Pori small and indistinct with no well defined circular edge and no annulus present. Porus demarcated merely by the cessation of reticulate sculpture: *Buxus*

31b Well-defined, more or less circular edge to each porus, annuli may or may not be present. Grains foveolate to eureticulate: 32

32a Annuli distinctly present, each seen as a thinning of the sexine, usually accompanied by a neat ring or band around the edge of each circular porus: some members of the Caryophyllaceae [see the special key on this page for separation of types within this family]

32b Annuli either indistinctly present or absent, porus edge neat but often irregular and not smoothly circular. Porus membrane always with large irregular granules or islands. Grain foveolate, or eureticulate with wide muri that approximately equal the width of the foveolae or lumina: *Liquidambar styraciflua**

SPECIAL KEY TO CARYOPHYLLACEAE − FROM KEY EXIT 8b

This key covers North West European species, being limited by the type material available to us. It is very tentative and to be used only in conjunction with type slides. Not every fossil grain will be identifiable. The key does not include *Herniaria*, *Illecebrum* and *Corrigiola* as these are rather distinct from other Caryophyllaceae and are separated in the main part of the polypantoporate key. The use of phase contrast microscopy is strongly advised for this key, wrong identification may occur without it. Counting of pori requires some skill; careful focusing allows the detection of the pori in face view on both sides of the grain, but the pori that are in optical section around the grain outline are easily missed, leading to an underestimate of the total number. Other works which have attempted distinctions within the Caryophyllaceae include Chanda (1962), Fredskild (1967), Birks (1973), Faegri and Iversen (1989). These works formed the basis of our investigations, and thus of the key that follows.

1a Grains foveolate or eureticulate in the mesoporium, foveolae or lumina wider than the width of a columella. Columellae in a reticulate pattern under the microechinate muri. Usually more than 20 pori: 2

1b Grains tectate-perforate, perforations <or = width of a columella in cross section. Columellae may or may not be in reticulate pattern (an infrareticulum):   3

2a Muri duplicolumellate or possibly represented by one row of branched columellae:
      Silene latifolia ssp. *alba*

2b Muri simplicolumellate: *Silene dioica* type
[includes *S. dioica*, *S. noctiflora*, *S. gallica* and possibly others]

3a Perforations more frequent (of greater density) than columellae in the mesoporium. Perforations tend to be more regularly arranged than columellae: 4

3b Perforations of equal frequency to columellae or less frequent than columellae in the mesoporium: 10

4a Pori in depressions on grain surface, grain outline obtusely angular with 5−6 sides due to the mesoporium rising up into a pattern of ridges between the pori. Less than 15 pori: 5

4b Pori not in depressions that cause the rest of the grain to appear in a pattern of ridges. Outline thus not obtusely angular with 5−6 sides. More or less than 15 pori: 7

5a Each porus edged by a **broad** (1.5−2.5 µm wide) **solid** ring or band (appears black in phase contrast):   *Gypsophila repens* type
[includes *G. repens* and *G. fastigiata*. Occasional grains of *G. paniculata* can key out here as they appear angular and have pori with rings that approach the lower end of the above width range]

5b No such broad solid band as part of the annulus to each porus. Either narrow bands or, if a diffuse ring-like structure is present, then it never appears solid black in phase contrast: 6

6a Columellae uniform, very coarse, cylindrical, and limited in distribution a few rows under the ridges only — leaving a wide area around each porus where no columellae occur: *Scleranthus*

6b Columellae a mixture of both coarse and fine, longest and coarsest under the ridges but becoming slimmer and shorter towards each porus margin. Prominent microechinae on the ridges:
      Stellaria holostea
[occasionally, other *Stellaria* species appear angular, but never to the same degree as *S. holostea*]

7a Grain with rather small, poorly defined pori that may not have even a narrow distinct band around the edge. Always >15 pori, and commonly >20. Columellae are irregularly distributed, but do not vary much in thickness. Pori with membranes intact may show a **solid** operculum. Grain size <35 µm and commonly <30 µm: *Sagina*
[microechinae vary from indistinct to very distinct. Fredskild, 1967, reports up to 67 pori in some Greenland *Sagina* species. He also indicates that an oval grain shape is common. Such grains are expected to be bigger than the sizes quoted here]

7b Grains with more clearly defined pori that are larger in relation to grain size. Columellae irregularly distributed but may or may not vary in thickness. Porus membranes (when present) always with several granules: 8

8a Columellae usually thinning towards their bases, varying to a greater or lesser degree in length and thickness. They are irregularly distributed in the following manner — coarse and sparsely arranged ones in the middle of the mesoporium; slimmer and more densely arranged ones in the region nearest the pori (may appear to be aggregated around the pori):
      Cerastium type
[includes *Cerastium*, most species of *Stellaria*, *Moenchia erecta*, *Holosteum umbellatum* and *Myosoton aquaticum*. Some *Stellaria* species appear slightly angular in outline. Porus number varies from 10 in *Myosoton* to >30 in some *Cerastium* species. Microechinae most distinct in *C. cerastioides* and some *Stellaria* species]

8b Columellae cylindrical, not varying much in length or in thickness and never aggregated around the pori. They are irregularly distributed

throughout the mesoporium, occasionally in a
faintly infrareticulate pattern: 9

9a With many pori (characteristically 30−40):
                               Lychnis flos-cuculi
[Birks, 1973, quotes this species with as low as
20 pori]

9b With fewer pori, 10−30:   *Arenaria* type
[includes *Arenaria ciliata*, *A. serpyllifolia*,
Moehringia trinerva, *Minuartia sedoides*,
M. verna, *M. hybrida*, *M. rubella*, *M. recurva*,
M. stricta. Note this type is very similar to the
Lychnis viscaria type − see exit 15b of this key.
For example grains of *Silene rupestris* may have
sufficiently dense perforations to key out under
9b]

10a Grain large, usually >45 μm, with >15 pori.
Columellae always rather crowded in the
mesoporium: *Agrostemma githago*
[Some very large grains of the *Silene vulgaris*
type may key out here, but they may be
distinguished by their coarse, spaced out
columellae]

10b Grains usually <45 μm, with more or less than
15 pori, columellae crowded or rather sparse: 11

11a Columellae arranged in a reticulate pattern at
low focus in the mesoporium (this pattern may
be very obscure and difficult to detect − the
reticulate pattern may break down so that the
columellae are more in irregular curving lines):   12

11b Columellae inordinately arranged (rather irregularly packed, not in a reticulate pattern
under the muri at low focus):   13

12a Perforations rather sparse, one small one in
the tectum covering each lumen of the infrareticulum. Columellae irregular or angular
in cross section, thinning towards the base.
Usually less than 16 pori. Grains up to 48 μm:
                               Dianthus type
[includes *Dianthus*, *Petrorhagia*, *Vaccaria* and
possibly *Saponaria* although the reticulate
arrangement is very obscure in this last. *Lychnis
alpina* may occasionally show an infrareticulate
pattern and key out here, however it is much
smaller than the others in the type]

12b Perforations rather more densely spaced, often
not directly over the lumina of the obscure
infrareticulum. More or less than 16 pori.
Grains up to 55 μm: *Silene vulgaris* type
[includes *S. vulgaris* ssp. *vulgaris*, *S. vulgaris*
ssp. *maritima*, *S. acaulis*, *S. otites*, *S. nutans*,
and *Cucubalus baccifer*. See also 15a below]

13a Grain with numerous pori (approximately
30−40):   14

13b Grains with <30 pori: 15

14a With coarse, usually sparse, columellae: *Silene*

conica type
[includes *S. conica*, *S. conoidea* and some grains
of *S. acaulis* with high porus number. *S. conica*
itself may be distinguished by the sculpture of
its operculum which takes the form of a ring
shape enclosing one or more islands]

14b With more slender columellae: *Lychnis flos-cuculi*

15a With coarse, usually sparse, columellae; and
prominent microechinae: *Silene vulgaris* type
[includes those species mentioned under 12b
and, additionally, *S. dichotoma*]

15b With more slender columellae, microechinae
may or may not be prominent:
                               Lychnis viscaria type
[includes *L. viscaria*, *L. alpina*, *Honkenya
peploides*, *Silene alpestris*, *S. rupestris*, *Gypsophila muralis*, *G. paniculata* and very occasional grains of *Lychnis flos-cuculi* with low
porus numbers. It should be noted that this
type is very similar to *Arenaria* type − see exit
9b. *G. muralis* and *G. paniculata* are distinct
because of their small size, and in addition,
G. paniculata is distinct because of its low
porus number of 12]

SYNCOLPATE (Plates 25, 28, 29)

Only grains which are very frequently syncolpate
are included here. Many trizonocolpate or trizonocolporate grains are occasionally syncolpate at the
poles.

1a Colpi running in spirals; or one or more circles,
curves or loops round the grain: 2

1b Colpi running relatively straight over the surface of the grain. More than one colpus: 5

2a Colpi always in neat, parallel spirals around the
grain; mesocolpia echinate: *Eriocaulon*

2b Colpi rarely in neat spirals, most commonly in
one or more regular or irregular circles or
loops: 3

3a One circular loop around the grain which
divides it neatly into two equal halves. Colpus
edges ragged, mesocolpia psilate:
                               Pedicularis palustris type
[includes *P. palustris* and *P. sylvatica*. *Hypecoum* may be syncolpate in this manner]

3b With one or more loops or spirals: 4

4a Usually with 2 colpi in rather neat concentric
circles on each side of the circular grain. Mesocolpia microreticulate: *Mimulus*

4b With colpi that are irregular, non-concentric,
elongated loops or in a loop which divides the
grain into two equal interlocking parts as do the

seams on a tennis ball. Mesocolpia tectate, psilate: *Berberis vulgaris* [see Blackmore and Heath, 1984] 6

5a Grain a flattened triangular-obtuse convex shape in polar view (the only view that it usually presents): 6

5b Grain circular or lobed in polar view, circular or elliptic in equatorial view. 10

6a Colpi fusing in each apocolpium to cut off an island of exine shaped like a triangle: 7

6b Colpi fusing in the apocolpia without cutting off islands of any sort. Surface psilate to scabrate-verrucate: *Myrtus* [*Primula scotica* and *P. stricta* may key out here but they have a tectate-perforate or microreticulate sculpturing according to Punt *et al.*, 1976]

7a Grain >30 μm in diameter, surface coarsely rugulate to striate with straight, abruptly truncated, muri which occasionally branch and are irregularly arranged. Muri simpli-columellate: *Nymphoides peltata*

7b Grain <30 μm in diameter, psilate or reticulate to tectate-perforate: 8

8a Psilate, no columellae detectable upon careful examination (×1000, phase contrast): *Eucalyptus** [often used as an 'exotic marker' grain in absolute pollen analysis. Frequently planted]

8b Microreticulate or tectate-perforate, fine columellae usually detectable upon careful examination (×1000, phase contrast): 9

9a Grains very small, <15 μm in diameter, microreticulate or tectate-perforate (sometimes psilate): *Primula farinosa* [The species shows heterostyly and two size classes are detectable. Data from Punt *et al.*, 1976]

9b Grains slightly larger, 15–21 μm: *Primula stricta, P. scotica* [data from Punt *et al.*, 1976 — see this reference for the distinction between these two species. The commonly planted *P. denticulata* has similar grains]

10a Grain 3- or 4-zonocolpate: 11

10b Grain with 5 or more colpi, pantocolpate: 13

11a With non-sunken colpi that have markedly ragged and granular edges, often split open. Surface psilate: *Pedicularis oederi*

11b With slightly sunken or inrolled colpi, grain size <22 μm: 12

12a Tectate, scabrate; with granulate colpus membranes. Colpi narrow in the mesocolpia and commonly widening in the sunken fusion area at each apocolpium: *Soldanella alpina* [data from Punt *et al.*, 1976. Our own observations indicate that some grains are not syncolpate and of the syncolpate ones, some do not show the widening of the colpi in the fusion area]

12b Microreticulate or tectate-perforate with nongranulate membranes: *Primula scotica, P. stricta* [data from Punt *et al.*, 1976 — see this reference for distinction between these species]

13a Microreticulate or tectate-perforate, <22 μm: *Primula stricta, P. scotica* [data from Punt *et al.* 1976 — see this reference for distinction between these species]

13b Echinate, verrucate-undulate, tectate-perforate or tectate-psilate. Usually >22 μm: 14

14a Surface tectate-perforate, echinate. Columellae coarse and possibly sitting in a reticulate pattern in the middle of the mesocolpia. Exine thinning towards the colpi: *Montia fontana*

14b Surface psilate to verrucate-undulate with no perforations. Colpi divide the surface into rounded or angular plates: 15

15a Surface irregularly verrucate-undulate. Consult a type slide: *Corydalis lutea* type [includes *C. lutea* and *C. ochroleuca*. Data from Kalis, 1979]

15b Surface psilate to faintly verrucate. Consult a type slide: *Mahonia aquifolia** [see Blackmore and Heath, 1984]

MONOCOLPATE (Plates 23–27, 60)

In this class are included all single-furrowed (monolete) spores as well as all single-furrowed pollen grains. This is because tetrad orientation is not apparent for any unknown sub-fossil grain or spore; and thus it is not possible to tell the difference between a distal polar position for the furrow (the colpus in monocolpate pollen grains) and a proximal polar position for the furrow (the laesura in monolete spores). Knox (1951) and Sorsa (1964) present detailed descriptions and/or keys of fern spores which will be useful to the specialist. Erdtman and Sorsa (1971) present some useful descriptions in relation to pteridophyte taxonomy. No attempt has been made to include North American pteridophyte spores in the present work, as our type material is inadequate for this area. McAndrews *et al.*, (1973) identify many fern taxa in their key covering pollen and spores of the Great Lakes region. It should be noted that the perine of many species is not very resistant to acetolysis (e.g. *Blechnum*) and perine features may be useful only if samples are unacetolysed or given very brief acetolysis times.

1a Grain with very long echinae (longest 5−12 μm) which are cylindrical-conical, solid, often with one or two constrictions along their length, and scattered over the surface. No columellae visible in the area between the echinae. The bases of the echinae project downwards to produce verrucae on the inner surface of the nexine. Colpus wide, obtuse-ended and often sunken or inrolled, but always with an operculum which bears echinae as large as those on the rest of the grain surface. Longest grain dimension >45 μm: *Nuphar*
[See Jones and Clark, 1981. *Cystopteris fragilis* may key out here but note that it has echinae hollow at the bases and a non-operculate furrow]

1b Grains elliptic, bean-shaped or D-shaped; with or without echinae but with colpi always narrow and non-operculate. Surfaces psilate, tectate-perforate, reticulate or with variously folded and wrinkled outer coats: 2

2a Grains elliptic, bean- or D-shaped, but always with some **columellate** structure in the wall (use ×1000 and view an optical section of the wall combined with LO analysis of the surface). Surface psilate, tectate-perforate, echinate or apparently eureticulate: 3

2b Grains usually bean-shaped, never with any columellate structure in the wall (viewed as in 2a). Surface psilate, granulate or possessing two coats; the outer (perine) variously folded, wrinkled or thrown into papillae, verrucae or echinae: 20
[note that columella-like structures are seen in *Asplenium* species in the outer coat, but these are not true columellae]

3a Grains semitectate, foveolate to eureticulate; at least in the centre of the mesocolpium (use phase contrast to clarify this feature):   4

3b Grains tectate-perforate to microreticulate (muri >or = lumina in width) in the centre of the mesocolpium; or microechinate; or with fine indistinct psilate-scabrate sculpturing (use phase contrast to clarify this feature):   12

4a Muri simplicolumellate, at least in the centre of the mesocolpium. Lumina may or may not decrease towards the colpus: 5

4b With duplicolumellate or pluricolumellate muri. Lumina decreasing in size gradually towards the colpus margins so that there is a wide area of tectate-perforate, or even tectate-psilate exine next to the colpus, where the columellae are in a more or less even carpet: 10

5a Grain surface foveolate in phase contrast. Lumina size decreases towards colpus so that margin appears tectate: *Tulipa sylvestris*

5b Grain surface reticulate, even under phase contrast. Lumina size may or may not decrease towards the colpus: 6

6a Grain very large (>50 μm in length, and often over 90 μm) Lumina size showing either no decrease, or only a slight decrease towards the colpus margin and grain ends. Columellae very coarse, heads (capita) appearing swollen in exine optical section: 7

6b Grain smaller, <40 μm. Lumina gradually decreasing in size towards the colpus margin and grain ends. Columellae coarse or fine: 8

7a Muri relatively thinner compared to grain size and not verrucate. Floors of lumina not sunken. Colpus a large irregular-edged rupture. Lumina size may vary from one part of the grain to another. Careful analysis under phase contrast may reveal the presence of a very thin, perforate tectum covering the lumina: *Iris pseudacorus* type
[includes *I. pseudacorus* and *I. foetidissima*. Some distinction may be made between these two on the basis of exine thickness. The latter has a thicker exine composed of longer columellae − consult type slides]

7b Muri thick and apparently verrucate; each composed of a single line of coarse structures which are oblong, quadrangular or triangular in surface view and narrow towards their bases. Columellae very short and slim in a line underneath these quadrangular or triangular structures. Floor of each lumen sunken and covered with more or less deciduous granules (some ring-shaped). Reticulum meshes very large compared to grain size. Grain neatly elliptic. Colpus membrane granulate if present:
                    Lilium martagon type
[includes *L. martagon* and *L. bulbiferum*. The structures on the muri are quadrangular or triangular in the former and more oblong in the latter]

8a Lumina small (approaching microreticulate) fairly uniform in size in the mesocolpium but showing a slight and very gradual decrease towards the colpus and the grain ends. Shape of grain usually narrowly elliptic, exine thin and with fine columellae − these may be clearly visible only under phase contrast:
                    Narthecium ossifragum

8b Lumina larger, uniform or varying in size in the mesocolpium but showing a **marked decrease** in size towards the colpus. Grain shape circular to elliptic. Columellae coarser and clearly visible: 9

9a Lumina in the mesocolpium decrease sharply to a broad colpus margin that is clearly tectate-perforate with columellae in a uniform carpet. Lumina >3 μm in diameter, rather rounded. Grain often circular in outline:

Butomus umbellatus

9b Lumina decrease more gradually to colpus margin. No tectate-perforate margin, i.e. still some small lumina detectable at the very edge. Largest lumina 3 μm or less, lumina rather irregular in shape (not rounded) with strong, black muri in phase contrast. Grain more often elliptic. Consult type slides: *Veratrum album* [the introduced *Erythronium dens-canis** may key out here]

10a Colpus very long — curving round the ends of the grain so that it is almost divided into two lobes like the halves of a bivalve mollusc shell:

Muscari

[our observations indicate that some separation of species may be possible, based on the size of reticulum meshes]

10b Colpus not curving around the ends of the grain in the above characteristic manner, but ending at or before the grain ends: 11

11a In the centre of the mesocolpium some lumina may be large (>2 μm) although there may be small ones immediately adjacent (consult type slides on this feature):   *Fritillaria* type [includes *Fritillaria meleagris*, *Anthericum* and *Ornithogalum*. It is possible that *Polygonatum* may key out here; see 17b and 18a below]

11b Largest lumina in the mesocolpium 1−2 μm only. Sizes of the lumina in the mesocolpium may or may not vary. Consult type slides:

Scilla type

[includes *Scilla*, *Hyacinthoides*, *Maianthemum*, *Lloydia* and perhaps also some grains of *Polygonatum* and *Gagea* — see exits 17b, 18a and 19b below. Some distinction within this type is possible if sufficient type slides are available. *Lloydia* has distinctive thick muri that appear solid black in phase contrast]

12a Grain with coarse, **very sparse** columellae visible under the tectum. Surface psilate, tectate-perforate and microechinate. Size usually >50 μm: *Gladiolus* type [includes *Gladiolus* and *Romulea*. *G. illyricus* has microechinae so small as to be almost undetectable]

12b Grain with denser, finer columellae; and surface never microechinate or echinate, being tectate-psilate or tectate-perforate: 13

13a Grain psilate, apparently non-perforate: 14

13b Grain tectate-perforate (observe ×1000, phase contrast may be needed to detect the perforations) or microreticulate: 18

14a Grain D-shaped in side view, in 3 dimensions shaped rather like an orange segment. Colpus set on curved face of the grain, i.e. the one opposite the straight edge of the 'D'. Colpus sometimes slightly sunken-in and in some species with clearly rounded or squared-off ends: 15

14b Grain elliptic in side view, in three dimensions not shaped like an orange segment: 16

15a Grains <50 μm in length. Consult type slides:

Allium type

[includes *Allium*, *Narcissus* and *Ruscus*. Some variation in the length of the colpus is observed on type slides of *Allium* species — a very long colpus that travels round to the other side of the grain is characteristic of *Allium carinatum*, whilst *A. triquetrum* and *A. fistulosum* have rather short colpi; other species have intermediate colpus lengths)

15b Grains usually >50 μm in length. Consult type slides: *Polygonatum*

16a Colpus long and travelling round the ends of the grain such that it sometimes splits open along the colpus into two lobes like an opened bivalve mollusc shell. Grain <35 μm:

Galanthus type

[includes *Galanthus* and *Leucojum*. Some *Muscari* sp. may key out here]

16b Colpus stops at or before the ends of the grain. Size usually >35 μm: 17

17a Mesocolpium concave, such that the grain appears slightly 'bone'-shaped in the view with colpus facing observer and 'boomerang'-shaped in side view. Colpus on the only convex side to the grain: *Sisyrinchium*

17b Mesocolpium flat or convex. This exit leads to the following three taxa which require comparison with type slides for separation:

(i) *Brasenia purpurea* [this species has few positive features under LM (has distinctive surface under SEM) but may be an important fossil taxon in European interglacial sediments (the plant is not found in the flora today). See account of morphology in Clarke and Jones, 1981c]

(ii) *Allium* type

(iii) *Polygonatum* [grains of these last two key out here if they are not clearly D-shaped or if they present in a position with colpus facing the observer]

18a Perforations quite **widely spaced** (view under phase contrast) and present only in the middle of the mesocolpium between large undulations which can look like irregular rugulae or reticulations. Grain commonly D-shaped in side view, usually >50 μm long: *Polygonatum*

18b Perforations rather closely spaced all over the grain surface (distance between the holes the same, or not much more than, the size of the holes). Perforations usually present in the meso-colpium and near the colpus margins, although they may be smaller in the latter position. Grains elliptic in side view: 19

19a Perforations regular in shape, very small and very close together all over the grain surface (i.e. about the same in density and distribution as the columellae):   *Convallaria* type
[includes *Convallaria* and *Asparagus*. The latter may be significantly smaller — consult type slides. See also *Narthecium*]

19b Perforations regular or irregular-elongate in shape, larger and less densely spaced than the columellae: *Paris* type
[includes *Paris* and some *Gagea* grains with small foveolae. *Paris* has a very wide, granulate-verrucate colpus membrane and distinctive surface in phase-contrast. *Gagea* may be distinct because the perforations in the membrane become gradually smaller and closer together towards the colpus]

20a Grain (spore) neatly bean-shaped and completely psilate:

    Polypodiaceae — spores of the monolete ferns
[spores minus their outer perine which has been lost during fossilization or acetolysis. Note that some grains of *Calla palustris* appear monocolpate and key out here. They are distinct because of their small size — usually less than 32 μm]

20b Grain (spore) bean-shaped but never completely smooth. Bean-shape often obscured by outer coat (perine) which is variously wrinkled, echinate, verrucate, foveolate or reticulate or any combination of these: 21

21a Sculptured outer perine closely attached to inner coat, or thrown up into **solid** elevations such as echinae or ridges (no hollows detectable in these structures):   22

21b Perine separated from inner coat either as a whole; or by **hollow** elevations such as sacci, ridges or echinae, between which it is quite closely adherent to the inner coat (note that the hollow part may be confined to the very base of the echinae):   26

22a Spore large, >60 μm long. Surface with large, undulating, solid verrucae. Individual large verrucae >3 μm high, elongated in the region of the laesura (furrow) and sometimes noticeably covered by granules or smaller verrucae: *Polypodium vulgare* type

[includes *P. vulgare*, *P. australe* and *P. interjectum*]

22b Spores <60 μm long. Surfaces various, but not with large, undulating, solid verrucae: 23

23a Perine extremely faint, forming quite a thick layer and appearing reticulate or with 'enclosed bubbles'. It is essential to compare with a type slide: *Blechnum spicant*
[it is thought extremely unlikely that the perine of this species would survive in a fossil state, it has not been seen to survive acetolysis in the making of type slides. If anything of the perine survives it may be as a very thin, slightly wrinkled layer — see picture in Sorsa 1964]

23b Perine echinate or with various sizes of granules: 24

24a Surface with small gemmae, verrucae or scabrae: *Athyrium filix-femina*
[note that spores of many other species may conform to this description when the perine is in the last stages of disintegration, so great care must be taken in the identification of *A. filix-femina* as a fossil]

24b Surface echinate: 25

25a With densely spaced, irregular-echinate processes: *Polystichum* species
[see also exit 35a]

25b With long, apparently solid, echinae: *Cystopteris fragilis*
cf. also *Thelypteris palustris* and exit 37a]

26a Perine either entirely detached from the bean-shaped inner coat, or thrown into sacci or ridges that have only slight contact with it, so that the overall perine outline only roughly conforms to that of the inner coat: 27

26b Perine closely attached to the bean-shaped inner coat **except** where it is thrown up into hollow ridges, echinae or irregularly pointed processes. Thus its outline conforms to that of the inner coat (the perine may appear falsely separate if the bean-shaped inner coat is collapsed or contracted):   34

27a Perine either unridged or thrown into a few loose wrinkles only (usually up to 4 ridges visible on one side of the spore). Perine loose, often apparently completely separated from the inner coat: 28

27b Perine thrown into many sacci (rounded protrusions, broader than high) or ridges. Usually >4 sacci or ridges visible on one side of the spore: 29

28a Perine not especially bulging out over the laesura (furrow). Surface densely covered in small echinae (1.5–12.0 μm long) which project from small rugulae: *Dryopteris dilatata*

[from Sorsa, 1964, it appears that *Matteuccia*, *Athyrium distentifolium* and *A. crenatum* may key out here if they have rather few folds on the perine. None of these have the clear echinae and rugulae]

28b Perine bulges out over the laesura (furrow) in a very characteristic manner although this may not be noticeable as the perine is often ruptured and ragged in this area. Perine surface psilate, granulate or verrucate: *Isoetes* [includes *I. lacustris*, *I. echinospora* and *I. hystrix*. *I. lacustris* has granules on the perine. *I. hystrix* has low faint verrucae or broad-based echinae. Birks, 1973, has shown how size criteria are useful in separation of *I. lacustris* and *I. echinospora* in a mixed fossil population]

29a Perine surface smooth or with only very faint granules, the whole surface being thrown into short curving and twisting ridges or bulging sacci which rarely intersect and anastomose: *Dryopteris filix-mas* type [includes *D. filix-mas*, *D. oreades*, *D. affinis*, *D. aemula*, *D. submontana*, *Gymnocarpium robertianum*, *Oreopteris limbosperma*, *Cystopteris dickieana*]

29b Perine surface echinate, granulate, microverrucate or foveolate or reticulate; surface thrown into sacci or ridges which may or may not intersect and anastomose: 30

30a With mainly sacci or short ridges which do not anastomose: 31

30b with mainly ridges or folds that are long and sinuous or tend to intersect and anastomose to form a very coarse reticulum: 33

31a With echinae and small rugulae on the sacci: *Dryopteris cristata* type [includes *D. cristata* and *D. carthusiana*. *Cystopteris dickeana* may key out here]

31b With granules, foveolae or reticulations on the sacci. Perine thick: 32

32a Perine surface granulate: *Phegopteris connectilis*

32b Perine surface foveolate-reticulate: *Gymnocarpium dryopteris* [*Polystichum braunii* may key out here — see Sorsa, 1964]

33a With fairly low ridges, whole perine surface may be covered with small granules, scabrae, microechinae or rugulae: *Woodsia* type [includes *W. ilvensis*, *W. glabella*, *W. alpina*, *Athyrium distentifolium*, and perhaps *A. crenatum* and *Matteuccia struthiopteris* — see Sorsa, 1964]

33b With perine thrown into fairly high, coarse ridges which may or may not anastomose. Small echinae are present on the tops of the ridges

and columella-like structures can be seen under them in most species: *Asplenium* type [includes *A. adiantum-nigrum*, *A. viride*, *A. billotii*, *A. trichomanes*, *A. ruta-muraria*, *A. septentrionale*, *A. ceterach*. They may be distinguished from *Woodsia* by the lack of columella-like structures in the latter]

34a With ridges being the most prominent gross feature. Ridges are fairly high, coarse, and may or may not anastomose. Between these ridges the perine adheres to the grain inner coat. Small echinae are present on top of the ridges and columella-like structures can be seen under the ridges in most species: *Asplenium* type [includes *A. adiantum-nigrum*, *A. viride*, *A. billotii*, *A. trichomanes*, *A. ruta-muraria*, *A. septentrionale*, *A. ceterach*. They may be distinguished from *Woodsia* by the lack of columella-like structures in the latter]

34b With **echinae** or **papillae** being the most prominent gross feature: 35

35a With densely spaced irregular-echinate processes which are joined at their bases by small ridges forming a reticulum-like structure: *Polystichum* type [includes *Polystichum setiferum*, *P. lonchitis*, *P. aculeatum*. *Asplenium marinum* and *Asplenium scolopendrium* may conform to this description but are distinct upon consulting type slides. According to Sorsa, 1964, *Polystichum braunii* appears to be rather different to the other *Polystichum* species]

35b Discrete echinae or papillae present: 36

36a With low, rounded papillae or processes shaped somewhere between papillae and hollow echinae: *Cystopteris montana* type [includes *C. montana* and *C. sudetica* — see Sorsa, 1964]

36b At least some long pointed papillae or echinae present: 37

37a Echinae usually densely spaced, long (up to 10 μm) and mostly solid — usually **hollow at the bases only**: *Cystopteris fragilis*

37b Echinae shorter, **hollow along more of their length** and with lines or radiating reticulate pattern between their bases. The reticulate pattern may remain even when the echinae are reduced or lacking — see Sorsa (1964): *Thelypteris palustris* [*Asplenium marinum*, *A. ruta-muraria* and *A. scolopendrium* may key out here if their echinae are very pronounced, but they are distinct from *Thelypteris palustris* — consult type slides]

DIZONOCOLPATE (Plate 25)

Very few pollen grains have two colpi as their normal complement, but the 2-colpate condition does sometimes occur in grains which are normally 3- (or more) colpate, as an aberrant state.

1a Grain psilate, microechinate or faintly scabrate; in the last case sometimes with faint foveolae that are elongated into crack-like lines in phase contrast: 2

1b Grain reticulate: 4

2a No columellae visible, even in phase contrast. Not syncolpate, although apocolpium may be very small. Surface psilate, scabrate, or with foveolae that may be elongated and crack-like:
Calla palustris
[consult a type slide. Grains may also appear irregularly monocolpate]

2b Columellae visible as a fine and dense carpet. Grain may or may not be syncolpate at the poles: 3

3a Surface completely psilate, grain syncolpate at the poles with the colpi often split open such that it appears in two halves:
Pedicularis palustris type
[includes *P. palustris* and *P. sylvatica*]

3b Surface densely microechinate, grain not syncolpate at the poles: *Hypecoum*

4a Grain reticulate, lumina becoming gradually and markedly smaller towards the colpi. Columellae visible (×1000) coarsest and longest in the area between the colpi. Greatest dimension >25 µm:
Tamus communis
[in this grain the two colpi are actually unlikely to be in an equatorial zonal position — see Clarke and Jones, 1981b; but in practical analysis it is impossible to determine this feature on a fossil grain. It is likely to be exceedingly rare as a fossil and workers may more often encounter aberrant dicolpate *Salix* grains. Consult a type slide]

4b Grain reticulate with lumina near the colpi only slightly smaller than those away from the colpi. Columellae minute to invisible (×1000, using phase contrast). Exine thin (1 µm). Grain <25 µm: *Tofieldia*

TRIZONOCOLPATE (Plates 29–40, 62, 65–67, 69–71)

This class covers very many grains and thus is subdivided into sections based on the sculpturing type as follows (see pp. 74–77 for details of sculpturing):

Trizonocolpate, psilate to perforate or microreticulate

Note: No key is entirely satisfactory for this very difficult section. Many of the important features are at the limits of resolution of the light microscope. The key was written using grains acetolysed and embedded in glycerine jelly. The increase in size associated with these two treatments, compared with simple KOH and glycerine jelly, or with acetolysis and silicone oil means that differences in very small structures such as columellae and perforations are more easily seen. Workers using methods other than the ones we have, may well find that they do not have enough resolution to detect the very fine details required by this key and may thus end up in the wrong branch, with consequent misidentification. Even if an identification seems clear, type slides should always be consulted.

The correct use of the term 'microreticulate' may present slight problems in this key. Generally a grain has been included under this heading if it could be classed as 'microreticulate' in any of the type collections at our disposal; even if some of the slides showed bigger reticulations due, perhaps, to longer acetolysis times.

1a With exine thickness varying over the surface of the grain, e.g clearly thickest at poles or thickest at various regions in the mesocolpia (careful observation of optical sections under ×1000 is necessary):   2

1b With exine thickness the same all over the grain surface: 10

2a Exine clearly thickest at the poles: 3

2b Exine thickest in the mesocolpia (or at least in **parts** of each one e.g. thickest in a ridge down the centre of the mesocolpium, or thickest in a ring around a thin-walled circular depression in the centre of each mesocolpium):   8

3a Colpi relatively short, occupying approximately $\frac{2}{3}$–$\frac{3}{4}$ of the distance between the poles (whole colpus length easily visible in equatorial view). Colpi narrow, slit-like, without granulate membranes. Grains elliptic to circular in equatorial view. Surface may be perforate, with each perforation **directly over a branched columella** (careful LO analysis of surface, ×1000 is needed). This is especially clear in the thickened apocolpia. Grains very variable in size: *Polygonum bistorta* type

[including *P. bistorta* and *P. viviparum*. Rare grains where the pori are not apparent. This pollen type is very variable. See also p. 142 *P. aviculare* may occasionally have the endoporus inconspicuous and key out here, although it is distinct because of its non-perforate tectum]

3b Colpi long, grain **narrowly elliptic** to rhombic-obtuse in equatorial view. Colpi narrow and slit-like or wide   4

4a Narrow slit-like colpi: 5

4b Wide colpi with **granulate** membranes: 6

5a Grains rhombic-obtuse, with flat or concave mesocolpia. In polar view, shape triangular-obtuse with colpi on the angles:
                 Bupleurum and *Pleurospermum* type
[grains where the colpi are unruptured and the endoporus is undetectable. See p. 138 for the taxa included and consult type slides for their differentiation]

5b Grains elliptic, mesocolpia always convex. Shape circular to lobed in polar view. Columellae long and prominent in surface view and optical section, being longest at the poles and present in a dense, even carpet in surface view. The thickened polar areas may give the appearance of a swollen 'cap' to each apocolpium, giving the equatorial view an outline similar to that of a lemon. Careful examination under phase contrast may reveal perforations and microechinae: *Adonis aestivalis* type
[according to W. Punt (personal communication of unpublished data) this includes *A. aestivalis*, *A. annua* and *A. flammea*; only the first of these showing the polar 'caps' well developed in most of its grains. It is our observation that *Aconitum* grains that are not well expanded may key out here, but they are easily distinguished by their short, indistinct columellae. It is noted that *Stachys officinalis* may key out here if the reticulum is indistinct, but this grain is clearly different if type slides are available]

6a Tectum thick, columellae **short and indistinct to invisible** in surface view under ×1000 and phase contrast (detectable in optical section). Colpi wide and gaping (if not inrolled) with markedly granulate membranes. Surface completely psilate or very faintly microechinate. Consult a type slide: *Aconitum*
[type slides should always be consulted before identifying a grain to this type, because it has a shape that is assumed by many other grains when in a semi-collapsed state. In this condition the grain folds-in along the lines of the colpi, making them appear longer and the grains thus

lobed in polar view. Such folding obscures pori and thus grains which fit the above description have been seen on slides of, for example, *Sorbus*, *Pyrus*, *Malus*, *Rubus*. Slight thickening of the poles may appear in such folded grains. They should, of course, be distinct from *Aconitum* because of their narrow colpi]

6b Columellae longer and thicker, being prominent in surface view and optical section, longest at the poles. Microechinae detectable on top of the tectum on close examination (×1000, phase contrast):   7

7a Columellae in a dense, even carpet in surface view. Microechinae less densely spaced than the columellae and not associated with them. Thicker exine at each pole sometimes developed into a polar 'cap', giving the grain a lemon-shaped outline in equatorial view. Consult type slide: *Consolida ambigua* type
[according to W. Punt (personal communication of unpublished data) this includes *C. ambigua* and *C. regalis* ssp. *regalis*]

7b Columellae coarse and rather spaced-out. Tectum layer clear in optical cross section. Tectum perforate and with a microechina on top of each coarse columella:
                 Cuscuta europaea type
[includes *C. europaea*, *C. epithymum*, *C. epilinum* and *C. campestris* — see Cronk and Clarke, 1981. Pantocolpate grains may also be produced]

8a With exine thickest, and columellae longest and coarsest, in a zone down the centre of each mesocolpium. Shape may be rectangular-obtuse in equatorial view and is markedly triangular-obtuse in polar view (colpi between the angles):   *Alchemilla* type
[includes *Alchemilla* and *Aphanes*. Some *Alchemilla* species produce only deformed grains. Some *Euphorbia* species may key out here if there are no indications of pori e.g. *E. exigua*. *Euphorbia* may be distinguished from *Alchemilla* type by its perforate tectum, coarse columellae and circular or lobed polar view]

8b With exine thinnest in the centre of each mesocolpium (see polar view, optical section):   9

9a With a circular area of short columellae and thinner (sometimes sunken) exine in the centre of each mesocolpium. This area is surrounded by a slightly projecting rim of thicker exine where the columellae may be coarser, longer and more spaced out. Colpi markedly narrow and slit-like. Grain shape triangular-obtuse to circular in polar view (colpi on the angles) octangular to elliptic in equatorial view.

Surface tectate, non-perforate: *Melampyrum*
[the depressed centre of each mesocolpium
varies in its prominence between the species of
the genus]

9b Exine thicker in a strip down the margins of
each colpus, slightly thinner in a wide zone
down the centre of each mesocolpium. Columellae coarse and rather irregularly spaced out
under the smooth tectum. Grain circular in
polar and equatorial views: *Anemone nemorosa*

10a Grain large, >45 μm long. Exine thick (>4 μm)
and columellae **very coarse** and **clearly branched**
at $\frac{2}{3}$ of their length. Tectum with very small,
closely spaced perforations (much more
numerous than columellae). Colpus membrane
sparsely covered with granules, gemmae or
bacula: *Convolvulus arvensis*

10b Grains usually <45 μm long, but some may be
larger. Exine usually <4 μm, columellae
unbranched. Perforations (if present) sparser,
i.e about as numerous, or less numerous, than
columellae; or grain microreticulate: 11

11a Grain triangular-obtuse or triangular-convex in
polar view, with short narrow colpi. P/E ratio
1.0 or less: 12

11b Grain circular to lobed in polar view, colpi
longer. P/E ratio usually **1.0 or more**:   13

12a Colpi sometimes split and gaping; each one
crossed by two nexinous ridges, one near each
of its ends. Columellae distinct under ×1000.
Grain markedly broader than long (P/E ratio
<1.0):   *Symphoricarpos**
[this grain is actually trizonocolporate with a
very wide transverse endocolpus, the margins
of which are represented by the nexinous ridges.
See also *Linnaea borealis*]

12b Colpi never gaping and with no nexinous ridges
crossing underneath. Each colpus so short as to
resemble an elongated porus; margins irregular.
Columellae fine to invisible. Surface totally
psilate. Grain as long as broad, or only slightly
broader than long. Consult type slide: *Monotropa*

13a Grains microreticulate with lumina as wide, or
wider than, the muri which separate them (at
least in the middle of the mesocolpium). Phase
contrast ×1000, may be needed to detect this
and even then, this is a very difficult feature to
apply because of the extreme smallness of the
sculpturing elements: 14

13b Grains psilate or tectate-perforate with perforations <or = 1 μm and smaller than the areas
of tectum which separate them. Phase contrast
×1000 may be needed to detect the perforations as separate entities from the columellae: 21

14a Colpi relatively wide, with granulate mem-

branes (if colpi ruptured then more or less
ragged edges):   15

14b Colpi narrow, not markedly granulate: 19

15a Colpi relatively short (whole length visible in
an equatorial view) with obtuse ends: *Platanus**

15b Colpi longer (whole length not usually visible
in equatorial view) with acute ends: 16

16a Suprareticulate — careful focusing down on a
surface view under ×1000 and phase contrast is
the best method for detecting this condition,
although an optical section of the grain wall is
useful as it shows the reticulum walls to be very
shallow:
  some members of *Stachys sylvatica* type on p. 122
[type slides should be consulted. Keying out
here are those species with the smallest reticulations e.g. *S. alpina*, *S. palustris*, *S. arvensis*,
S. germanica, *S. sylvatica*, *S. annua*, *Lamium
purpureum*, *L. hybridum*, *L. amplexicaule*,
Leonurus cardiaca. No subdivision of the type
is possible because there is size overlap in the
lumina size — some preparations of these
species show lumina >1 μm wide]

16b Eureticulate — careful examination as under
16a is required. In optical section the reticulum
walls may be seen to be deeper: 17

17a Exine thins slightly towards colpi — columellae
can be seen to become gradually shorter
towards margins in polar view, equatorial
optical section (similar to, but not as pronounced as in, *Artemisia*). Consult type slide:
                                      Frankenia laevis

17b Exine not thinning towards the colpi — columellae of the same length right up to the edge
of the colpi (polar view, equatorial optical
section):   18

18a Grain most commonly circular in equatorial
view, microreticulations do not vary in size
over grain surface. Columellae clearly detectable under ×1000 and phase contrast. Consult
type slide: *Chelidonium majus*
[cf. also *Helleborus foetidus*].

18b Grain most commonly elliptic with obtusely
pointed apocolpia in equatorial view, microreticulations detectably larger in the centre
of each mesocolpium. Columellae not visible
under ×1000 and phase contrast — only the
microreticulum appears visible as a sharp black
network: *Reseda lutea* type
[includes *R. lutea*, *R. luteola* and *R. phyteuma*.
Some distinction between these may be possible, based on lumina size]

19a Sexine relatively thick — thicker than the
nexine (see optical section of wall, ×1000).
Columellae long, making up most of the sexine

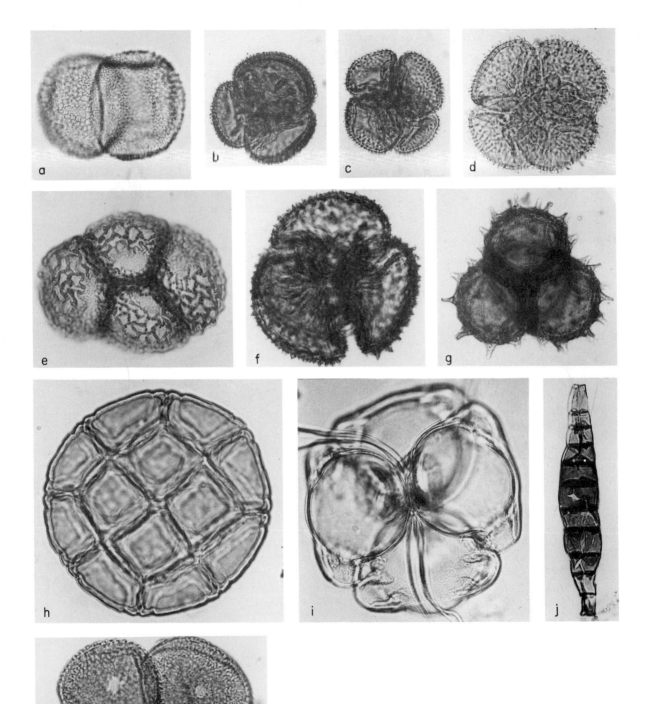

PLATE I (×1000 unless stated): (a) *Scheuchzeria palustris*; (b, c) *Drosera intermedia* (×500); (d, f) *Drosera rotundifolia* (d × 500); (e) *Listera* type (e.g. *L. cordata*); (g) *Selaginella selaginoides* microspores (×500); (h) Mimosaceae (e.g. *Acacia saligna*); (i) *Epilobium* type (e.g. *E. hirsutum*); (j) Fungal material (×750); (k) *Typha latifolia*.

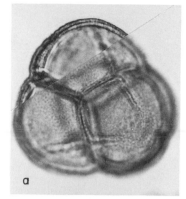

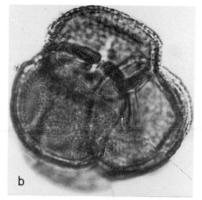

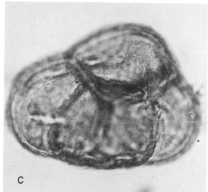

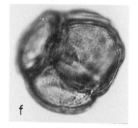

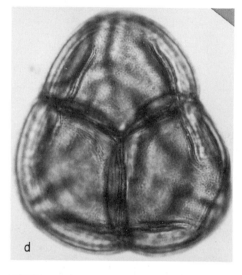

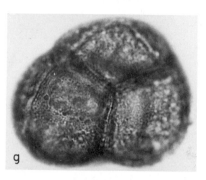

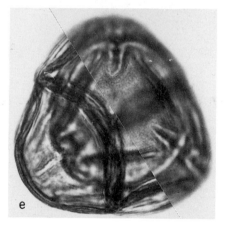

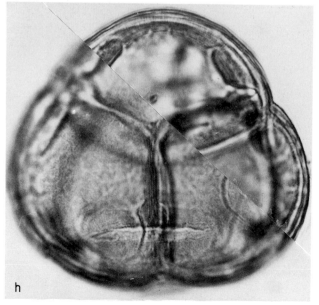

PLATE 2 (×1000) Ericaceae: (a) *Pyrola* type (e.g. *P. minor*); (b, c) *Calluna vulgaris* (see SEM Plate 59); (d, e) *Andromeda polifolia*; (f, g) *Phyllodoce caerulea*; (h) *Rhododendron ponticum*.

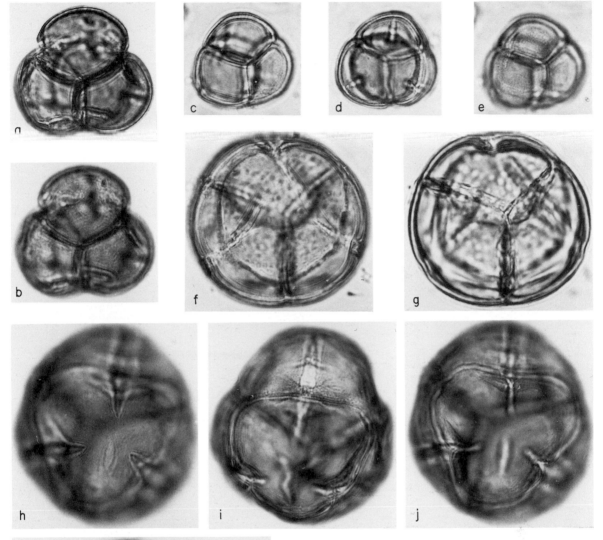

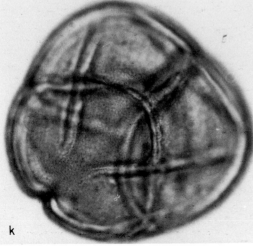

PLATE 3 (×1000) Ericaceae: (a, b) *Loiseleuria procumbens*; (c–e) *Empetrum nigrum*; (f, g) *Arctostaphylos alpinus*; (h–j) *Arbutus unedo*; (k) *Arctostaphylos uva-ursi*.

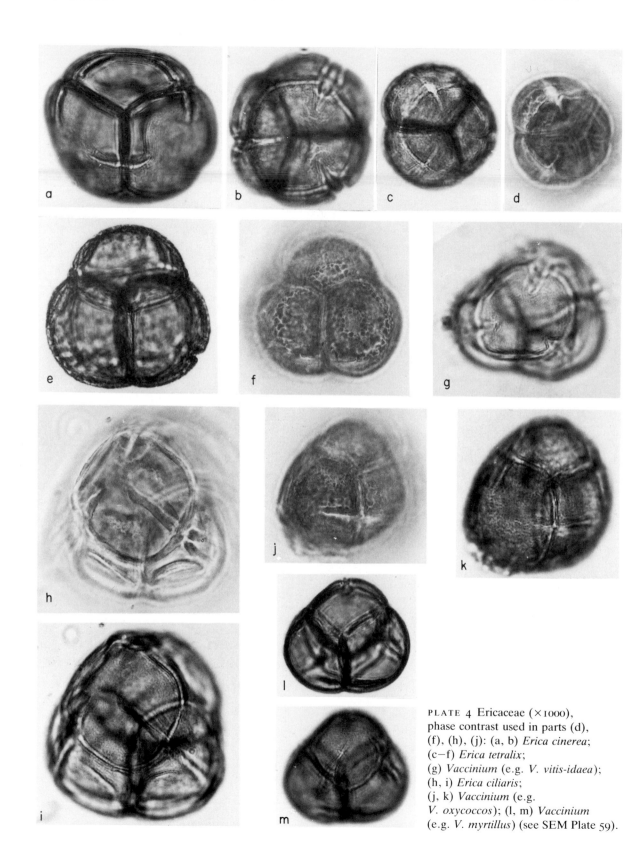

PLATE 4 Ericaceae (×1000),
phase contrast used in parts (d),
(f), (h), (j): (a, b) *Erica cinerea*;
(c–f) *Erica tetralix*;
(g) *Vaccinium* (e.g. *V. vitis-idaea*);
(h, i) *Erica ciliaris*;
(j, k) *Vaccinium* (e.g.
V. oxycoccos); (l, m) *Vaccinium*
(e.g. *V. myrtillus*) (see SEM Plate 59).

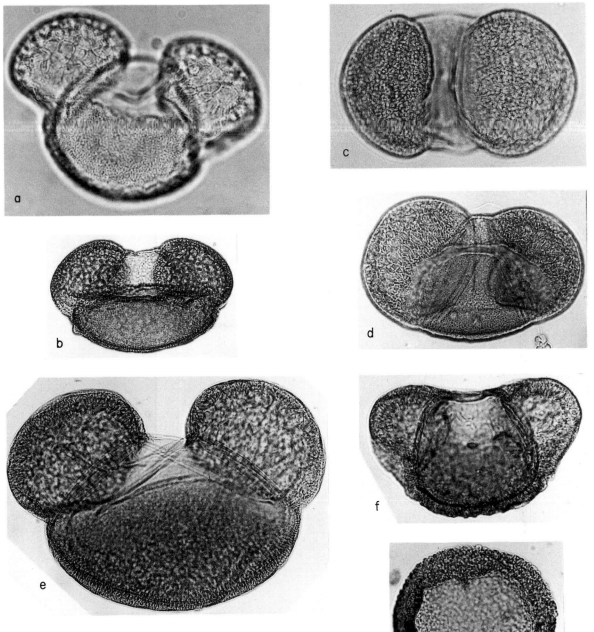

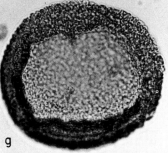

PLATE 5 (×1000 unless stated): (a) *Pinus* subgenus *Diploxylon* (e.g.
P. sylvestris — SEM Plate 59); (b) *Pinus* subgenus *Haploxylon* (e.g.
P. wallichiana); (c, d) *Picea* (e.g. *P. glauca*) (×500); (e) *Abies* (e.g.
A. pinsapo) (×400); (f) *Cedrus* (e.g. *C. deodara*) (×500); (g) *Tsuga* (e.g.
T. canadensis) (×500).

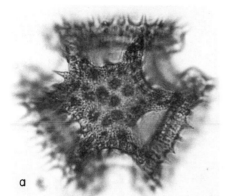

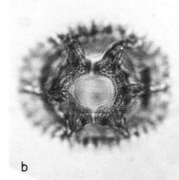

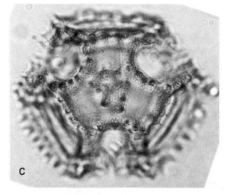

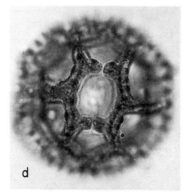

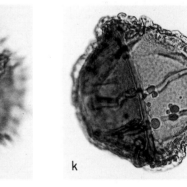

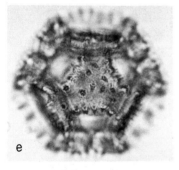

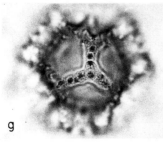

PLATE 6 (×1000 unless stated)
See SEMs in Plate 61:
(a, b) Lactuceae (e.g. *Taraxacum officinale*); (c, d) Lactuceae (e.g. *Sonchus arvensis*); (e, f) Lactuceae (e.g. *Tragopogon pratense*); (g, h) Lactuceae (e.g. *Lactuca sativa*); (i–k) *Trapa natans* (×500).

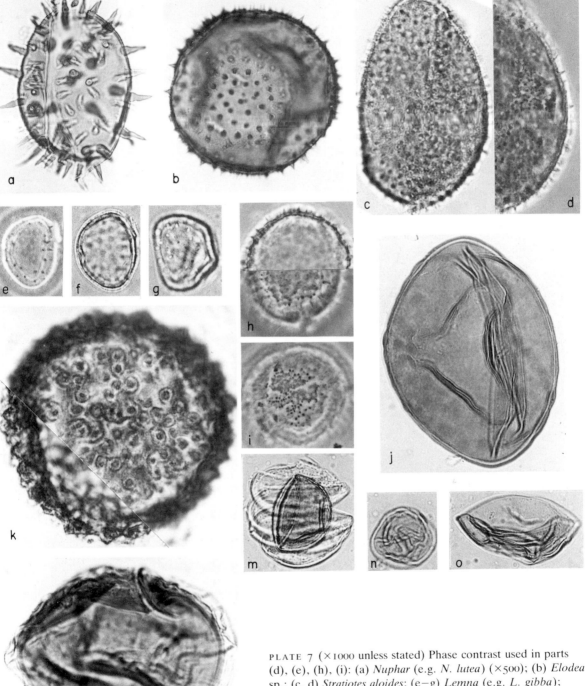

PLATE 7 (×1000 unless stated) Phase contrast used in parts
(d), (e), (h), (i): (a) *Nuphar* (e.g. *N. lutea*) (×500); (b) *Elodea*
sp.; (c, d) *Stratiotes aloides*; (e−g) *Lemna* (e.g. *L. gibba*);
(h, i) *Hydrocharis morsus-ranae*; (j) *Larix* type (e.g. *L. decidua*)
(× 500); (k) *Ranunculus arvensis*; (l−o) *Equisetum* (e.g.
E. arvense); (m−o) show variation in crumpled spores;
(m) with elaters evident (×500).

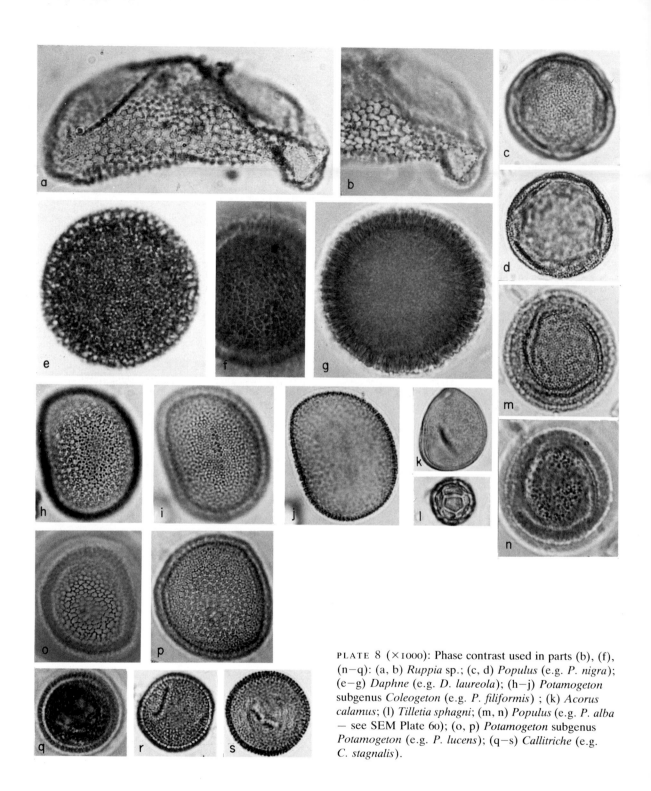

PLATE 8 (×1000): Phase contrast used in parts (b), (f), (n–q): (a, b) *Ruppia* sp.; (c, d) *Populus* (e.g. *P. nigra*); (e–g) *Daphne* (e.g. *D. laureola*); (h–j) *Potamogeton* subgenus *Coleogeton* (e.g. *P. filiformis*) ; (k) *Acorus calamus*; (l) *Tilletia sphagni*; (m, n) *Populus* (e.g. *P. alba* – see SEM Plate 60); (o, p) *Potamogeton* subgenus *Potamogeton* (e.g. *P. lucens*); (q–s) *Callitriche* (e.g. *C. stagnalis*).

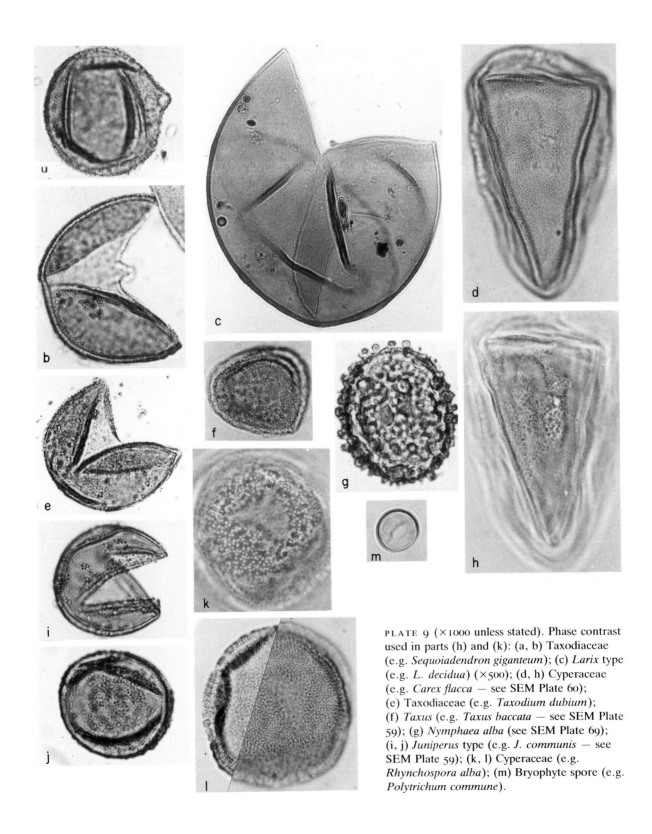

PLATE 9 (×1000 unless stated). Phase contrast used in parts (h) and (k): (a, b) Taxodiaceae (e.g. *Sequoiadendron giganteum*); (c) *Larix* type (e.g. *L. decidua*) (×500); (d, h) Cyperaceae (e.g. *Carex flacca* — see SEM Plate 60); (e) Taxodiaceae (e.g. *Taxodium dubium*); (f) *Taxus* (e.g. *Taxus baccata* — see SEM Plate 59); (g) *Nymphaea alba* (see SEM Plate 69); (i, j) *Juniperus* type (e.g. *J. communis* — see SEM Plate 59); (k, l) Cyperaceae (e.g. *Rhynchospora alba*); (m) Bryophyte spore (e.g. *Polytrichum commune*).

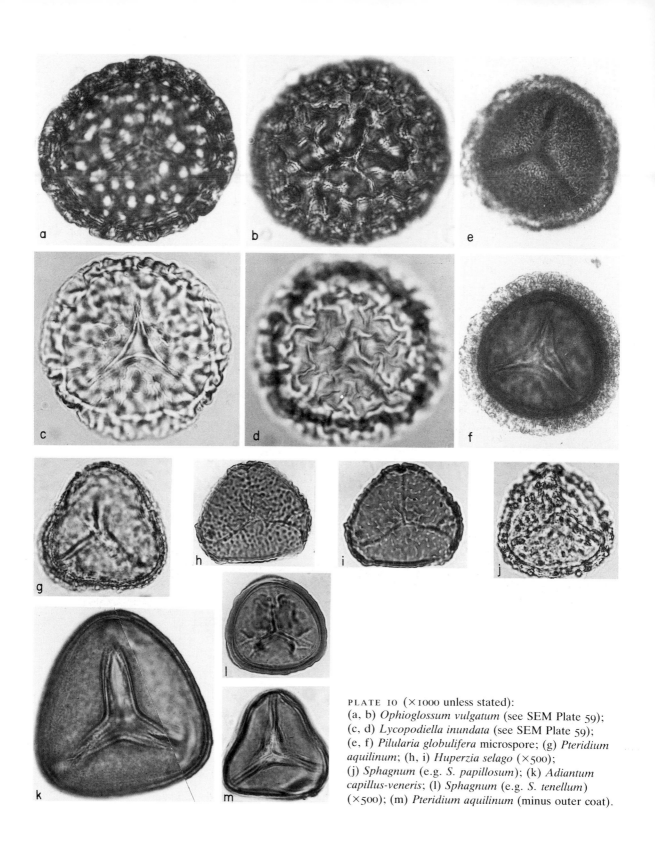

PLATE 10 (×1000 unless stated):
(a, b) *Ophioglossum vulgatum* (see SEM Plate 59);
(c, d) *Lycopodiella inundata* (see SEM Plate 59);
(e, f) *Pilularia globulifera* microspore; (g) *Pteridium aquilinum*; (h, i) *Huperzia selago* (×500);
(j) *Sphagnum* (e.g. *S. papillosum*); (k) *Adiantum capillus-veneris*; (l) *Sphagnum* (e.g. *S. tenellum*) (×500); (m) *Pteridium aquilinum* (minus outer coat).

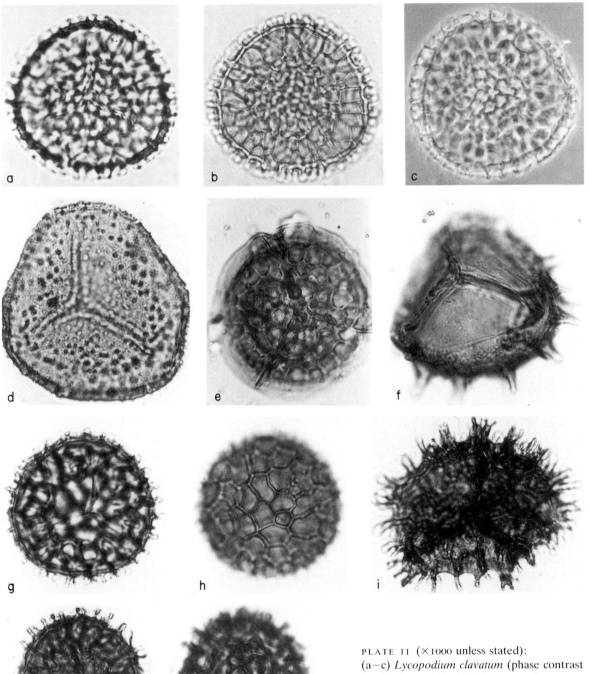

PLATE 11 (×1000 unless stated):
(a–c) *Lycopodium clavatum* (phase contrast used for part (c)); (d) *Phaeoceros laevis*;
(e) *Riccia* type (e.g. *R. glauca*) (×500);
(f) *Selaginella selaginoides* microspore;
(g, h) *Diphasiastrum* type (e.g. *D. alpinum*);
(i) *Anthoceros punctatus*; (j, k) *Lycopodium annotinum*.

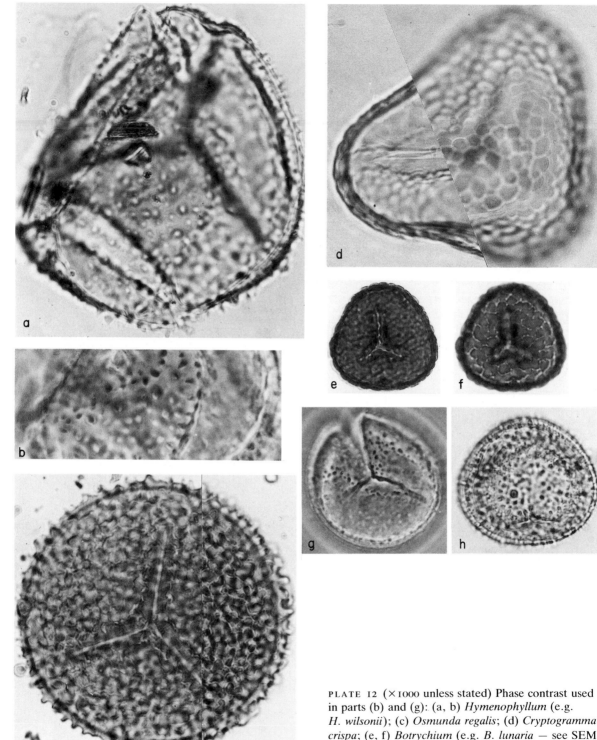

PLATE 12 (×1000 unless stated) Phase contrast used in parts (b) and (g): (a, b) *Hymenophyllum* (e.g. *H. wilsonii*); (c) *Osmunda regalis*; (d) *Cryptogramma crispa*; (e, f) *Botrychium* (e.g. *B. lunaria* — see SEM Plate 59) (×500); (g, h) *Trichomanes speciosum*.

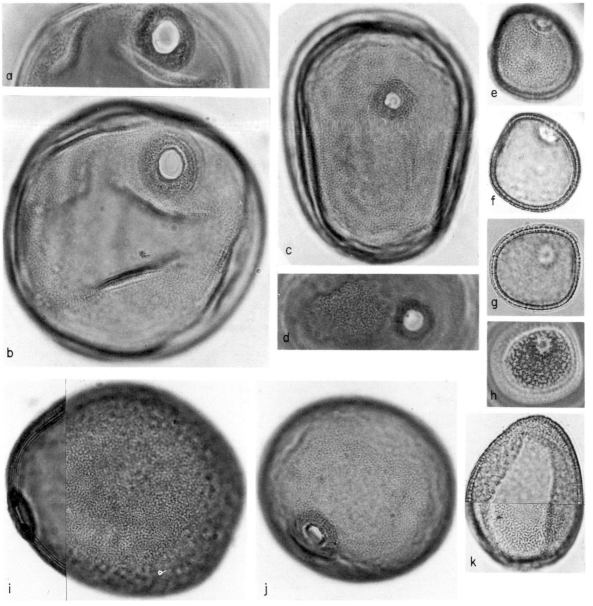

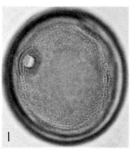

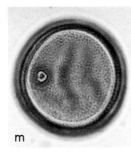

PLATE 13 (×1000). Phase contrast used for parts (a), (d), (h). See SEM Plate 60: (a, b) Gramineae (*Triticum aestivum*); (c, d) Gramineae (*Secale cereale*); (e, f) *Sparganium erectum*; (g, h) *Typha angustifolia*; (i) Gramineae (*Avena sativa*); (j) Gramineae (*Triticum turgidum*); (k) *Cephalanthera* (e.g. *C. damasonium*); (l) Gramineae (*Festuca rubra*); (m) *Gramineae* (*Poa annua*).

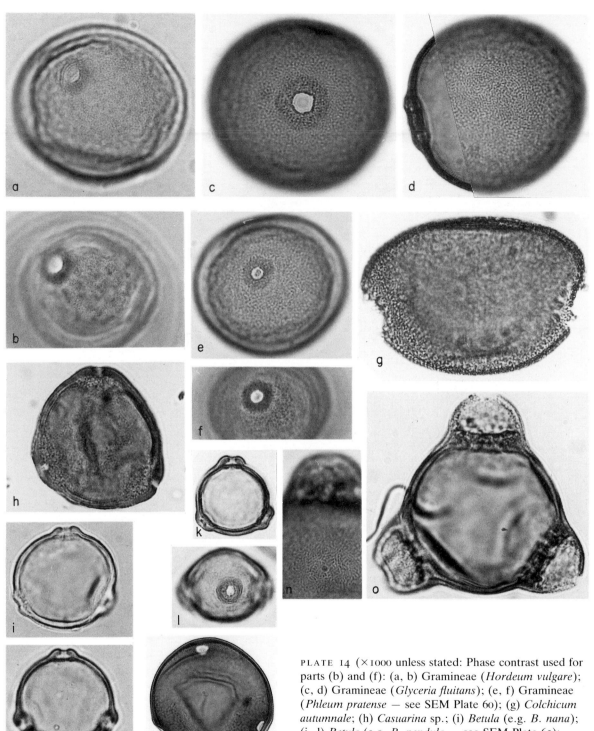

PLATE 14 (×1000 unless stated: Phase contrast used for parts (b) and (f): (a, b) Gramineae (*Hordeum vulgare*); (c, d) Gramineae (*Glyceria fluitans*); (e, f) Gramineae (*Phleum pratense* — see SEM Plate 60); (g) *Colchicum autumnale*; (h) *Casuarina* sp.; (i) *Betula* (e.g. *B. nana*); (j–l) *Betula* (e.g. *B. pendula* — see SEM Plate 62); (m) *Carya* (e.g. *C. aquatica*) (×500); (n, o) *Circaea* (e.g. *C. lutetiana*).

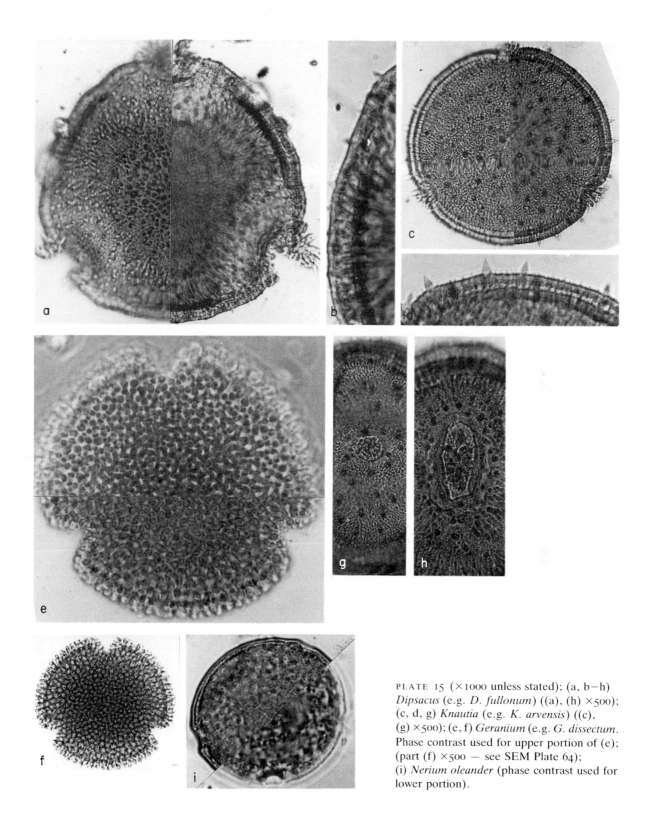

PLATE 15 (×1000 unless stated); (a, b–h)
Dipsacus (e.g. *D. fullonum*) ((a), (h) ×500);
(c, d, g) *Knautia* (e.g. *K. arvensis*) ((c),
(g) ×500); (e, f) *Geranium* (e.g. *G. dissectum*.
Phase contrast used for upper portion of (e);
(part (f) ×500 — see SEM Plate 64);
(i) *Nerium oleander* (phase contrast used for
lower portion).

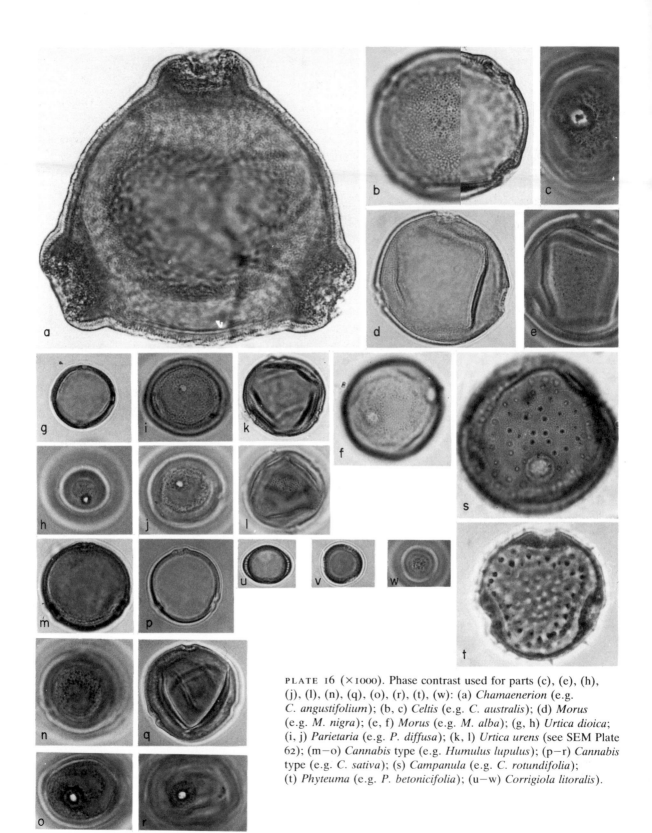

PLATE 16 (×1000). Phase contrast used for parts (c), (e), (h), (j), (l), (n), (q), (o), (r), (t), (w): (a) *Chamaenerion* (e.g. *C. angustifolium*); (b, c) *Celtis* (e.g. *C. australis*); (d) *Morus* (e.g. *M. nigra*); (e, f) *Morus* (e.g. *M. alba*); (g, h) *Urtica dioica*; (i, j) *Parietaria* (e.g. *P. diffusa*); (k, l) *Urtica urens* (see SEM Plate 62); (m–o) *Cannabis* type (e.g. *Humulus lupulus*); (p–r) *Cannabis* type (e.g. *C. sativa*); (s) *Campanula* (e.g. *C. rotundifolia*); (t) *Phyteuma* (e.g. *P. betonicifolia*); (u–w) *Corrigiola litoralis*.

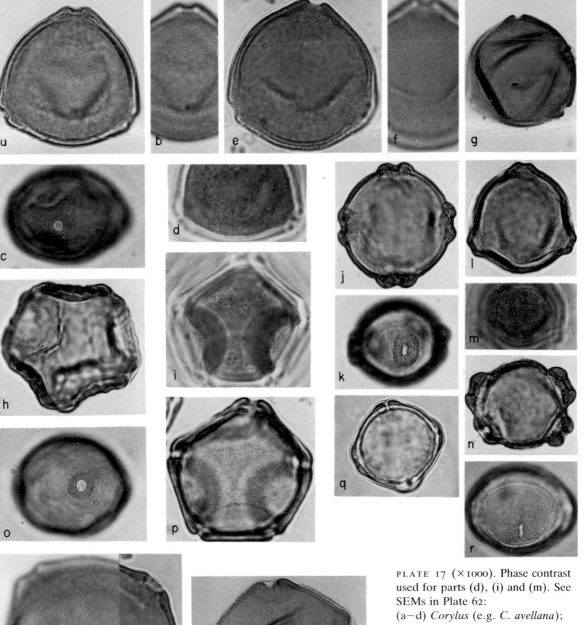

PLATE 17 (×1000). Phase contrast used for parts (d), (i) and (m). See SEMs in Plate 62:
(a−d) *Corylus* (e.g. *C. avellana*);
(e, f) *Myrica* (e.g. *M. gale*);
(g) *Comptonia* (e.g. *C. peregrina*);
(h) *Ulmus* (e.g. *U. glabra*);
(i, p) *Alnus* (e.g. *A. glutinosa*);
(j, k) *Myriophyllum spicatum*;
(l, m) *Ostrya* (e.g. *O. carpinifolia*);
(n) *Myriophyllum alterniflorum*;
(o, s) *Carpinus* (e.g. *C. betulus*);
(q, r) *Myriophyllum verticillatum*;
(t) *Pterocarya* (e.g. *P. fraxinifolia*).

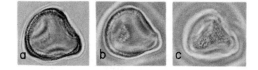

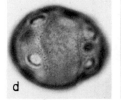

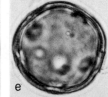

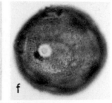

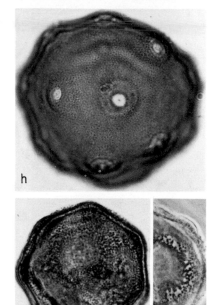

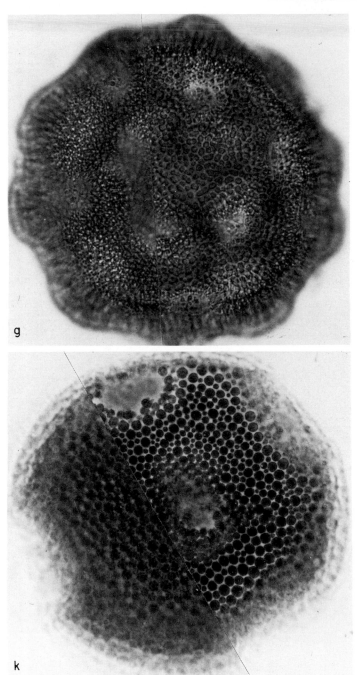

PLATE 18 (×1000). Phase contrast used for parts (c), and (j), and left portion (k): (a–c) *Herniaria* type (e.g. *H. glabra*); (d–f) *Ribes rubrum*; (g) *Calystegia* (e.g. *C. sepium* — see SEM Plate 63); (h) *Juglans* type (e.g. *J. nigra*); (i, j) *Alisma* type (e.g. *A. plantago-aquatica*); (k) *Linum perenne* ssp. *anglicum* (see SEM Plate 68).

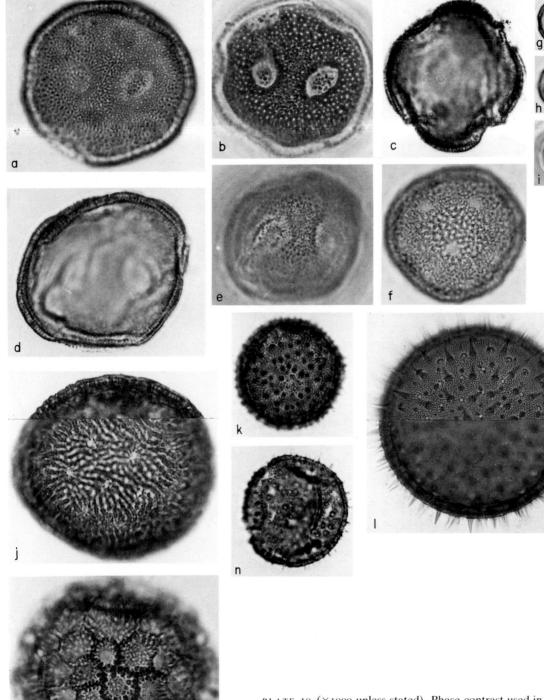

PLATE 19 (×1000 unless stated). Phase contrast used in parts (b), (e), (i): (a, b) *Liquidambar styraciflua*; (c–e) *Papaver argemone*; (f) *Buxus* (e.g. *B. sempervirens*); (g–i) *Illecebrum verticillatum*; (j) *Polemonium* (e.g. *P. caeruleum*); (k) *Sagittaria* (e.g. *S. sagittifolia*); (l) *Malva* type (e.g. *M. sylvestris* ×400); (m) *Polygonum persicaria*; (n) *Koenigia islandica*.

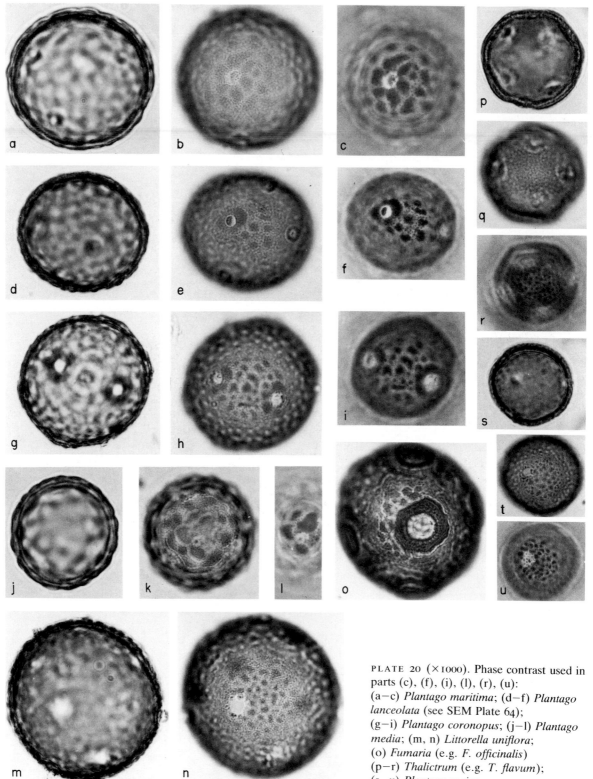

PLATE 20 (×1000). Phase contrast used in parts (c), (f), (i), (l), (r), (u):
(a–c) *Plantago maritima*; (d–f) *Plantago lanceolata* (see SEM Plate 64);
(g–i) *Plantago coronopus*; (j–l) *Plantago media*; (m, n) *Littorella uniflora*;
(o) *Fumaria* (e.g. *F. officinalis*)
(p–r) *Thalictrum* (e.g. *T. flavum*);
(s–u) *Plantago major*.

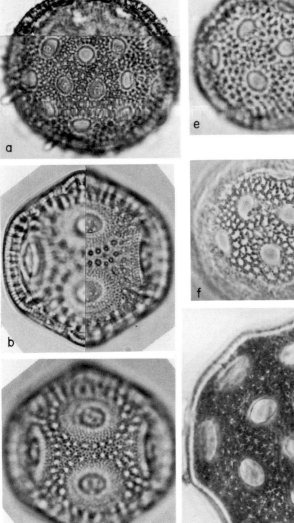

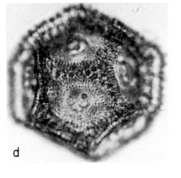

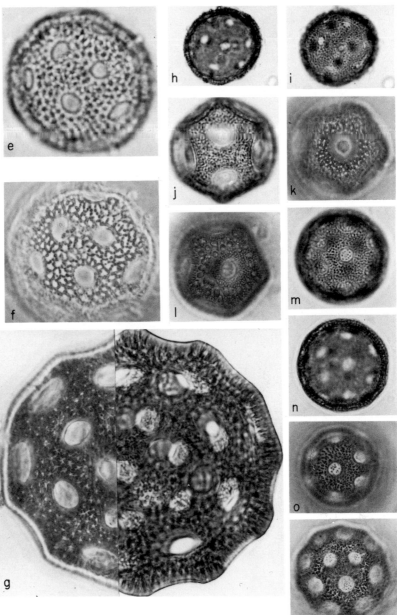

PLATE 21 (×1000) see SEMs in Plates 63 & 64. Phase contrast used
for the left portion of part (f) and for parts (g), (k), (l), (o), (p).
Caryophyllaceae: (a) *Silene dioica* type (e.g. *S. noctiflora*);
(b, c) *Scleranthus* (e.g. *S. perennis*); (d) *Stellaria holostea*;
(e, f) *Silene dioica*; (g) *Agrostemma githago*; (h, i) *Sagina*
(e.g. *S. procumbens*); (j, k) *Gypsophila repens*; (l) *Scleranthus*
(e.g. *S. annuus*).
Chenopodiaceae: (m−o) *Chenopodium polyspermum*; (p) *Salsola
kali*.

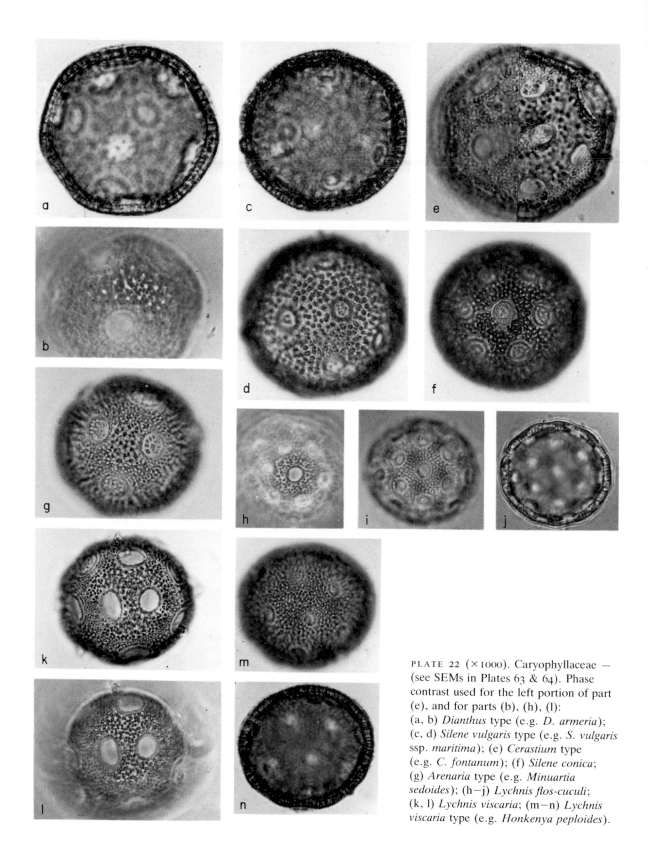

PLATE 22 (×1000). Caryophyllaceae —
(see SEMs in Plates 63 & 64). Phase
contrast used for the left portion of part
(e), and for parts (b), (h), (l):
(a, b) *Dianthus* type (e.g. *D. armeria*);
(c, d) *Silene vulgaris* type (e.g. *S. vulgaris*
ssp. *maritima*); (e) *Cerastium* type
(e.g. *C. fontanum*); (f) *Silene conica*;
(g) *Arenaria* type (e.g. *Minuartia
sedoides*); (h–j) *Lychnis flos-cuculi*;
(k, l) *Lychnis viscaria*; (m–n) *Lychnis
viscaria* type (e.g. *Honkenya peploides*).

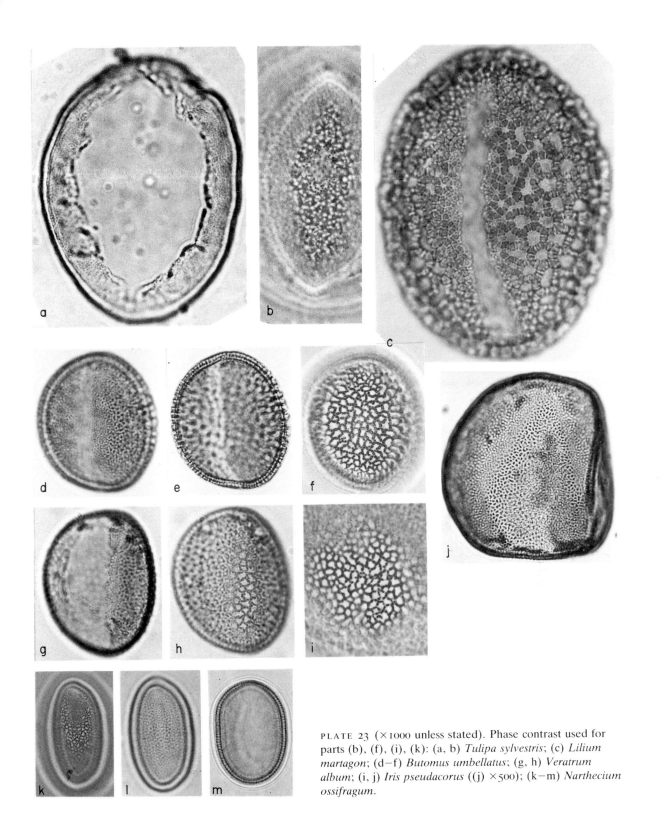

PLATE 23 (×1000 unless stated). Phase contrast used for parts (b), (f), (i), (k): (a, b) *Tulipa sylvestris*; (c) *Lilium martagon*; (d–f) *Butomus umbellatus*; (g, h) *Veratrum album*; (i, j) *Iris pseudacorus* ((j) ×500); (k–m) *Narthecium ossifragum*.

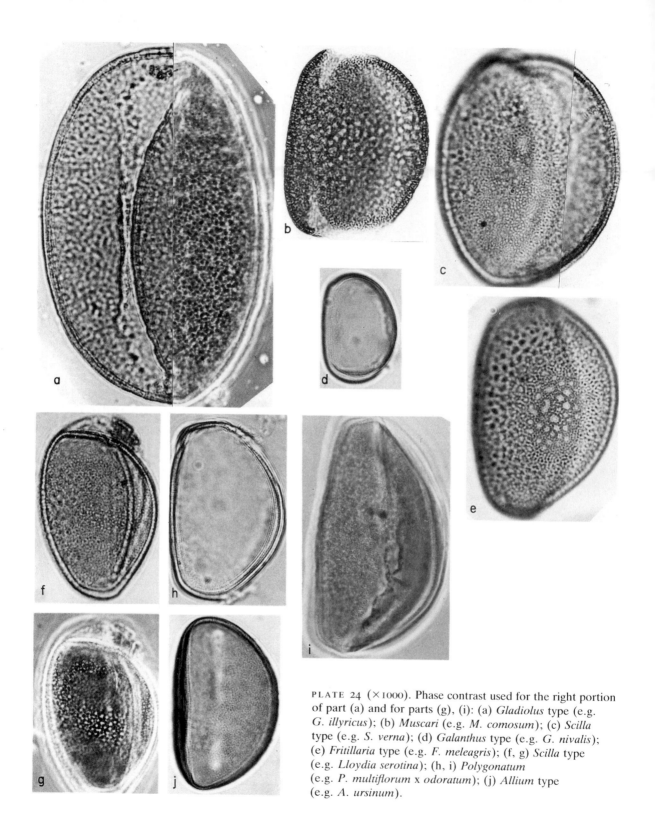

PLATE 24 (×1000). Phase contrast used for the right portion of part (a) and for parts (g), (i): (a) *Gladiolus* type (e.g. *G. illyricus*); (b) *Muscari* (e.g. *M. comosum*); (c) *Scilla* type (e.g. *S. verna*); (d) *Galanthus* type (e.g. *G. nivalis*); (e) *Fritillaria* type (e.g. *F. meleagris*); (f, g) *Scilla* type (e.g. *Lloydia serotina*); (h, i) *Polygonatum* (e.g. *P. multiflorum* x *odoratum*); (j) *Allium* type (e.g. *A. ursinum*).

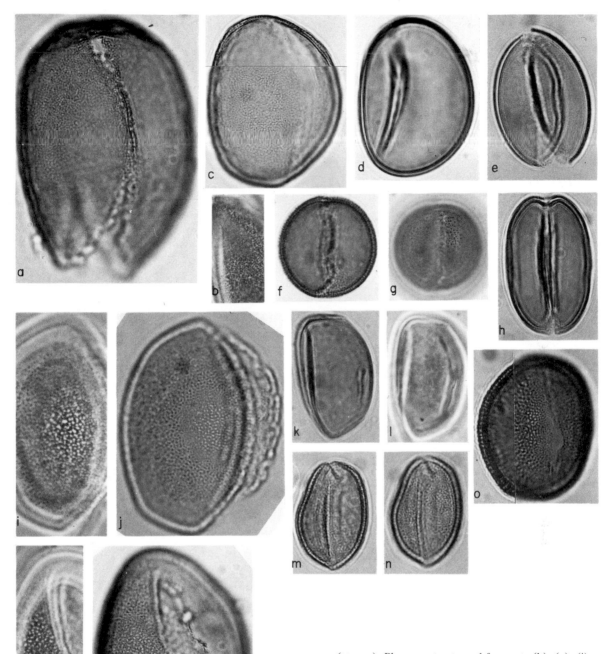

PLATE 25 (×1000). Phase contrast used for parts (b), (g), (i), (l), (p): (a, b) *Brasenia purpurea* (see SEM Plate 60); (c) *Convallaria* type (e.g. *C. majalis*); (d) Polypodiaceae spore minus outer coat (e.g. *Blechnum spicant*); (e) *Pedicularis palustris*; (f, g) *Hypecoum* (e.g. *H. pendulum*); (h) *Pedicularis sylvatica*; (i, j) *Paris* type (e.g. *P. quadrifolia*); (k, l) *Calla palustris*; (m, n) *Tofieldia* (e.g. *T. pusilla*); (o) *Tamus communis*; (p, q) *Paris* type (e.g. *Gagea chlorantha*).

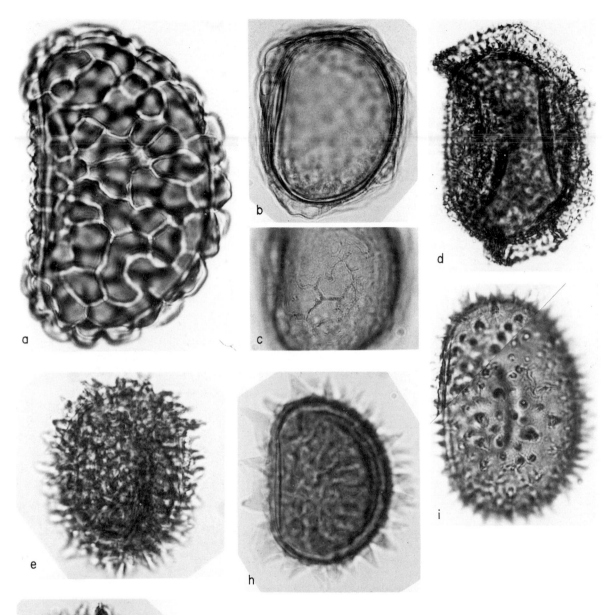

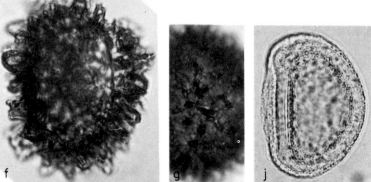

PLATE 26 (×1000) (a) *Polypodium vulgare*; (b, c) *Blechnum spicant*; (d) *Dryopteris dilatata* (see SEM Plate 60); (e–g) *Polystichum* (e.g. *P. lonchitis*); (h) *Cystopteris fragilis* (see SEM Plate 60); (i) *Thelypteris palustris*; (j) *Athyrium filix-femina*.

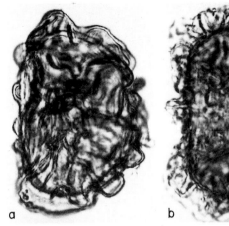

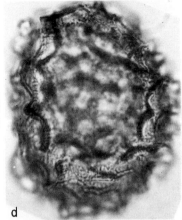

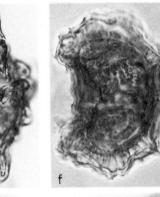

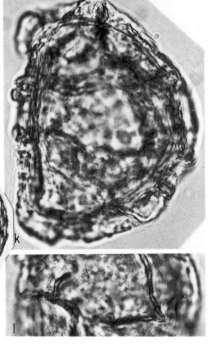

PLATE 27 (×1000). Phase contrast used for part (f): (a) *Dryopteris filix-mas* (see SEM Plate 60); (b, c) *Dryopteris cristata* type (e.g. *D. carthusiana*); (d) *Asplenium* (e.g. *A. adiantum-nigrum*); (e) *Gymnocarpium dryopteris*; (f, g) *Asplenium* (e.g. *A. ruta-muraria*); (h, i) *Woodsia* (e.g. *W. alpina*); (j) *Isoetes* microspore (e.g. *I. histrix*); (k, l) *Phegopteris connectilis*.

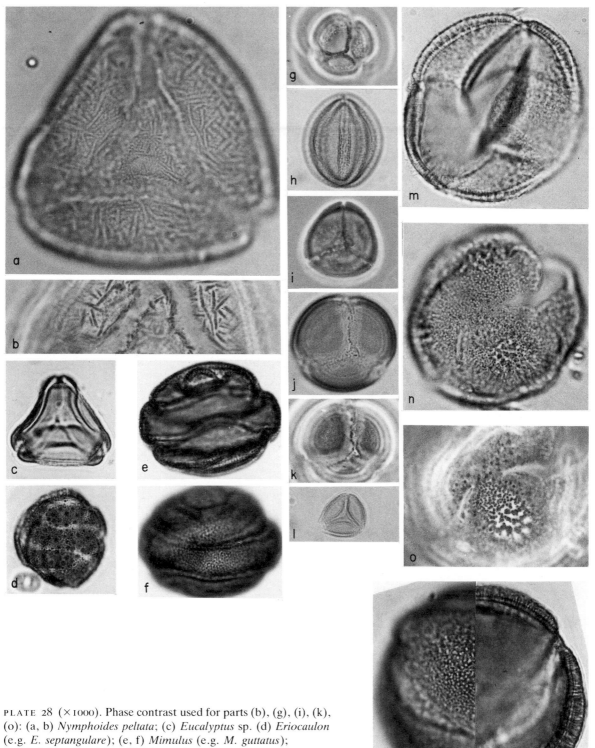

PLATE 28 (×1000). Phase contrast used for parts (b), (g), (i), (k), (o): (a, b) *Nymphoides peltata*; (c) *Eucalyptus* sp. (d) *Eriocaulon* (e.g. *E. septangulare*); (e, f) *Mimulus* (e.g. *M. guttatus*); (g, h) *Soldanella alpina*; (i) *Myrtus* (e.g. *M. communis*); (j, k) *Pedicularis oederi*; (l) *Primula farinosa*; (m−o) *Montia fontana*; (p) *Montia perfoliata*.

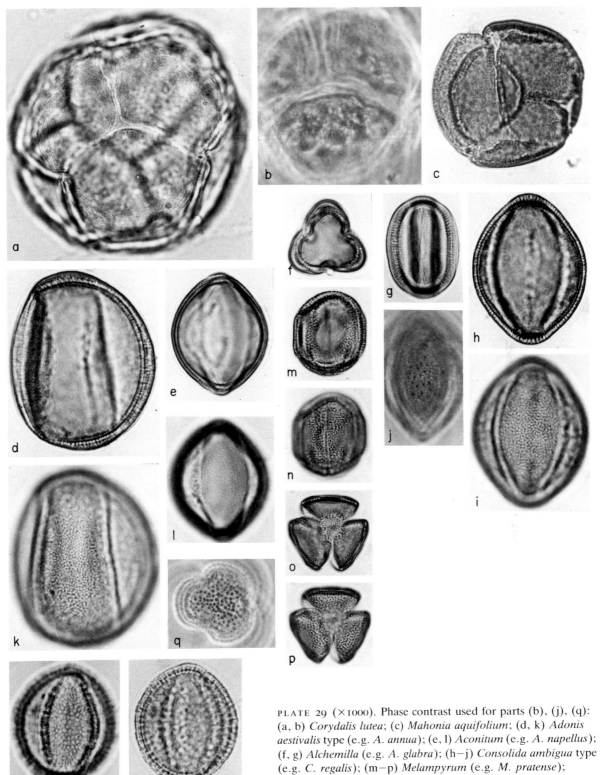

PLATE 29 (×1000). Phase contrast used for parts (b), (j), (q): (a, b) *Corydalis lutea*; (c) *Mahonia aquifolium*; (d, k) *Adonis aestivalis* type (e.g. *A. annua*); (e, l) *Aconitum* (e.g. *A. napellus*); (f, g) *Alchemilla* (e.g. *A. glabra*); (h–j) *Consolida ambigua* type (e.g. *C. regalis*); (m–p) *Melampyrum* (e.g. *M. pratense*); (q–s) *Cuscuta europaea* type (e.g. *C. epithymum*).

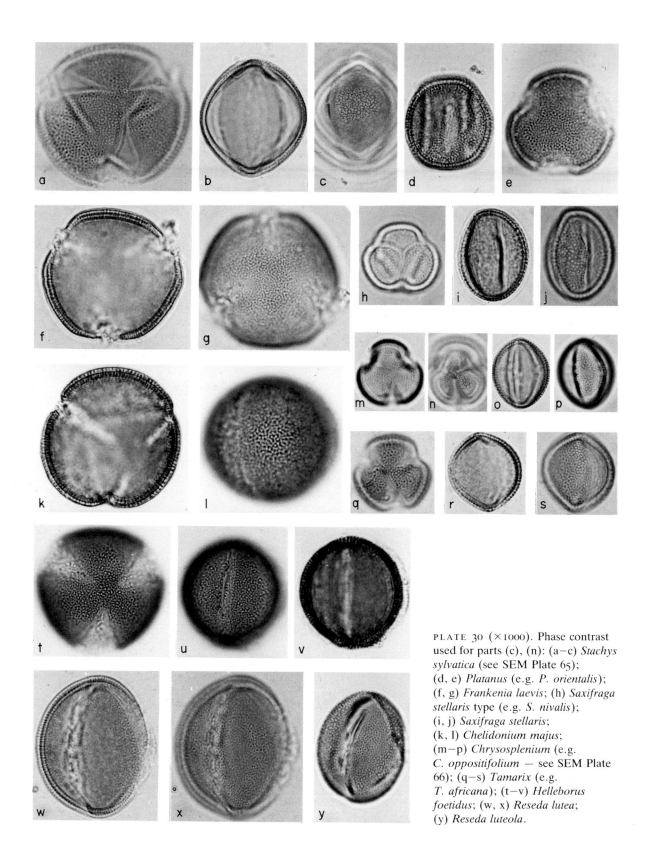

PLATE 30 (×1000). Phase contrast used for parts (c), (n): (a–c) *Stachys sylvatica* (see SEM Plate 65); (d, e) *Platanus* (e.g. *P. orientalis*); (f, g) *Frankenia laevis*; (h) *Saxifraga stellaris* type (e.g. *S. nivalis*); (i, j) *Saxifraga stellaris*; (k, l) *Chelidonium majus*; (m–p) *Chrysosplenium* (e.g. *C. oppositifolium* — see SEM Plate 66); (q–s) *Tamarix* (e.g. *T. africana*); (t–v) *Helleborus foetidus*; (w, x) *Reseda lutea*; (y) *Reseda luteola*.

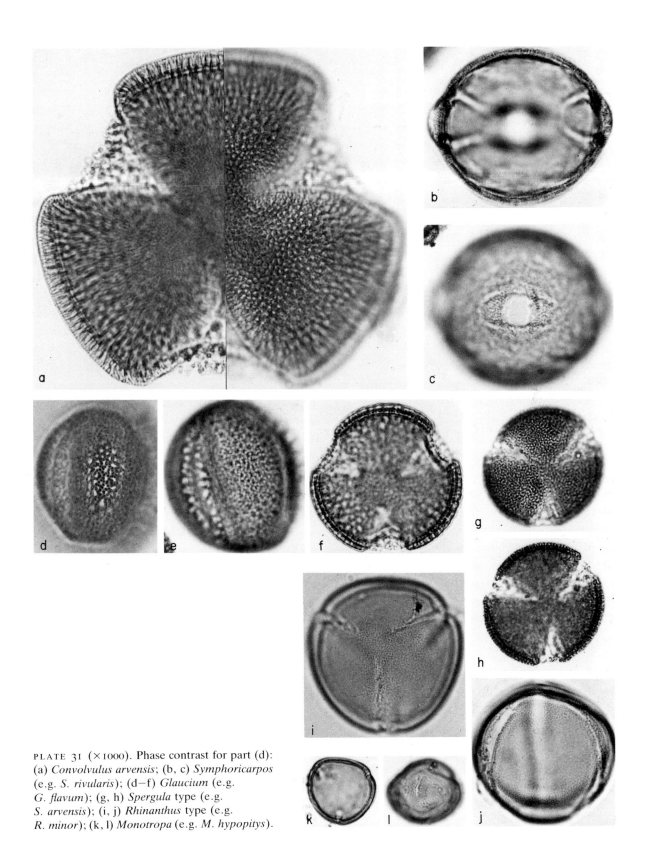

PLATE 31 (×1000). Phase contrast for part (d):
(a) *Convolvulus arvensis*; (b, c) *Symphoricarpos*
(e.g. *S. rivularis*); (d–f) *Glaucium* (e.g.
G. flavum); (g, h) *Spergula* type (e.g.
S. arvensis); (i, j) *Rhinanthus* type (e.g.
R. minor); (k, l) *Monotropa* (e.g. *M. hypopitys*).

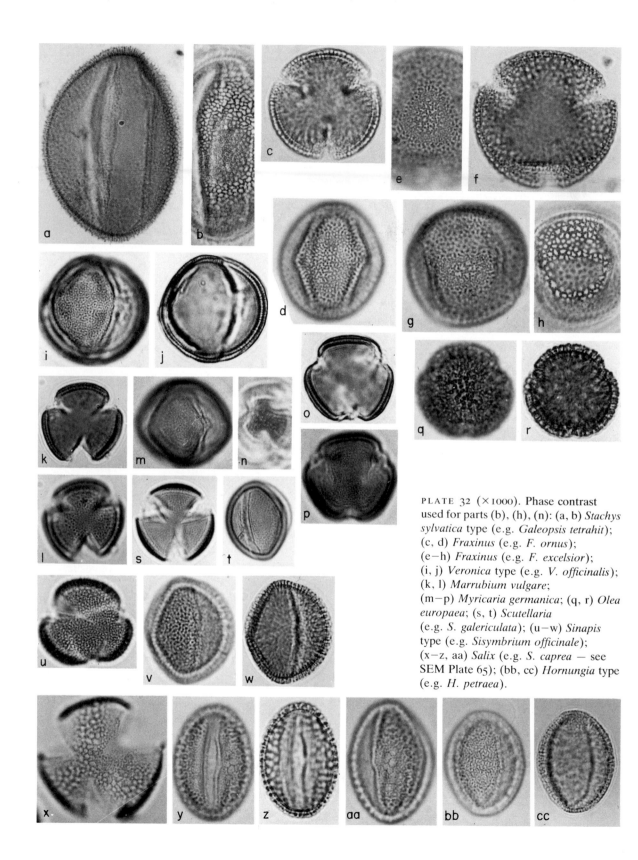

PLATE 32 (×1000). Phase contrast used for parts (b), (h), (n): (a, b) *Stachys sylvatica* type (e.g. *Galeopsis tetrahit*); (c, d) *Fraxinus* (e.g. *F. ornus*); (e–h) *Fraxinus* (e.g. *F. excelsior*); (i, j) *Veronica* type (e.g. *V. officinalis*); (k, l) *Marrubium vulgare*; (m–p) *Myricaria germanica*; (q, r) *Olea europaea*; (s, t) *Scutellaria* (e.g. *S. galericulata*); (u–w) *Sinapis* type (e.g. *Sisymbrium officinale*); (x–z, aa) *Salix* (e.g. *S. caprea* — see SEM Plate 65); (bb, cc) *Hornungia* type (e.g. *H. petraea*).

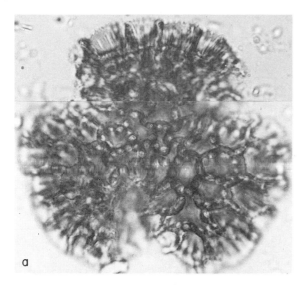

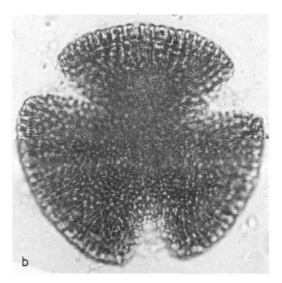

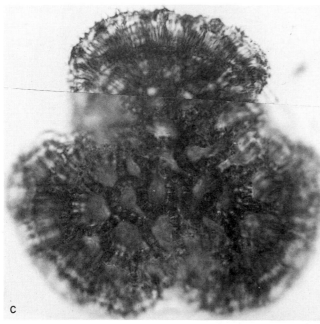

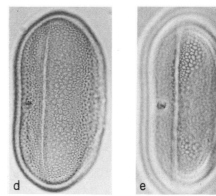

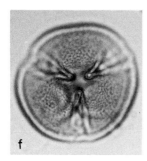

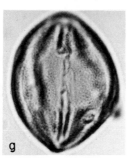

PLATE 33 (×1000). Phase contrast for part
(e): (a) *Limonium vulgare* type A (e.g.
Limonium vulgare — see SEM Plate 67);
(b) *Armeria maritima* type B (e.g. *Limonium
vulgare*); (c) *Armeria maritima* type A (e.g.
Armeria maritima — see SEM Plate 67);
(d, e) *Onobrychis* type (e.g. *O. viciifolia*);
(f, g) *Ulex* type (e.g. *U. europaeus*).

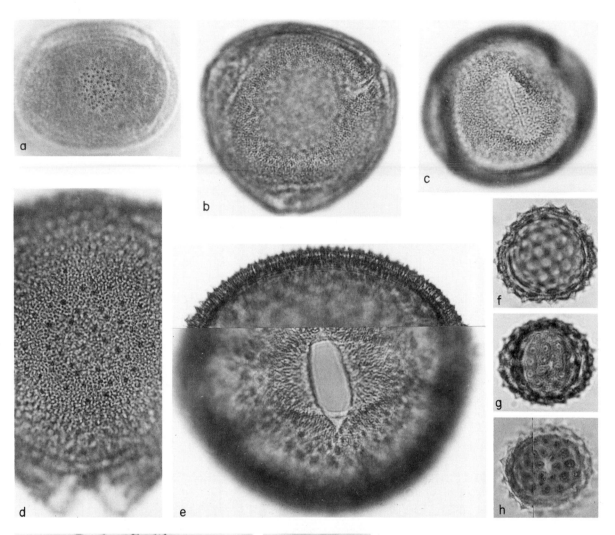

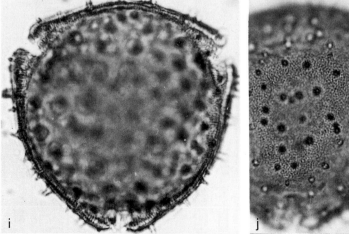

PLATE 34 (×1000). Phase contrast used for parts (a), (h): (a–c) *Linnaea borealis*; (d, e) *Lonicera periclymenum*; (f–h) *Ambrosia* type (e.g. *A. trifida*); (i, j) *Lonicera xylosteum*.

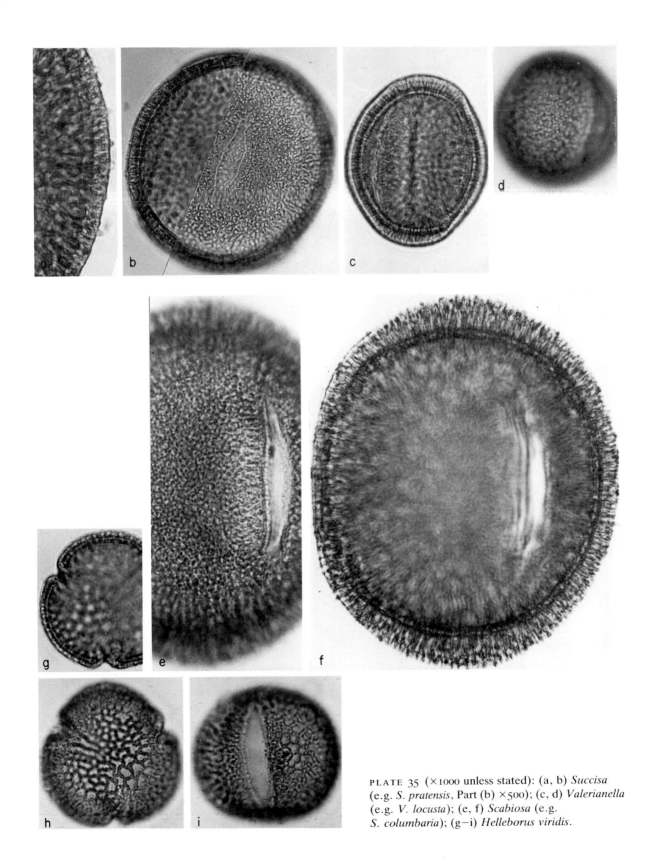

PLATE 35 (×1000 unless stated): (a, b) *Succisa* (e.g. *S. pratensis*, Part (b) ×500); (c, d) *Valerianella* (e.g. *V. locusta*); (e, f) *Scabiosa* (e.g. *S. columbaria*); (g–i) *Helleborus viridis*.

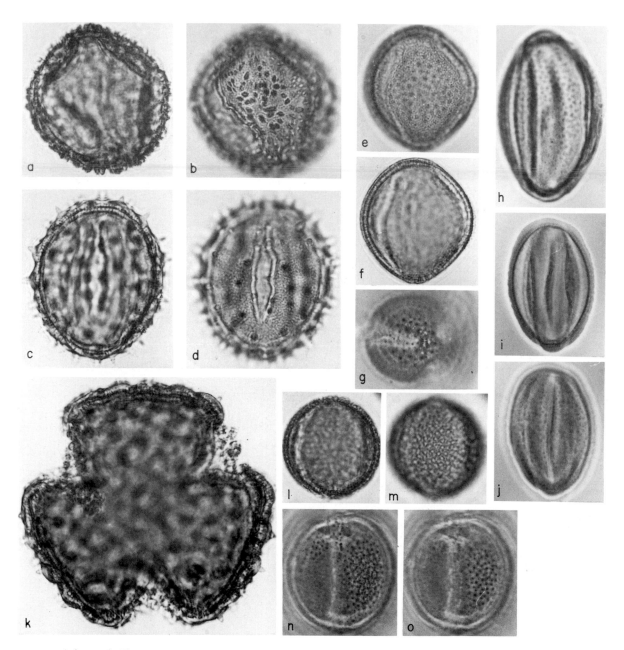

PLATE 36 (×1000). Phase contrast used for parts (g), (i), (j), (n), (o): (a, b) *Rubus chamaemorus*; (c, d) *Valeriana dioica*; (e–g) *Caltha* type (e.g. *C. palustris*); (h) *Teucrium* (e.g. *T. botrys*); (i, j) *Teucrium* (e.g. *T. scorodonia*); (k) *Valeriana officinalis*; (l–o) *Papaver rhoeas* type (e.g. *P. dubium*).

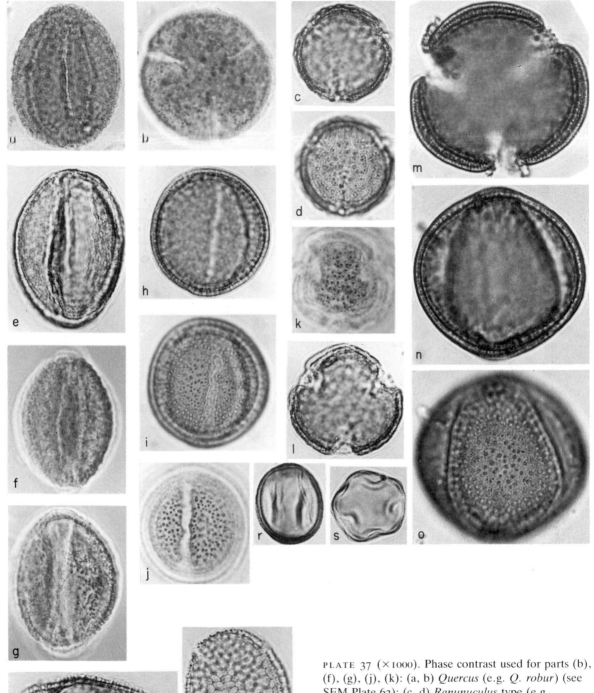

PLATE 37 (×1000). Phase contrast used for parts (b), (f), (g), (j), (k): (a, b) *Quercus* (e.g. *Q. robur*) (see SEM Plate 62); (c, d) *Ranunculus* type (e.g. *R. sardous*) (see SEM Plate 67); (e, f) *Saxifraga granulata*; (g) *Saxifraga cernua*; (h–j) *Anemone nemorosa*; (k, l) *Ranunculus* type (e.g. *R. flammula*); (m–o) *Ranunculus* type (e.g. *Pulsatilla vulgaris*); (p, q) *Impatiens* (e.g. *I. parviflora*); (r, s) *Hippuris vulgaris*.

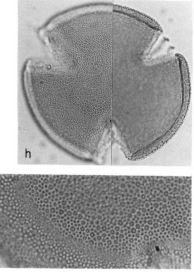

PLATE 38 (×1000 unless stated). Phase contrast used for parts (c), (l): (a–c) *Viscum* (e.g. *V. album* — see SEM Plate 69); (d) *Linum austriacum* type, long-styled (e.g. *L. perenne* ssp. *perenne*); (e, f) *Ilex* type (e.g. *I. aquifolium* — see SEM Plate 69); (g) *Linum catharticum*; (h, i) *Linum bienne* type (e.g. *L. usitatissimum*. Part (h) ×500 — also see SEM in Plate 68); (j, k) *Linum austriacum* type, short-styled (e.g. *L. viscosum*. Part (j) ×500); (l, m) *Radiola* (e.g. *R. linoides*).

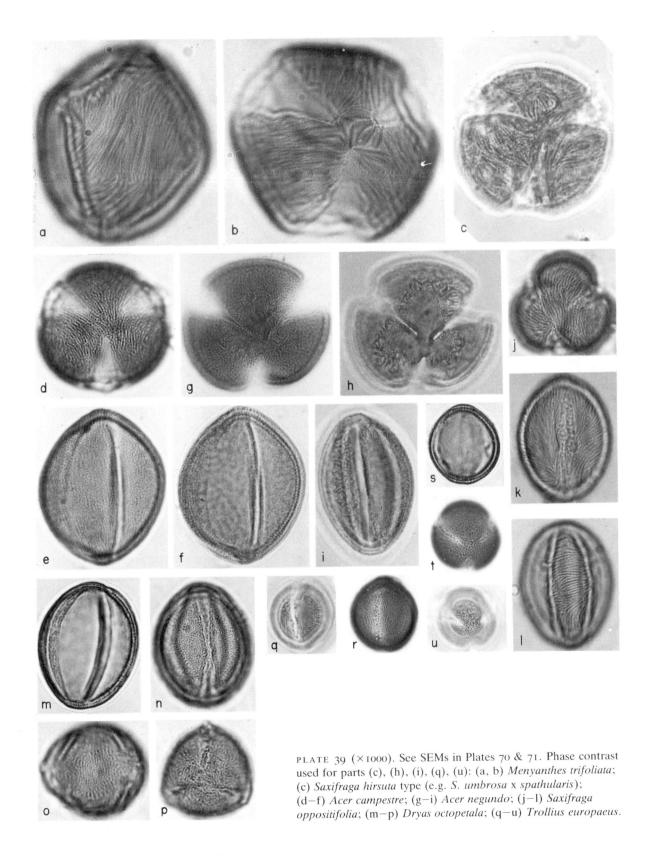

PLATE 39 (×1000). See SEMs in Plates 70 & 71. Phase contrast used for parts (c), (h), (i), (q), (u): (a, b) *Menyanthes trifoliata*; (c) *Saxifraga hirsuta* type (e.g. *S. umbrosa* x *spathularis*); (d–f) *Acer campestre*; (g–i) *Acer negundo*; (j–l) *Saxifraga oppositifolia*; (m–p) *Dryas octopetala*; (q–u) *Trollius europaeus*.

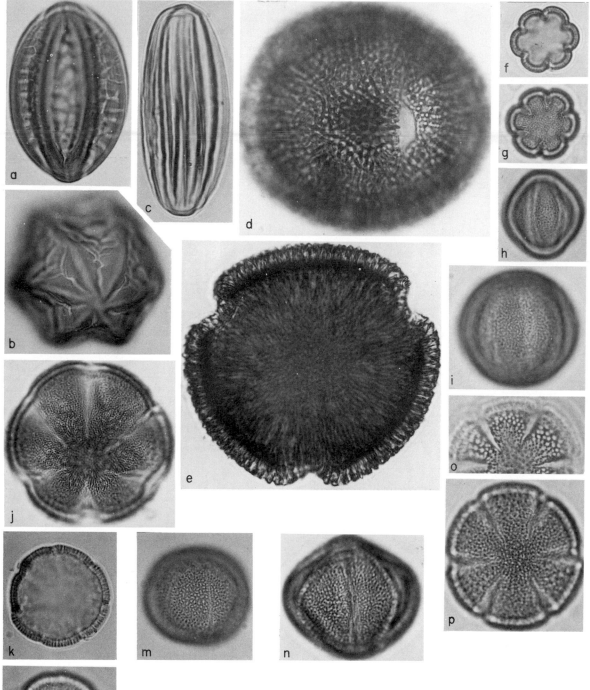

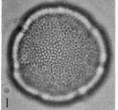

PLATE 40 (×1000). Phase contrast used for part (o): (a, b) *Ephedra distachya*; (c) *Ephedra fragilis*; (d, e) *Erodium* (e.g. *E. guttatum* − see SEM in Plate 71); (f−h) *Galium* type (e.g. *G. cruciata*); (i) *Galium* type (e.g. *Rubia peregrina*); (j) *Prunella* type (e.g. *P. vulgaris*); (k−m) *Primula veris* type (e.g. *P. vulgaris*); (n−p) *Mentha* type (e.g. *Lycopus europaeus*).

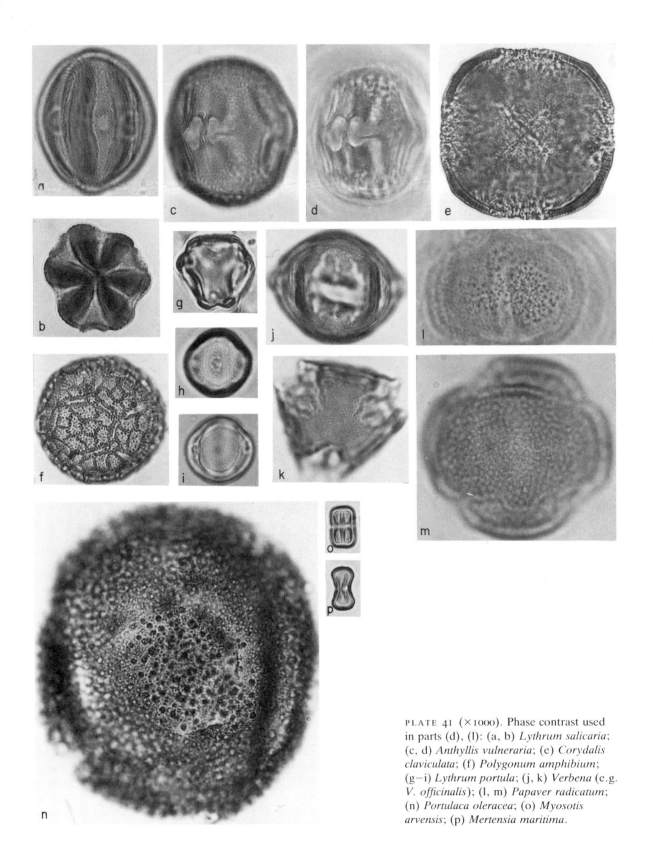

PLATE 41 (×1000). Phase contrast used in parts (d), (l): (a, b) *Lythrum salicaria*; (c, d) *Anthyllis vulneraria*; (e) *Corydalis claviculata*; (f) *Polygonum amphibium*; (g–i) *Lythrum portula*; (j, k) *Verbena* (e.g. *V. officinalis*); (l, m) *Papaver radicatum*; (n) *Portulaca oleracea*; (o) *Myosotis arvensis*; (p) *Mertensia maritima*.

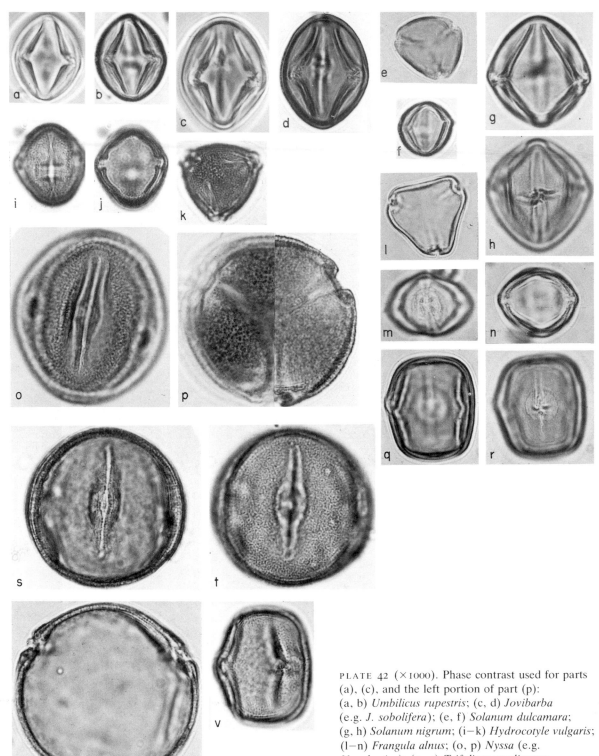

PLATE 42 (×1000). Phase contrast used for parts (a), (c), and the left portion of part (p): (a, b) *Umbilicus rupestris*; (c, d) *Jovibarba* (e.g. *J. sobolifera*); (e, f) *Solanum dulcamara*; (g, h) *Solanum nigrum*; (i–k) *Hydrocotyle vulgaris*; (l–n) *Frangula alnus*; (o, p) *Nyssa* (e.g. *N. sylvatica*); (q–r) *Trifolium spadiceum*; (s–u) *Fagus* (e.g. *F. sylvatica*); (v) *Ornithopus perpusillus*.

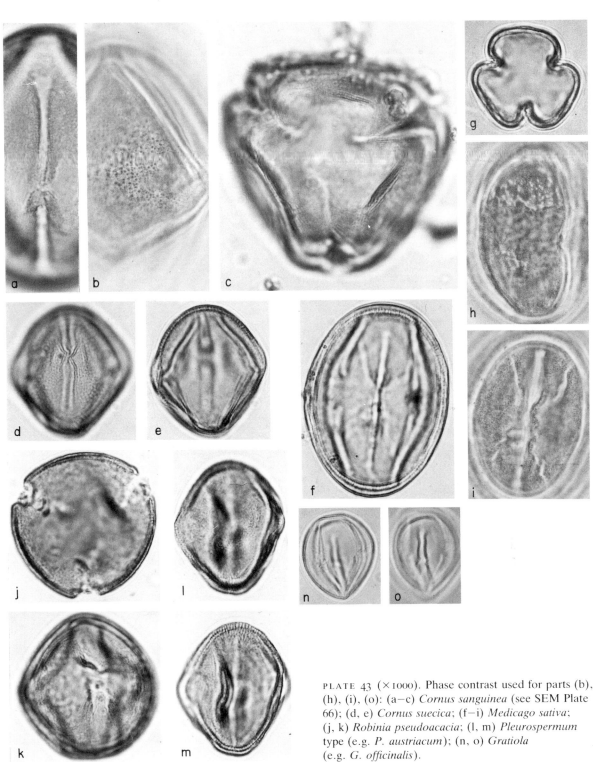

PLATE 43 (×1000). Phase contrast used for parts (b), (h), (i), (o): (a–c) *Cornus sanguinea* (see SEM Plate 66); (d, e) *Cornus suecica*; (f–i) *Medicago sativa*; (j, k) *Robinia pseudoacacia*; (l, m) *Pleurospermum* type (e.g. *P. austriacum*); (n, o) *Gratiola* (e.g. *G. officinalis*).

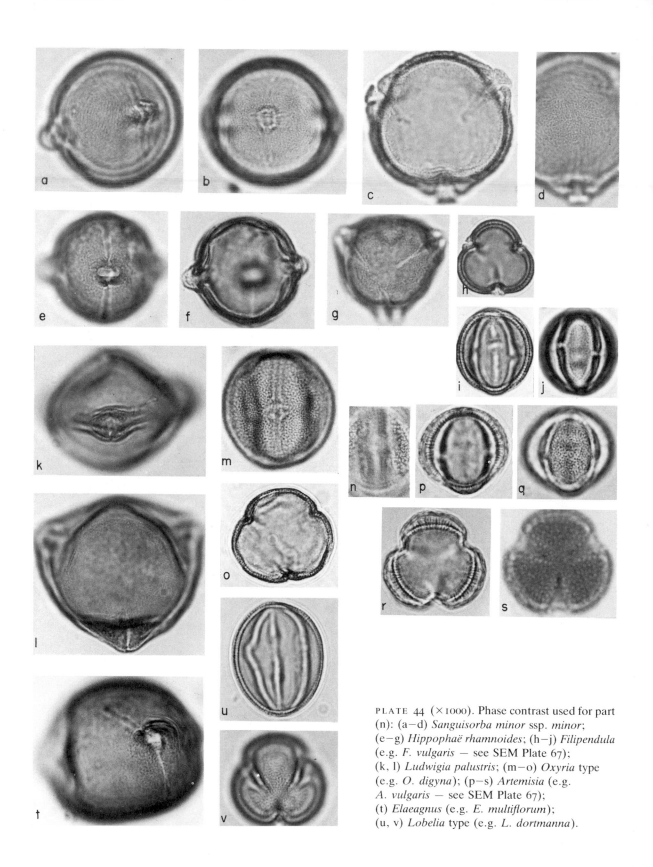

PLATE 44 (×1000). Phase contrast used for part (n): (a–d) *Sanguisorba minor* ssp. *minor*; (e–g) *Hippophaë rhamnoides*; (h–j) *Filipendula* (e.g. *F. vulgaris* – see SEM Plate 67); (k, l) *Ludwigia palustris*; (m–o) *Oxyria* type (e.g. *O. digyna*); (p–s) *Artemisia* (e.g. *A. vulgaris* – see SEM Plate 67); (t) *Elaeagnus* (e.g. *E. multiflorum*); (u, v) *Lobelia* type (e.g. *L. dortmanna*).

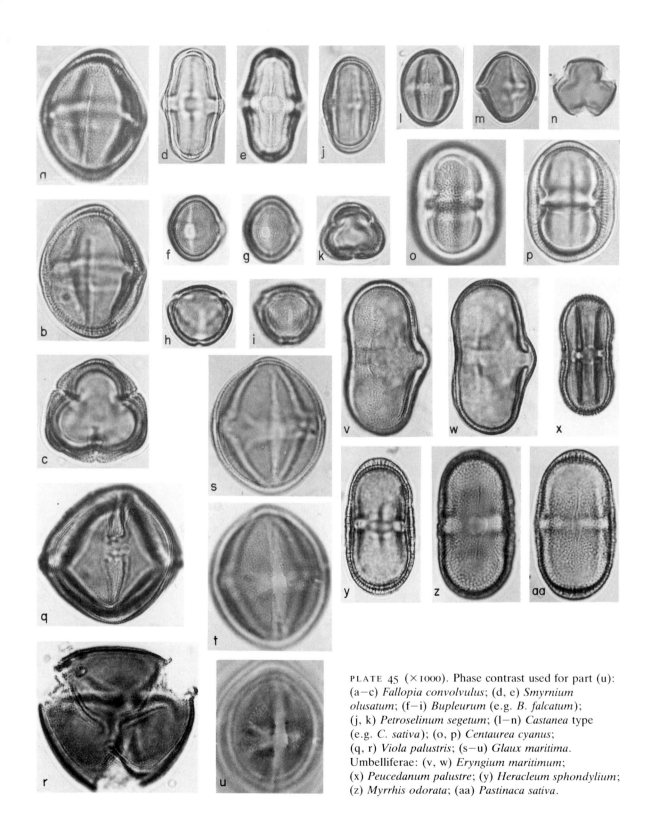

PLATE 45 (×1000). Phase contrast used for part (u):
(a–c) *Fallopia convolvulus*; (d, e) *Smyrnium
olusatum*; (f–i) *Bupleurum* (e.g. *B. falcatum*);
(j, k) *Petroselinum segetum*; (l–n) *Castanea* type
(e.g. *C. sativa*); (o, p) *Centaurea cyanus*;
(q, r) *Viola palustris*; (s–u) *Glaux maritima*.
Umbelliferae: (v, w) *Eryngium maritimum*;
(x) *Peucedanum palustre*; (y) *Heracleum sphondylium*;
(z) *Myrrhis odorata*; (aa) *Pastinaca sativa*.

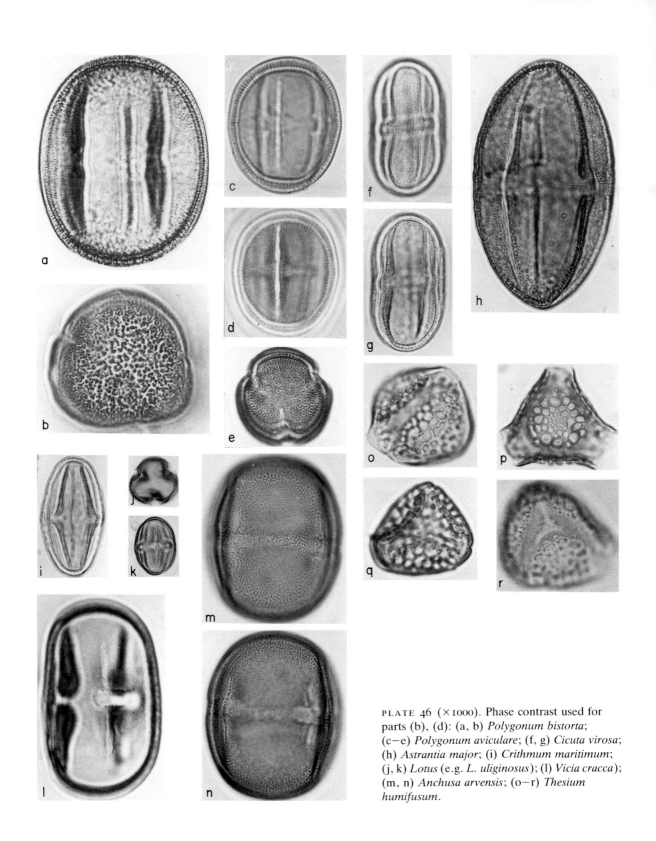

PLATE 46 (×1000). Phase contrast used for parts (b), (d): (a, b) *Polygonum bistorta*; (c–e) *Polygonum aviculare*; (f, g) *Cicuta virosa*; (h) *Astrantia major*; (i) *Crithmum maritimum*; (j, k) *Lotus* (e.g. *L. uliginosus*); (l) *Vicia cracca*); (m, n) *Anchusa arvensis*; (o–r) *Thesium humifusum*.

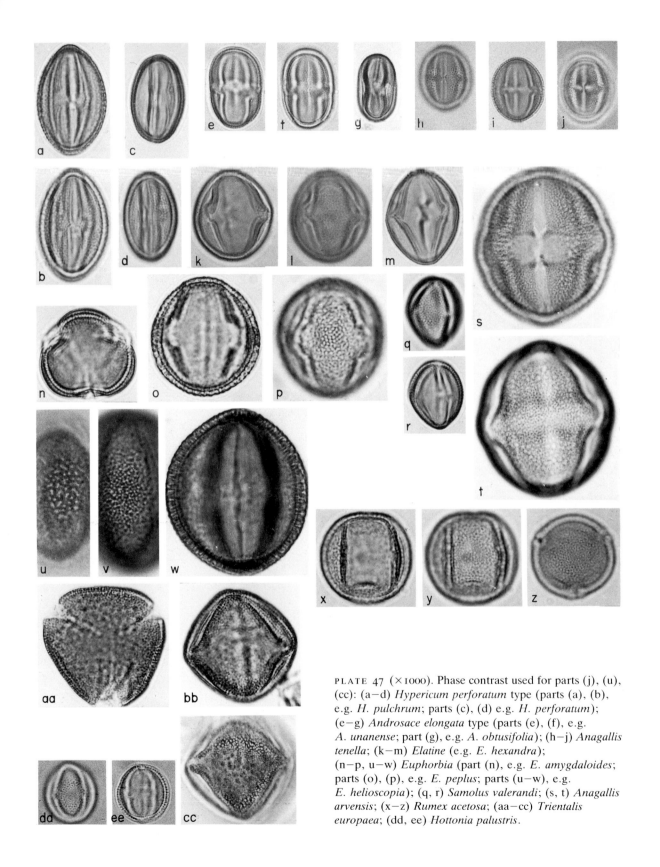

PLATE 47 (×1000). Phase contrast used for parts (j), (u), (cc): (a–d) *Hypericum perforatum* type (parts (a), (b), e.g. *H. pulchrum*; parts (c), (d) e.g. *H. perforatum*); (e–g) *Androsace elongata* type (parts (e), (f), e.g. *A. unanense*; part (g), e.g. *A. obtusifolia*); (h–j) *Anagallis tenella*; (k–m) *Elatine* (e.g. *E. hexandra*); (n–p, u–w) *Euphorbia* (part (n), e.g. *E. amygdaloides*; parts (o), (p), e.g. *E. peplus*; parts (u–w), e.g. *E. helioscopia*); (q, r) *Samolus valerandi*; (s, t) *Anagallis arvensis*; (x–z) *Rumex acetosa*; (aa–cc) *Trientalis europaea*; (dd, ee) *Hottonia palustris*.

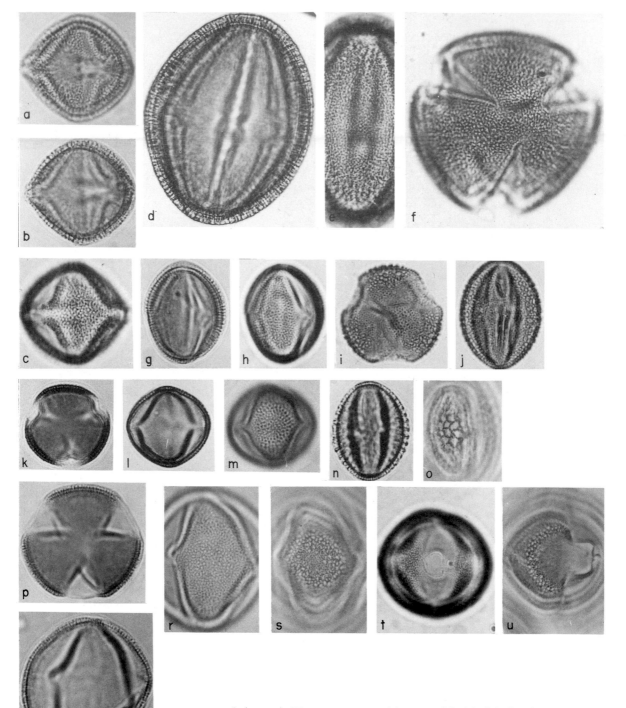

PLATE 48 (×1000). Phase contrast used for parts (o), (s), (u): (a−c) *Mercurialis* (e.g. *M. perennis* − see SEM Plate 65); (d−f) *Gentianella campestris* type (e.g. *G. anglica*); (g, h) *Sambucus nigra*; (i, j) *Adoxa moschatellina*; (k−m) *Sambucus racemosa*; (n, o) *Sambucus ebulus*; (p−u) *Digitalis purpurea*.

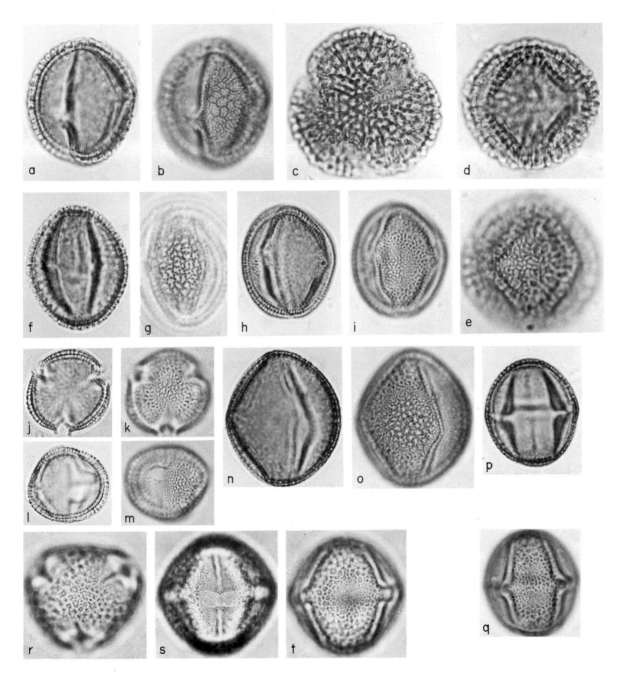

PLATE 49 (×1000). Phase contrast for part (g): (a, b) *Viburnum opulus*; (c–e) *Ligustrum vulgare*; (f, g) *Viburnum lantana*; (h, i) *Scrophularia* type (e.g. *Verbascum thapsus*); (j–m) *Euonymus europaeus*; (n, o) *Diapensia lapponica*; (p, q) *Lysimachia vulgaris*; (r–t) *Hedera helix* (see SEM Plate 65).

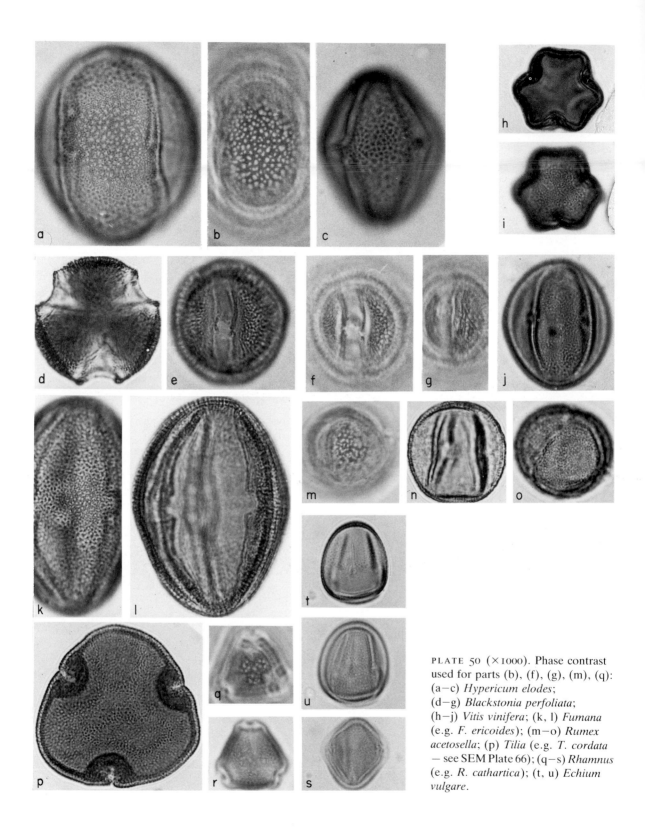

PLATE 50 (×1000). Phase contrast used for parts (b), (f), (g), (m), (q): (a–c) *Hypericum elodes*; (d–g) *Blackstonia perfoliata*; (h–j) *Vitis vinifera*; (k, l) *Fumana* (e.g. *F. ericoides*); (m–o) *Rumex acetosella*; (p) *Tilia* (e.g. *T. cordata* — see SEM Plate 66); (q–s) *Rhamnus* (e.g. *R. cathartica*); (t, u) *Echium vulgare*.

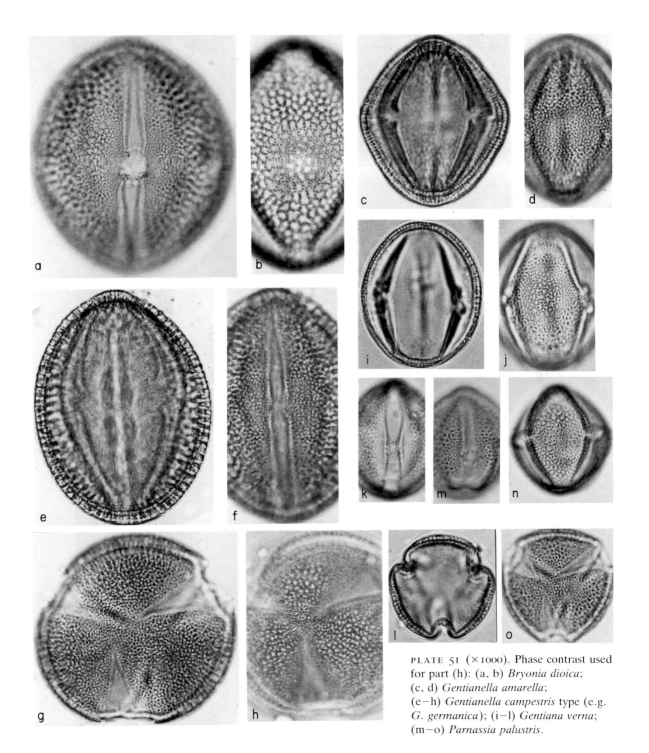

PLATE 51 (×1000). Phase contrast used for part (h): (a, b) *Bryonia dioica*; (c, d) *Gentianella amarella*; (e–h) *Gentianella campestris* type (e.g. *G. germanica*); (i–l) *Gentiana verna*; (m–o) *Parnassia palustris*.

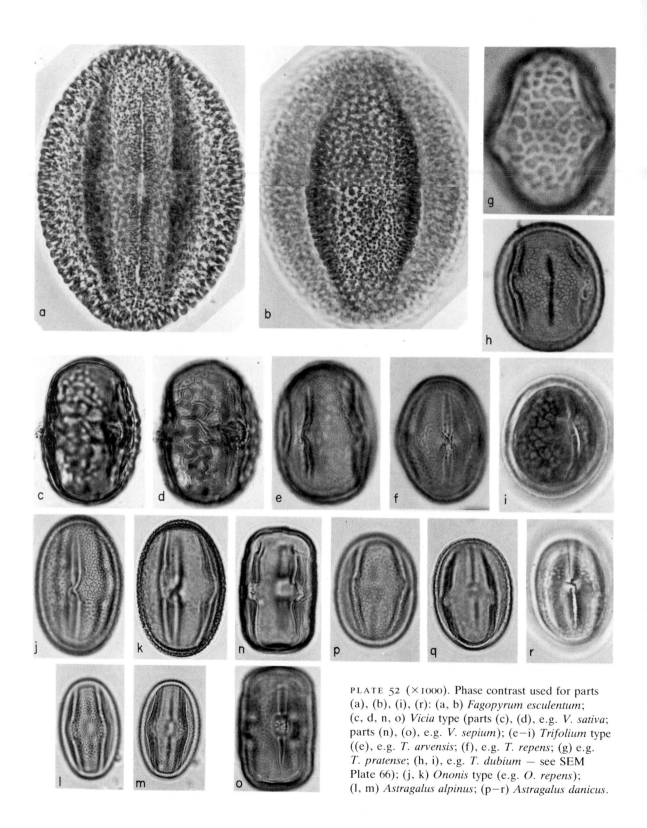

PLATE 52 (×1000). Phase contrast used for parts (a), (b), (i), (r): (a, b) *Fagopyrum esculentum*; (c, d, n, o) *Vicia* type (parts (c), (d), e.g. *V. sativa*; parts (n), (o), e.g. *V. sepium*); (e–i) *Trifolium* type ((e), e.g. *T. arvense*; (f), e.g. *T. repens*; (g) e.g. *T. pratense*; (h, i), e.g. *T. dubium* — see SEM Plate 66); (j, k) *Ononis* type (e.g. *O. repens*); (l, m) *Astragalus alpinus*; (p–r) *Astragalus danicus*.

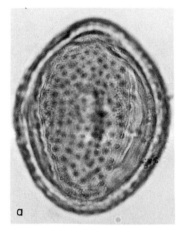

a

b

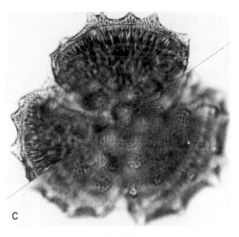

c

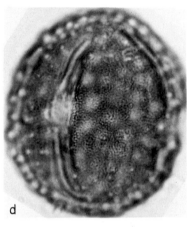

d

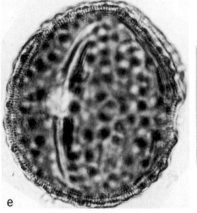

e

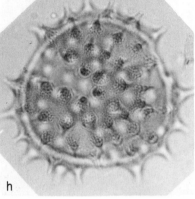

h

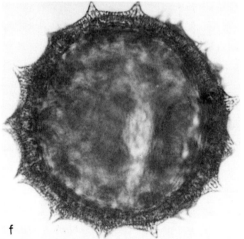

f

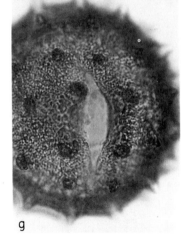

g

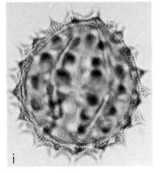

i

PLATE 53 (×1000):
(a, b) *Centaurea scabiosa*;
(c) *Serratula* type
(e.g. *S. tinctoria*);
(d, e) *Centaurea nigra*;
(f, g) *Cirsium* (e.g.
C. dissectum); (h, i) *Aster*
type (e.g. *A. tripolium* —
(see SEM Plate 67).

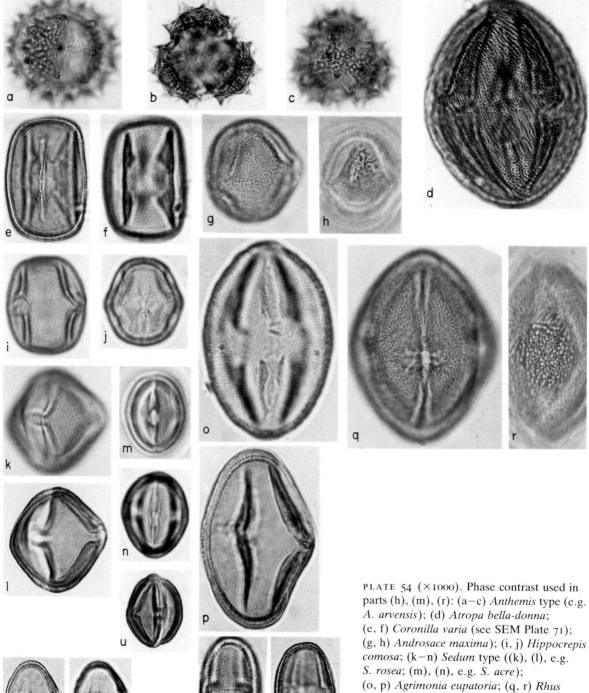

PLATE 54 (×1000). Phase contrast used in parts (h), (m), (r): (a–c) *Anthemis* type (e.g. *A. arvensis*); (d) *Atropa bella-donna*; (e, f) *Coronilla varia* (see SEM Plate 71); (g, h) *Androsace maxima*); (i, j) *Hippocrepis comosa*; (k–n) *Sedum* type ((k), (l), e.g. *S. rosea*; (m), (n), e.g. *S. acre*); (o, p) *Agrimonia eupatoria*; (q, r) *Rhus* (e.g. *R. typhina*); (s, t) *Anthriscus caucalis*; (u) *Crassula* (e.g. *C. tillaea*); (v, w) *Apium inundatum* type ((v), e.g. *A. nodiflorum*; (w) e.g. *A. graveolens*) (see SEM Plate 71).

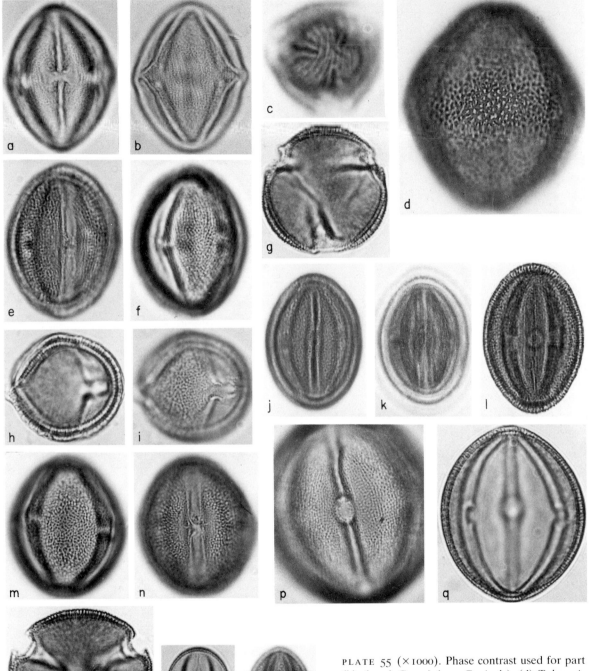

PLATE 55 (×1000). Phase contrast used for part (k): (a–c) *Geum* (e.g. *G. rivale*); (d) *Tuberaria guttata*; (e–g) *Centaurium* (e.g. *C. erythraea*); (h, i) *Lomatogonium rotatum*; (j–l) *Gentiana pneumonanthe*; (m–o) *Swertia perennis*; (p, q) *Helianthemum* (e.g. *H. canum*); (r, s) *Aesculus* (e.g. *A. hippocastanum*).

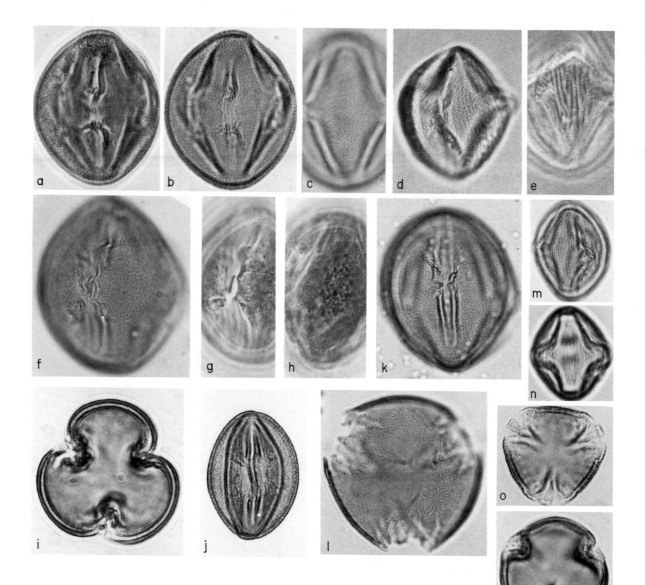
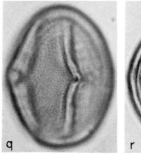
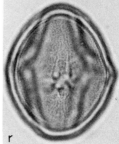

PLATE 56 (×1000). Phase contrast used for parts (e), (g), (h): (a–c) *Sorbus aria*; (d, e) *Potentilla* type (e.g. *Sibbaldia procumbens*); (f–i) *Crataegus* ((f), (g), (i) e.g. *C. monogyna*; (h), e.g. *C. laevigata*); (j) *Sorbus aucuparia* (see SEM Plate 70); (k, l) *Malus sylvestris*; (m–o) *Potentilla* type ((m) e.g. *P. sterilis*; (n), (o) e.g. *P. erecta*); (p–r) *Rosa* (e.g. *R. arvensis*).

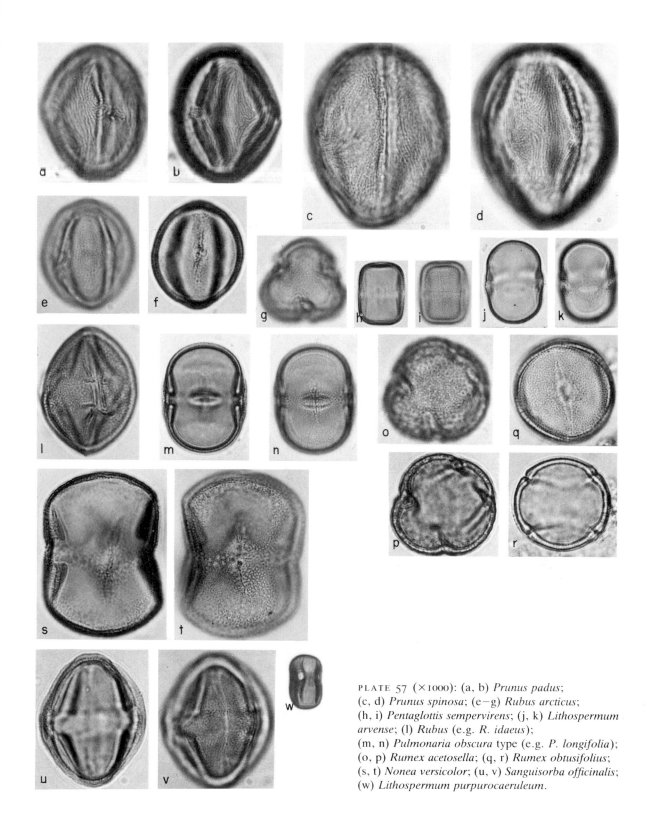

PLATE 57 (×1000): (a, b) *Prunus padus*;
(c, d) *Prunus spinosa*; (e–g) *Rubus arcticus*;
(h, i) *Pentaglottis sempervirens*; (j, k) *Lithospermum arvense*; (l) *Rubus* (e.g. *R. idaeus*);
(m, n) *Pulmonaria obscura* type (e.g. *P. longifolia*);
(o, p) *Rumex acetosella*; (q, r) *Rumex obtusifolius*;
(s, t) *Nonea versicolor*; (u, v) *Sanguisorba officinalis*;
(w) *Lithospermum purpurocaeruleum*.

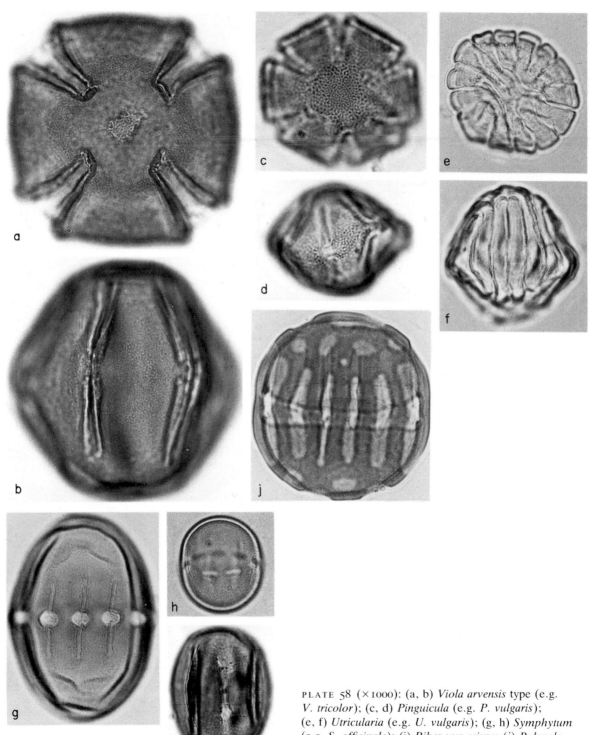

PLATE 58 (×1000): (a, b) *Viola arvensis* type (e.g. *V. tricolor*); (c, d) *Pinguicula* (e.g. *P. vulgaris*); (e, f) *Utricularia* (e.g. *U. vulgaris*); (g, h) *Symphytum* (e.g. *S. officinale*); (i) *Ribes uva-crispa*; (j) *Polygala* (e.g. *P. vulgaris*).

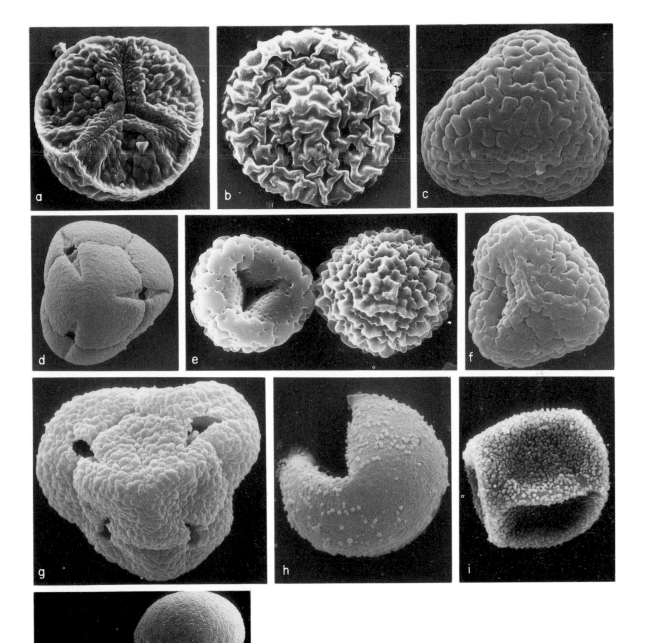

PLATE 59 SEM: (a, b) *Lycopodiella inundata* (×1000);
(c, f) *Botrychium lunaria* (×1000); (d) *Vaccinium myrtillus* (×1000);
(e) *Ophioglossum vulgatum* (×1000); (g) *Calluna vulgaris* (×2000);
(h) *Juniperus communis* (×2000); (i) *Taxus baccata* (×2000);
(j) *Pinus sylvestris* (×1000).

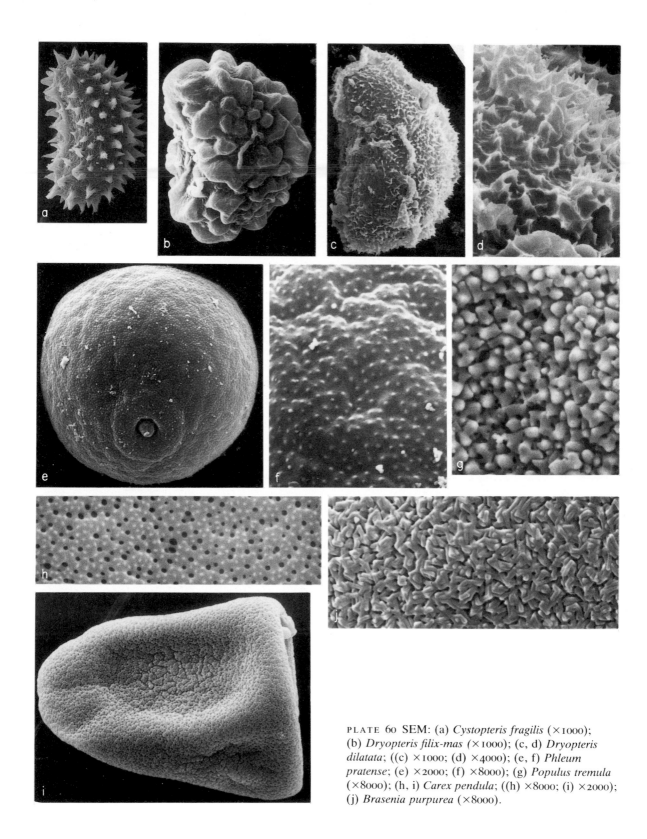

PLATE 60 SEM: (a) *Cystopteris fragilis* (×1000); (b) *Dryopteris filix-mas* (×1000); (c, d) *Dryopteris dilatata*; ((c) ×1000; (d) ×4000); (e, f) *Phleum pratense*; (e) ×2000; (f) ×8000); (g) *Populus tremula* (×8000); (h, i) *Carex pendula*; ((h) ×8000; (i) ×2000); (j) *Brasenia purpurea* (×8000).

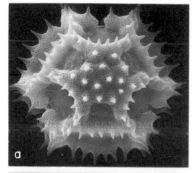

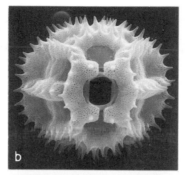

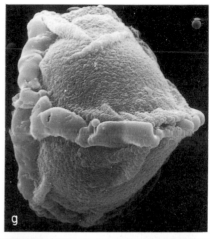

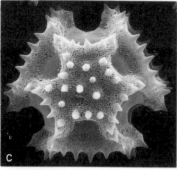

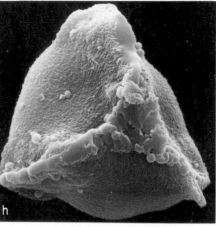

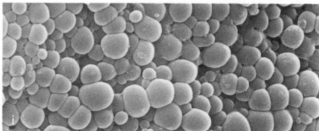

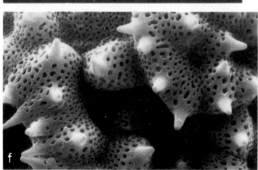

PLATE 61 SEM: (a, b) Lactuceae (e.g. *Sonchus arvensis*.
×1000); (c, d) Lactuceae (e.g. *Tragopogon pratensis*. ×1000);
(e) Lactuceae (e.g. *Lactuca saligna*. ×2000); (f) Lactuceae
(*Taraxacum officinale*. ×4000); (g–i) *Trapa natans* ((g),
(h) ×1000; (i) ×8000).

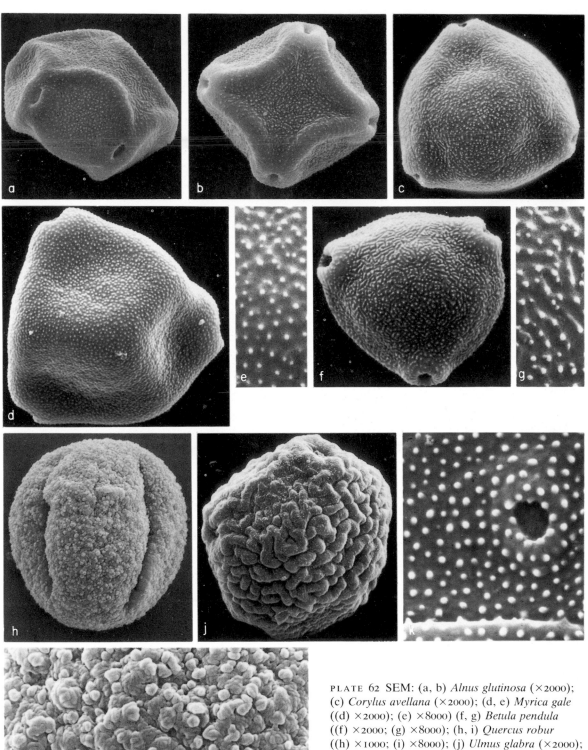

PLATE 62 SEM: (a, b) *Alnus glutinosa* (×2000); (c) *Corylus avellana* (×2000); (d, e) *Myrica gale* ((d) ×2000); (e) ×8000) (f, g) *Betula pendula* ((f) ×2000; (g) ×8000); (h, i) *Quercus robur* ((h) ×1000; (i) ×8000); (j) *Ulmus glabra* (×2000); (k) *Urtica urens* (×8000).

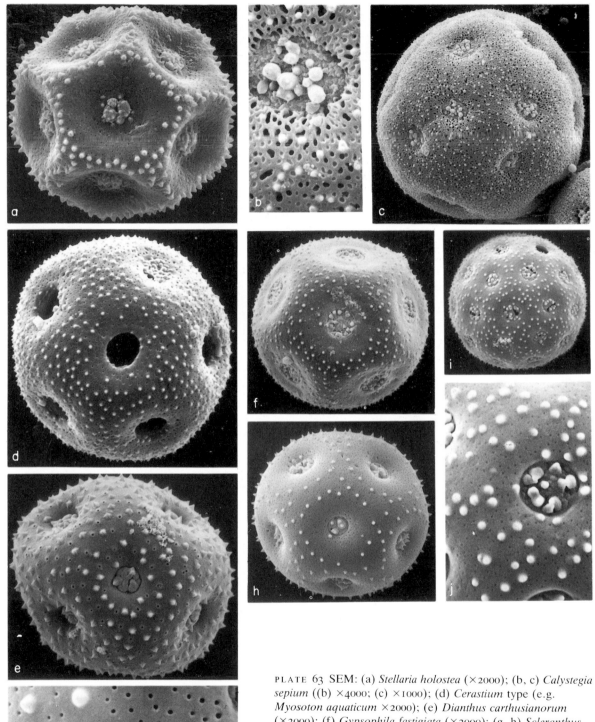

PLATE 63 SEM: (a) *Stellaria holostea* (×2000); (b, c) *Calystegia sepium* ((b) ×4000; (c) ×1000); (d) *Cerastium* type (e.g. *Myosoton aquaticum* ×2000); (e) *Dianthus carthusianorum* (×2000); (f) *Gypsophila fastigiata* (×2000); (g, h) *Scleranthus perennis* ((g) ×8000; (h) ×2000); (i, j) *Chenopodium rubrum* ((i) ×2000; (j) ×8000).

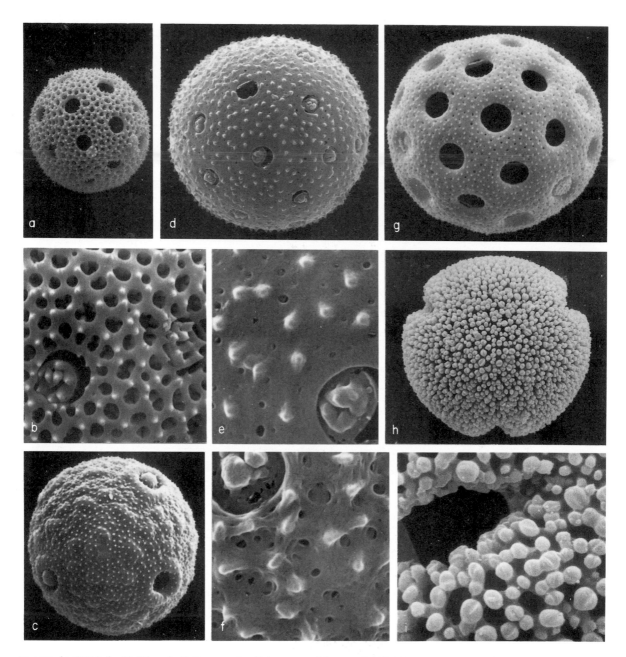

PLATE 64 SEM: (a, b) *Silene latifolia* ssp. *alba* ((a) ×1000; (b) ×4000); (c) *Plantago lanceolata* (×2000); (d–f) *Sagina procumbens*; ((d) ×2000; (e), (f) ×8000); (g) *Lychnis flos-cuculi* (×2000); (h, i) *Geranium dissectum*; ((h) ×1000; (i) ×4000).

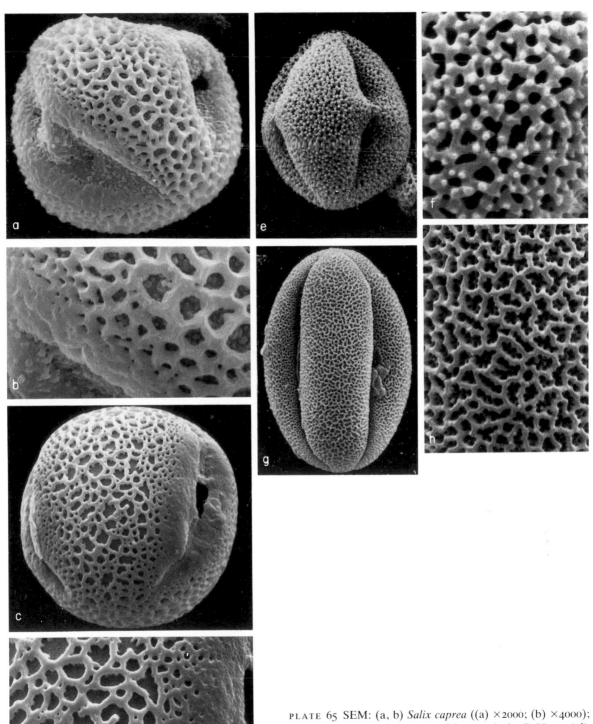

PLATE 65 SEM: (a, b) *Salix caprea* ((a) ×2000; (b) ×4000); (c, d) *Hedera helix* ((c) ×2000; (d) ×4000); (e, f) *Mercurialis perennis* ((e) ×2000; (f) ×8000); (g, h) *Stachys sylvatica* ((g) ×2000; (h) ×8000).

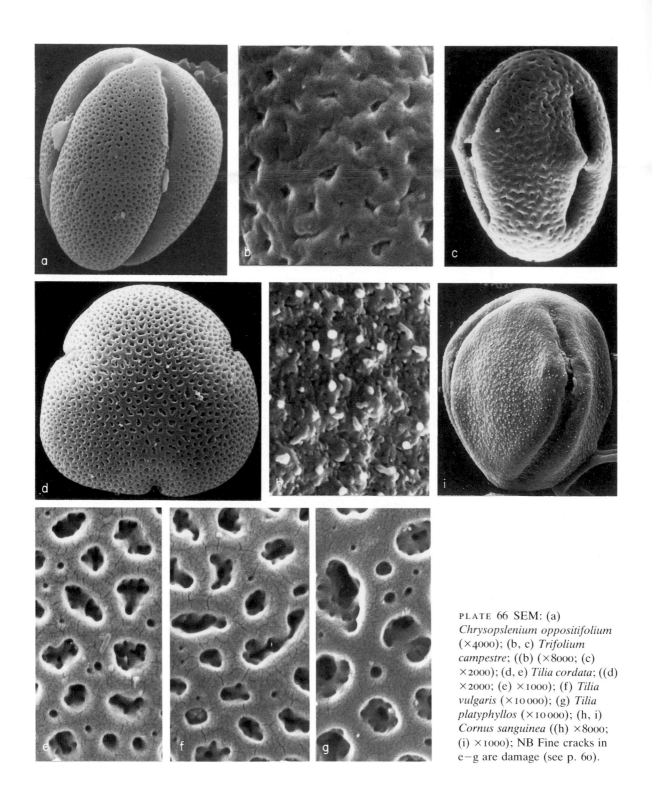

PLATE 66 SEM: (a) *Chrysopslenium oppositifolium* (×4000); (b, c) *Trifolium campestre*; ((b) (×8000; (c) ×2000); (d, e) *Tilia cordata*; ((d) ×2000; (e) ×1000); (f) *Tilia vulgaris* (×10000); (g) *Tilia platyphyllos* (×10000); (h, i) *Cornus sanguinea* ((h) ×8000; (i) ×1000); NB Fine cracks in e−g are damage (see p. 60).

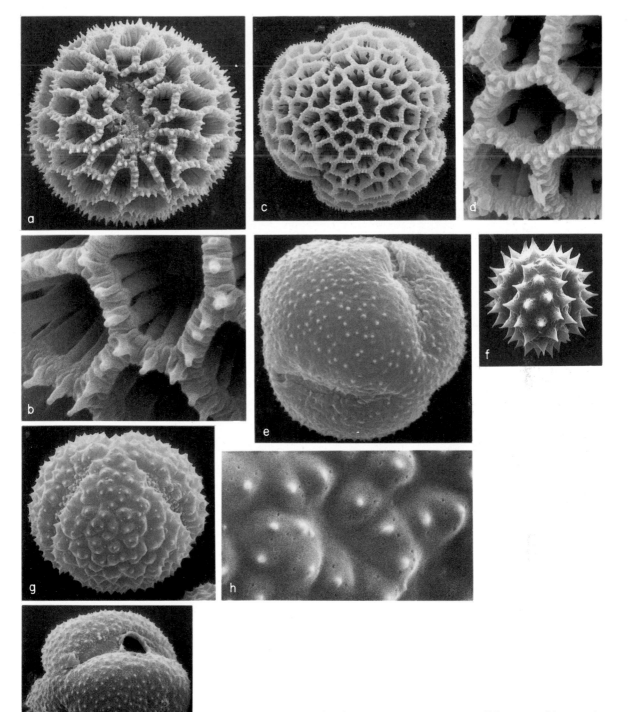

PLATE 67 SEM: (a, b) *Armeria maritima* type A ((a) ×1000; (b) ×4000); (c, d) *Limonium vulgare* type A ((c) ×1000; (d) ×4000); (e) *Filipendula vulgaris* (×4000); (f) *Aster tripolium* (×1000); (g, h) *Ranunculus bulbosus* ((g) ×2000); (h) ×8000); (i) *Artemisia vulgaris* ×2000.

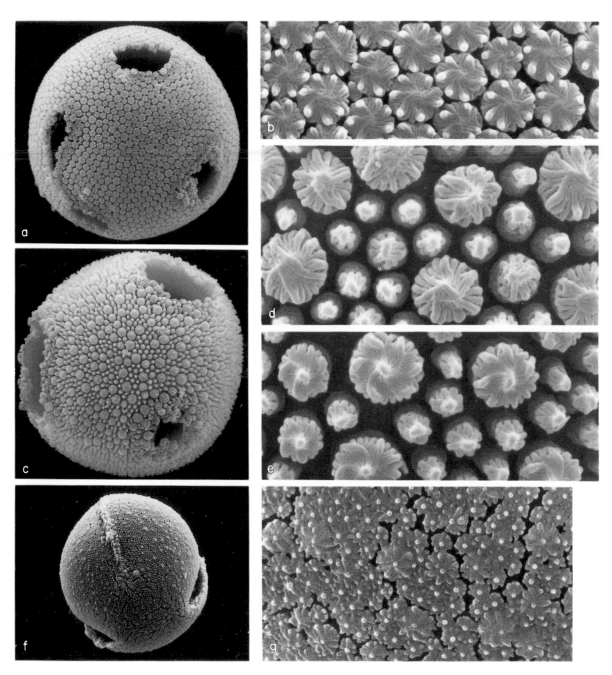

PLATE 68 SEM: (a, b) *Linum perenne* ssp. *anglicum* short-styled ((a) ×1000; (b) ×8000); (c–e) *Linum perenne* ssp. *anglicum* long-styled ((c) ×1000; (d), (e), ×8000); (f, g) *Linum bienne*; ((f) ×1000; (g) ×8000).

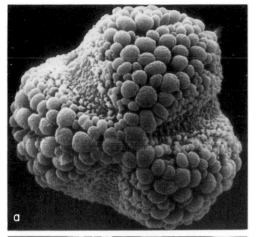
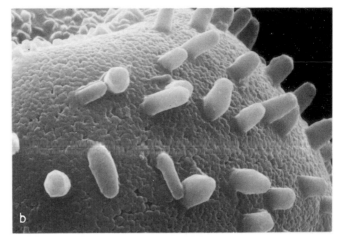
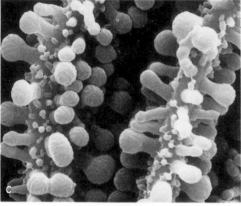
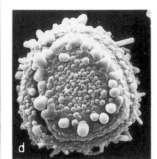
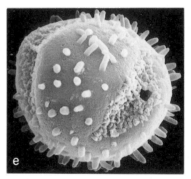
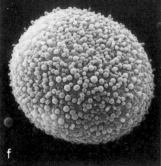
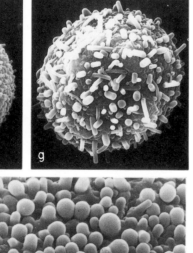

PLATE 69 SEM: (a, c) *Ilex aquifolium*
((a) ×2000; (c) ×4000); (b, e) *Viscum album*
((b) ×4000; (e) ×1000); (d, g) *Nymphaea*
alba (×1000); (f, h) *Nymphaea candida*
((f) ×1000; (h) ×4000).

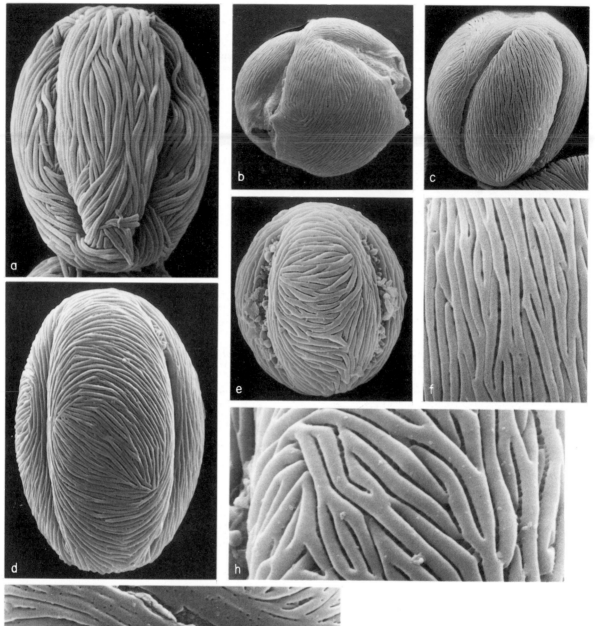
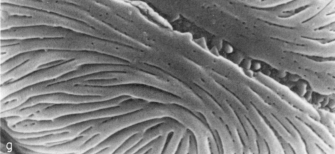

PLATE 70 SEM: (a) *Menyanthes trifoliata*
(×2000); (b) *Sorbus aucuparia* (×2000);
(c, f) *Acer pseudoplatanus* ((c) ×2000;
(f) ×8000); (d, g) *Trollius europaeus*
((d) ×4000; (g) ×8000); (e, h) *Saxifraga
oppositifolia* ((e) ×2000; (h) ×8000).

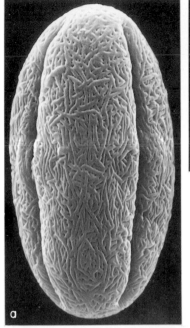

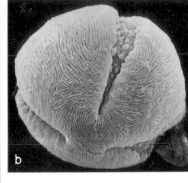

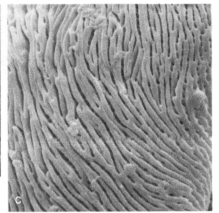

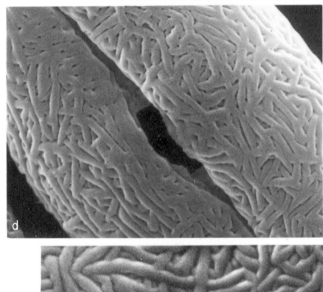

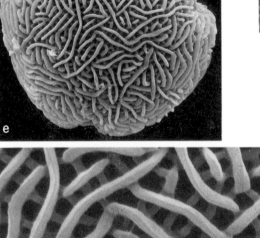

PLATE 71 SEM: (a, d) *Apium nodiflorum* ((a) ×4000; (d) ×8000) (b, c) *Dryas octopetala* ((b) ×2000; (c) ×8000); (e, f) *Erodium cicutarium* ((e) ×1000; (f) ×4000) (g) *Coronilla varia* (×8000).

thickness, thus muri (sexine 2) relatively shallow (depth of muri <length of columellae):

Tamarix

19b Sexine relatively thin — only as thick as, or slightly thicker than, nexine (see optical section of wall, ×1000) Columellae short, often indistinct, muri shallow. Expanded grains may show a few transverse endocracks across each colpus in the equatorial region: 20

20a Grain usually <18 μm long. Apocolpia not reticulate — having a few perforations only. Columellae often indistinct in wall cross section. Consult a type slide: *Chrysosplenium* [see Verbeek-Reuvers, 1977a]

20b Grains usually >18 μm long. Apocolpia reticulate, although some grains may be syncolpate at poles. Columellae distinct in wall cross section. The muri (sexine 2) relatively wide and deep (in wall optical section, depth of muri = length of columellae). Consult a type slide:

Saxifraga stellaris type

[this type is difficult to distinguish from *Chryso-splenium* and type slides should always be consulted. According to our observations and data in Verbeek-Reuvers, 1977a, this type includes *S. stellaris*, *S. nivalis*, *S. tenuis*, *S. clusii*, *S. hieracifolia* and *S. foliolosa*. Some preparations of species in this type show lumina of >1 μm diameter]

21a With coarse columellae arranged in a more or less clear infraticulum in the mesocolpia. Tectum relatively thin and with one perforation over each 'lumen' of the infraticulum. Colpi wide, with very coarse granules or bacula that are thicker than the columellae adjacent. Microechinae on the tectum situated over the muri only (×1000, phase contrast): *Glaucium* [includes the large, and possibly the medium, grain sizes of *G. flavum* and *G. corniculatum*. Kalis, 1979, reports that there are three sizes of pollen grain in each specimen of these species. Our observations indicate that the smallest size does not show an infraticulum. See 25b below]

21b Without any infraticulum visible — columellae evenly spaced or randomly irregularly spaced. Colpus membranes without such coarse granules: 22

22a Grains usually circular in polar and equatorial views. Long, narrow (and usually split) colpi with markedly **ragged**, **granular** edges and remnants of a granular membrane usually visible (colpi never inrolled or sunken-in). Slender, short columellae present in a very **fine and even carpet**, exine usually thin (may appear **pilate** in

optical cross section). Grain size >25 μm and most >30 μm: *Rhinanthus* type [includes *Rhinanthus*, *Bartsia*, *Parentucellia*, *Euphrasia*, *Lathraea*, some *Veronica* species, some *Orobanche* species e.g. *O. purpurea*; some *Pedicularis* species e.g. *P. flammea*, *P. sceptrum-carolinum*, *P. oederi*. *P. oederi* is syncolpate at the poles]

22b Without this exact combination of characters, grains either with thicker exines or coarser columellae or smaller sized or with non-ragged colpi: 23

23a Exine gradually thins towards each colpus (see polar view, grain optical section). Columellae can be seen to become gradually shorter towards each colpus (similar to, but not as pronounced as in *Artemisia*). Colpi granulate and rather sunken-in, so that a lobed shape is apparent in polar view. Grains tectate-perforate with densely spaced perforations. Consult a type slide: *Frankenia laevis*

23b Exine not thinning towards the colpus. The surface may or may not be perforate and the colpi may or may not be granulate. Shape lobed or circular in polar view: 24

24a Columellae coarse compared to grain size: 25

24b Columellae finer compared to grain size: 26

25a Narrow slit-like, non-granulate colpi, never gaping. No tectum perforations, no microechinae: *Anemone nemorosa*

25b With wide, granulate colpi, sometimes gaping. Tectum perforate and **commonly microechinate**. This description leads to the following three taxa which require type slides for differentiation: (i) *Spergula* type [includes *Spergula*, *Spergularia* and *Polycarpon*. This is probably the commonest type to key out here. Microechinae and perforations about equal in density and not associated with columellae. Colpi parallel-sided and **obtuse ended**, with finely granulate membranes] (ii) *Cuscuta europaea* type [includes *C. europaea*, *C. epithymum*, *C. epilinum* and *C. campestris* — see ref. under 7b. One microechina on top of each columella, perforations on tectum between columellae. Colpi coarsely granulate] (iii) *Glaucium* type [the smallest of the three grain sizes only — see Kalis (1979). Frequent microechinae, columellae very densely crowded, sparser perforations. Very coarsely granulate colpi which are often sunken-in, giving a lobed shape in polar view. Size <26 μm]

26a Colpus membranes psilate. Colpi with straight, non-ragged edges that are often inrolled as they widen towards the equator, where each shows a very slight constriction or some undulations. Columellae short, tectum layer distinct over them. Consult a type slide: *Myricaria germanica*

26b   Colpus membranes granulate. Colpi often split open with ragged edges: *Veronica* type [includes *most Veronica* species, *Odontites* and *Lamium album. Odontites* has the smallest grains. Note that some *Crataegus, Rubus* or *Sorbus* grains as well as *Viola palustris* type (p. 140) and *Digitalis* type (p. 146) may key out here if the equatorial constrictions to the colpi are not apparent. Distinctions may be possible if a very comprehensive type collection is available]

Trizonocolpate, reticulate

1a   Grains **suprareticulate**, i.e. the muri of the reticulum are walls which sit on top of a tectum. The reticulum is thus always independent of the pattern of the columellae, if the latter are visible (to detect columellae, observe the grain wall in optical section ×1000 and carry out careful LO analysis of the grain surface, whilst studying Fig 5.8):   2

1b   Grains **eureticulate**, i.e. the muri of the reticulum are walls which join the heads of the columellae. Distribution of columellae (if visible) commonly dependent on the pattern of the reticulum (view as in 1a and see Fig. 5.9):   5

2a   Grains rectangular-obtuse in equatorial view, with mesocolpium sometimes flattened or concave in the equatorial region only. Colpi narrow, slit-like (but can appear gaping if split open) with very straight edges. Lumina of the reticulum decrease sharply in size at the extreme margin of each colpus: *Onobrychis* type [includes *Onobrychis viciifolia* and *Hedysarum hedysaroides*. Separation of these two may be achieved by consulting type slides — *Hedysarum* grains are considerably smaller than those of *Onobrychis*]

2b   Grains circular, elliptic or rhombic-obtuse in equatorial view; mesocolpia always convex. Colpi wide or narrow, with granulate or smooth membranes: 3

3a   Colpi narrow, with out granulate membranes and rarely split or ruptured along their entire length (may be ruptured at the equator). Grain shape circular, elliptic or rhombic-obtuse. Phase contrast, ×1000, may reveal a very characteristic black blotches in a 'holly leaf' or 'star' pattern on the surface: *Ulex* type [includes *Ulex, Genista* and *Cytisus*. Some dis-

tinction may be possible between these by examination of type slides]

3b   Colpi wide and boat-shaped or gaping, with granulate membranes (may be represented as a torn and ragged granular edge to the colpus if rupture has occurred, may not be visible if inrolled):   4

4a   With lumina largest in the centre of each mesocolpium (muri here may appear winding). Lumina size decreasing **gradually** towards each colpus: *Scutellaria* type [includes *Scutellaria* and *Ajuga. A. chamaepitys* is distinct in its very coarse reticulum]

4b   Lumina more or less uniform in size all over the grain (may be so small as to be microreticulate):   *Stachys sylvatica* type [includes *S. sylvatica, S. palustris, S. arvensis, S. annua, S. germanica, S. alpina, S. recta, S. officinalis, Galeopsis, Lamium purpureum, L. amplexicaule, L. hybridum, Melittis melissophyllum, Leonurus cardiaca*. Some distinction may be possible if type slides are available. The largest lumina are present in *Galeopsis* and *S. recta. S. officinalis* is distinct because of its large grains and very faint, coarse reticulum. Some forms of *Myricaria germanica* may key out here]

5a   With very small reticulum, bordering on microreticulate i.e lumina <or = 1 μm in width: 6

5b   With bigger reticulum, lumina >1 μm in width: 12

6a   Colpi wide, with granulate membranes: 7

6b   Colpi narrow, no colpus membrane visible: 10

7a   Colpi relatively **short** (whole length easily visible in equatorial view) parallel sided, with diffuse edges and obtuse ends. Grain elliptic to circular in equatorial view: *Platanus* [*Reseda alba* may key out here. Consult type slides]

7b   Colpi **longer**, curving round the grain: 8

8a   Grains elliptic to rhombic-obtuse in equatorial view. Colpi boat-shaped, i.e widest at equator and narrowing to acute ends at the poles: *Reseda lutea* type [includes *R. lutea, R. luteola* and *R. phyteuma*. Some distinction may be possible between these, based on size of lumina. See also *Myricaria germanica* and *Acer saccharum* — if they are likely to be in the area]

8b   Grains circular in equatorial and polar view. Colpi may or may not be boat shaped, usually obtuse-ended: 9

9a   Grains usually <29 μm. Colpi always obtuse-ended. Lumina rather variable in size: *Helleborus foetidus* [consult type slides]

9b Grains 27 μm or more. Colpi may or may not be obtuse ended. Lumina uniform in size:
Chelidonium majus
[consult type slides]

10a Columellae distinct in surface view. Exine can usually be seen to be thickest and columellae longest in the centre of each mesocolpium, columellae shorter at poles and colpus margins.
Hornungia type
[a group of Crucifers which include many common small weedy species as well as rarer plants — data from Erdtman, *et al.*, 1963, and our own investigations, e.g. *Hornungia petraea, Capsella bursa-pastoris, Cardamine amara, C. impatiens, Cardaria draba, Descurainia sophia, Lepidium campestre, L. densiflorum, Arabidopsis thaliana, Arabis glabra, Berteroa incana, Alyssum, Arabis hirsuta, Arabidopsis suecica, Thlaspi arvense, T. alpestre*. Probably many more species belong to this type. Erdtman et al., 1963 separate *Arabis hirsuta, A. glabra* and *Arabidopsis suecica* from the above group on the basis of the exine being 2 μm or more thick, whereas the rest have exines <2 μm thick. It is noted that the small crucifers of *Hornungia* type can look very similar to grains of e.g. *Samolus, Hedysarum, Linaria* type, *Hottonia, Chrysosplenium, Tamarix* and *Saxifraga nivalis* type. Type slides should always be consulted]

10b Columellae indistinct, even in phase contrast only a fine black network is seen. Exine of the same thickness and columellae of the same length all over the grain surface: 11

11a Grains usually <18 μm in length. Apocolpia tectate-perforate with only a few perforations. Consult a type slide: *Chrysosplenium*

11b Grains usually >18 μm in length. Apocolpia reticulate. Sometimes syncolpate at the poles. Consult a type slide: *Saxifraga nivalis* type [includes *S. nivalis, S. clusii* and *S. tenuis* and possibly some grains of *S. stellaris* and *S. hieracifolia*. See Verbeek-Reuvers, 1977a. Compare also *Hedysarum, Samolus, Hottonia, Linaria* type, *Myricaria germanica* and *Tamarix*]

12a Exine >4 μm thick, grains usually >40 μm in size: 13

12b Exine <4 μm thick, grains usually <40 μm in size: 16

13a Colpi very short — resembling elongated pori, apocolpia consequently large. Columellae branching and anastomosing to form the reticulum. Reticulum is partially obscured by clavae-like projections on top of the muri (careful LO analysis is needed): *Geranium*

13b Colpi long. Muri supported by one row of unbranched columellae. Either no projections or echinae on top of the muri: 14

14a Reticulum very coarse, muri much narrower than lumina and with a row of echinae or microechinae on top of each. Columellae elliptical in cross section: 15

14b Reticulum less coarse, muri equal to or wider than lumina, no echinae or microechinae on muri. Columellae circular in cross section:
Armeria maritima type B grains
[from individuals with papillate stigmas. According to Turner and Blackmore, 1984, this includes half of the individuals of the following species: *A. maritima* ssp. *maritima, A. maritima* ssp. *halleri, A. maritima* ssp. *purpurea, A. alliacea, A. pseudoarmeria, Limonium vulgare, L. bellidifolium*; and all of the individuals of *L. auriculae-ursinifolium*. These authors report that the *Armeria* species are >60 μm in equatorial diameter whilst the *Limonium* species are <60 μm in equatorial diameter]

15a Muri with clear **echinae**, >2 μm long, pollen grains larger and with thicker exines (equatorial diameter >60 μm, exine >9 μm thick):
Armeria maritima type A grains
[from individuals with 'cob' or smooth stigmas. Data from Turner and Blackmore, 1984. Apparently includes half the individuals of *A. maritima* ssp. *maritima, A. maritima* ssp. *halleri, A. maritima* ssp *elongata, A. maritima* ssp. *purpurea, A. alliacea, A. pseudoarmeria*; and all the individuals of *A. maritima* ssp. *sibirica*]

15b Muri with **microechinae** (sometimes so small as to be invisible with the light microscope) pollen grains smaller and with thinner exines (equatorial diameter <77 μm, exine <11.5 μm):
Limonium vulgare type A grains or
Limonium humile type
[from individuals with 'cob' or smooth stigmas. Data from Turner and Blackmore, 1984. Apparently includes half of the individuals of *Limonium bellidifolium* and *L. vulgare*, all individuals of *L. humile, L. binervosum, L. recurvum* and *L. transwallianum*]

16a Colpi relatively short (whole length visible in equatorial view) narrow and non-granulate. Apocolpium large, grain circular in both polar and equatorial views: 17

16b Colpi longer, being either wide or narrow and inrolled or not inrolled along their length. Grain circular to elliptic in equatorial view: 18

17a Reticulum finer, muri simplicolumellate but sometimes with duplicolumellate areas at the intersections of the muri. Muri depth less than

length of columellae (see optical section of exine ×1000):   *Fraxinus*
[includes *Fraxinus excelsior* and *F. ornus*. The latter may be distinct because of its thicker sexine relative to the nexine, and longer colpi in comparison to the polar axis. Type slides should be consulted. *Hedera* may just possibly key out here if the endoporus is not detectable. It is more commonly duplicolumellate, with lumina more variable in size than in *Fraxinus*, and has a different shape see p. 148]

17b Reticulum very coarse. Muri always simplico-lumellate and very thick (deep) — depth of muri >length of the thin, short columellae (see optical section of exine, ×1000):   *Olea europaea*
[*Phillyrea angustifolia* has very similar grains, but with more lumina across a mesocolpium than *Olea*. Consult type slides and see discussion on pp. 79–80. Note that *Ligustrum* may key out here if the colpi do not exhibit constrictions]

18a Lumina size and exine thickness **markedly decreasing** towards the colpus edge such that a clear margo is present. The immediate edge of the colpus is actually **tectate** without lumina, but this is only visible on very expanded grains where the colpi are not so inrolled. Colpi fairly wide and sometimes showing granulate membranes. Edges of colpi usually inrolled: *Salix*
[some distinction within the genus has been attempted by other authors e.g. Erdtman *et al.* 1963, separate *S. pentandra* on the observation that adjacent lumina vary markedly in size in the middle of the mesocolpium, and *S. herbacea* on the observation that the muri can be more or less fragmentary. This last observation is not supported by our type material]

18b No **marked** decrease in lumina size and exine thickness towards the colpi (a slight decrease in lumina size may be shown, exine thickness may slightly decrease or increase towards the colpi). Edge of colpus **reticulate**, never tectate. Colpi wide or narrow: 19

19a Colpi narrow and slit-like; if split open and gaping, then without granulate membranes visible: 20

19b Colpi widening from acute or obtuse ends to the equator, granulate membranes always visible: 21

20a Muri simplicolumellate and narrower than lumina. Columellae longest in the centre of the mesocolpium, becoming shorter towards the poles and the edges of the colpi. Sexine always thicker than nexine, columellae joined only at their extreme apices into muri (reticulum only

clear at highest focus):   *Sinapis* type
[includes the remainder of the Cruciferae not covered under exit 10a. Data from Erdtman *et al.*, 1963, and our own observations i.e. *Sinapis, Matthiola, Bunias, Cakile, Teesdalia, Isatis, Crambe, Armoracia, Brassica, Arabis alpina, Cardamine bulbifera, C. pratensis, C. flexuosa, Alliaria, Braya linearis, Diplotaxis, Draba, Erophila, Erucastrum, Lepidium latifolium, L. ruderale, Lunaria, Raphanus, Rorippa, Sisymbrium, Barbarea, Cochlearia, Erysimum, Subularia, Cardaminopsis arenosa, Camelina microcarpa* and possibly more. Some division of the above list has been made by Erdtman *et al.*, 1963, on the basis of exine thickness and diameter of the largest lumen. Some *Matthiola* species are reported to have inaperturate grains (Andrew, 1984) and some *Rorripa* species to have inaperturate or irregularly polypantoporate grains. Note that *Ligustrum* has a very similar reticulum to some of the above grains and may key out here if the constriction to the colpi is not detectable. It would appear also that *Gentianella detonsa* keys out here — see Punt and Neinhuis, 1976]

20b Muri commonly as wide as, or wider than, the lumina and mostly duplicolumellate. Columellae the same length all over the grain surface, or very slightly longer at the poles. Lumina rather variable in size. Careful focusing shows that the muri are acute in cross section. Grain small, usually <25 μm:
                              Marrubium vulgare

21a With a coarse reticulum underlain by columellae which sit in a reticulate pattern in both apocolpia and mesocolpia. Lumina size decreasing slightly and gradually towards the margins of the colpi but not towards the apocolpia. Colpus margins edged by a narrow solid band. Exine in marginal area appears slightly thicker than exine elsewhere on the grain:
                              Helleborus viridus

21b Columellae very coarse and often irregular in cross section; inordinately (irregularly) arranged except in the centre of each mesocolpium, where they are roughly in a reticulate pattern. Lumina of reticulum actually nearly covered by a thin layer of tectum, with a foveola or a perforation visible over the centre of each lumen (surface view, ×1000, phase contrast helps) Microechinae may be detectable on the muri. Colpus membrane with granules as coarse, or coarser than, the columellae; colpus margin very diffuse: *Glaucium flavum* type
[a wide size range is visible on type slides of

Glaucium and the above description covers only the large and medium grains. According to Kalis, 1979, the type exhibits trimorphism in pollen grain size and includes *G. flavum* and *G. corniculatum*]

Trizonocolpate, echinate or microechinate

1a Two curved nexinous ridges (costae) cross under the polar ends of each ectocolpus (these are actually the margins of a very wide, transverse endoporus or endocolpus). Grain triangular-convex (with colpi on the angles) to circular in polar view. Colpi short and narrow, apocolpium consequently large in relation to grain size. Grain usually broader than long (P/E <1.0):   2

1b No such curved nexinous ridges passing under the ends of each ectocolpus. Grain circular or lobed in polar view. Colpi long or short: 4

2a Echinae <or = 1.5 μm (mostly microechinae) and densely clothing the surface of the grain. The transverse nexinous ridges are short, and cross at the extreme ends of each ectocolpus. Grain 50 μm or less in longest dimension: *Linnaea borealis*

2b Largest echinae >or = 1.5 μm, densely or sparsely clothing the surface. Echinae may be dimorphic (long and short mixed). Grains mostly >50 μm in longest dimension: 3

3a With long echinae, >1.5 μm and up to 3 μm, widely spaced on the surface and often thickened at the base: *Lonicera xylosteum* type [includes *L. xylosteum*, *L. alpigena*, *L. nigra* and some grains of *L. coerulea* that are not 4-aperturate. Separation of types is apparently possible, for example, *L. coerulea* has densely spaced small echinae as well as the sparser larger ones. See Punt *et al.*, 1976]

3b With shorter echinae, up to 1.5 μm long, more densely spaced on the exine:
                    Lonicera periclymenum type [includes *L. periclymenum* and *L. caprifolium*. Separation is apparently possible. See ref. under 3a]

4a Small rods traverse the tectum and bases of the echinae. These infratectal rods may appear to be columellae but no true columellae are present, the tectum thus lies adjacent to the nexine. Colpi long, or short and resembling elliptic pori. Echinae <2.0 μm, grain <30 μm:
                    *Ambrosia** type [includes *Ambrosia*, *Iva* and *Xanthium*. Distinction may be made within this type depending

on the length of the colpi — see McAndrews *et al.*, 1973. Note that other members of the Compositae may key out here if their pori are not detectable. See also such members of the Ranunculaceae as *Pulsatilla* and *Nigella*]

4b No infratectal rods traversing tectum and echinae. Always with columellae present on the nexine. Echinae may or may not be <2 μm: 5

5a Exine markedly thickest in the apocolpia, where the columellae are longest and sometimes branched: 6

5b Exine of the same thickness and columellae of the same length all over the grain surface: 9

6a Colpi very short, resembling sunken elliptical pori, each with a distinct smooth band (margo) on the edge, surrounded by a wide 'halo' or pale zone around each colpus where both columellae and nexine are absent. Columellae dimorphic — thick ones under the echinae, slimmer ones elsewhere. Colpus membrane with long branched echinae forming an operculum: *Dipsacus* [see Clarke and Jones, 1981a]

6b Colpi longer, membranes covering them with or without long branched echinae. Columellae not dimorphic, although they may show some degree of coalescence. Each colpus margin with or without a wide, pale 'halo' area where both columellae and nexine are missing: 7

7a Grain 40 μm or less along polar axis. Shape elliptic to circular or rhombic-obtuse in equatorial view, poles often slightly flattened. With microechinae (i.e. all <or = 1 μm long). Columellae showing some coalescence at the poles:
                    Valerianella [see Clarke and Jones, 1977b, for some differentiation between the species]

7b Grain >50 μm along the polar axis. Shape elliptic to rhombic-obtuse in equatorial view, with mixed echinae and microechinae: 8

8a Colpi rather wide, with margin surrounded by a narrow 'halo' or pale zone where columellae are absent and nexine is disrupted. Colpus membrane with long coarse echinae. Columellae coarse, being thicker and more distinct than the echinae. At the poles the columellae coalesce to form larger, scattered, irregular, linear or C-shaped units. Exine <7 μm thick in the apocolpia: *Centranthus* [see Clarke and Jones, 1977b]

8b Colpi very narrow, each with an extremely narrow, pale 'halo' zone which may not be detectable. Colpus membrane irregularly granulate (however it is hardly visible as the colpus is so narrow and slightly sunken). Columellae long, thin and densely packed all over the surface of the grain but showing a slight degree of coalescence at the poles. Columellae about as distinct as the echinae. Exine >7 µm thick in the apocolpia: *Scabiosa*

9a Echinae >1 µm long and rather sparsely spaced on the grain surface: 10

9b Echinae <or = 1 µm long (microechinae) rather densely covering the grain surface: 14

10a Grain with polar axis >60 µm: 11

10b Grain with polar axis <60 µm: 12

11a Colpi very short, resembling sunken elliptical pori, each with a distinct smooth band (margo) on the edge, surrounded by a wide 'halo' or pale zone around each colpus where both columellae and nexine are absent. Columellae dimorphic — thick ones under the echinae, slimmer ones elsewhere. Colpus membrane with long branched echinae forming an operculum: *Dipsacus*
[see Clarke and Jones, 1981a]

11b Colpi longer and narrower, margin irregular with a narrow margo, no 'halo'. Columellae very variable in thickness and often coalesced into irregular, linear or C-shaped units, but not clearly dimorphic. Tectum thick, minutely perforate. Colpus membrane with only a few irregular granules or short echinae: *Succisa*

12a Echinae constricted at their bases and cylindrical for most of their lengths. They may be acute or obtuse, thick or thin, and sometimes curved or bent over, but never situated on verrucae:
Rubus chamaemorus

12b Echinae not constricted at their bases, may or may not be cylindrical, but always each one set upon a low verruca: 13

13a Echinae long, length approaching that of thickness of the rest of the wall. Each echina cylindrical: *Valeriana dioica*

13b Echinae shorter, length less than the thickness of the rest of the wall. Each echina conical:
Valeriana officinalis
[includes *V. officinalis*, *V. tuberosa* and *V. pyrenaica*. See Clarke and Jones, 1977b]

14a A faint suggestion of lines joining the bases of the echinae (actually an infrareticulum):
Papaver rhoeas type
[includes *P. rhoeas*, *P. dubium*, *P. hybridum*, *P. orientale*, *P. somniferum*, *P. strigosum*, *Meconopsis cambrica* and occasional grains of *Papaver* species that are usually pantocolpate, e.g. *P. nudicaule*. See Kalis, 1979]

14b No such faint suggestion of lines joining the bases of the echinae. Columellae a dense, even carpet under the tectum: *Caltha* type
[includes *Caltha* and *Aquilegia*. Consult type slides. *Caltha* has finer and more densely spaced columellae than *Aquilegia* does]

Trizonocolpate, scabrate-verrucate to microechinate

It is sometimes very difficult to distinguish between scabrae and microechinae, so grains with either are included here.

1a Grain with thick exine (>3.5 µm) and tectum clearly traversed by infratectal bacula which are slimmer and more densely packed than the columellae underlying them. Colpi long and wide, with membranes coarsely granulate: *Nigella*
[data from W. Punt, personal communication of unpublished work. It is our opinion that similar, but fainter, exine structure is detectable in *Pulsatilla*]

1b Grain with exine usually thinner and never with any indication of infratectal bacula under the LM: 2

2a Grain tectate with **regularly spaced** verrucae, scabrae or microechinae on the tectum. These are underlain (use careful LO analysis and phase contrast if possible) by columellae which are regular or irregular in shape and distribution: 3

2b Tectum processes more or less **irregular in distribution** and sometimes in shape as well. Underlying columellae regular or irregular in shape and distribution: 6

3a Columellae thin, of uniform thickness and length in the mesocolpia, visible as a dense, uninterrupted carpet underneath the tectum which bears the verrucae, scabrae or microechinae: 4

3b Columellae never forming an even, uninterrupted carpet in the mesocolpia. They are either connected by a very faint reticulum-like pattern of lines; or varying in thickness and in length, and more or less aggregated into groups: 5

4a Grains elliptic to narrowly elliptic in equatorial view, with regularly spaced small verrucae or echinae. Tectum **conspicuously thickened** in the region of the mesocolpium **adjacent** to the apocolpium (thinner actually at each apocolpium). Exine markedly thinning from the mesocolpia towards the colpi. Each colpus with a narrow operculum: *Teucrium*
[some distinction between the species may be possible by reference to type slides. Grain size is the main difference. Grains of *Adonis* or *Consolida* may key out here if microechinae are

detectable, but note that these grains have the exine thickest actually at the poles and not in the region outside them]

4b Grains elliptic to circular in equatorial view, with regularly spaced microechinae on top of the tectum. Sexine more or less the same thickness all over the surface of the grain. Colpi wide, obtuse-ended and with granulate membranes:

Caltha type

[includes *Caltha* and *Aquilegia*. *Caltha* is distinct upon comparison with type slides because it possesses numerous very fine columellae whereas *Aquilegia* has fewer, coarser columellae. It is noted that some *Artemisia* species may key out here if the pori are not apparent; they may be distinguished by the non-granulate colpi and the fact that close examination shows that the exine is marginally thickest in the mesocolpia]

5a Columellae nearly uniform in thickness and connected by a faint reticulum (infrareticulum) underneath the tectum and microechinae. The microechinae are situated over the intersections of the reticulum: *Papaver rhoeas* type

[includes *P. rhoeas*, *P. dubium*, *P. hybridum*, *P. orientale*, *P. somniferum*, *P. strigosum*, *Meconopsis cambrica* and occasional grains of *Papaver* species that are usually pantocolpate e.g. *P. nudicaule*. Data on included species from Kalis, 1979. We note that some collections of *Meconopsis* show only inaperturate grains]

5b Columellae varying markedly in thickness, very coarse columellae (which are often irregular in cross section) occupy the centre of the meso-colpia, whilst finer, shorter columellae are seen near the colpus margins. Columellae may appear aggregated: *Artemisia*

[grains with such varying columellae are notice-able in a few species but they are most marked on type slides of some *A. maritima* subspecies. See Webb and Moore, 1982]

6a Tectum processes a mixture of scabrae and irregular, more or less elongated verrucae which are unevenly scattered over the tectum (careful LO analysis or phase contrast is needed to detect the difference between these processes and the underlying carpet of more regularly spaced, finer columellae): *Quercus*

[cf. also type slides of *Cornus* and *Medicago sativa* type, as these can look similar. *Oxalis acetosella* may key out here, but it may be distinguished by: a perforate tectum; wide, granulate, colpi and a thicker exine. The ever-green oaks e.g. *Q. ilex* may possibly be dis-tinguished if type slides are available]

6b Tectum either with microechinae or with scabrae and verrucae which are rounded and never elongate. Columellae either all uniform; or dimorphic, coarse and fine mixed: 7

7a Columellae varying in thickness i.e. dimorphic in the following fashion: sparser thick ones irregularly scattered amongst more numerous fine columellae (observe a surface view, ×1000; the slender columellae may be at the limit of visibility, although phase contrast helps). In some species the coarse columellae are each surrounded by a light ring (space) and then by a ring of slender columellae, which may or may not appear fused. The tectum may appear more or less undulating, rising up over the coarse columellae and diving down between them. Microechinae are sometimes detectable on the peaks, over the coarse columellae. Colpi usually wide and coarsely granulate, exine sometimes thicker along the colpus margins: *Ranunculus* type

[includes *Ranunculus*, *Clematis vitalba*, *Actaea spicata*, *Myosurus minimus*, *Eranthis hyemalis* and *Pulsatilla vulgaris*. Possibly *Adonis vernalis* may key out here (personal communication by W. Punt of unpublished data). See note at end of key on p. 128 for further distinctions within this type. Some species that key out here also have a high proportion of pantocolpate grains]

7b Columellae monomorphic (or apparently so): 8

8a Columellae coarse compared to grain size and characteristically irregularly spaced out in the mesocolpia (consult a type slide). Columellae may appear to be verrucae at high focus (white dots) but in fact the tectum is smooth, with no undulations — see optical section of wall. Exine adjacent to colpus slightly thicker than the exine in the middle of the mesocolpium. Colpi sparsely granulate, obtuse-ended and with parallel-sided, clearly defined edges. No perforations or micro-echinae visible with the LM: *Anemone nemorosa*

[other species of *Anemone* e.g. *A. apennina*, *A. sylvestris* and *A. ranunculoides*, are reportedly similar, but lack the characteristic thickening of the exine adjacent to the colpi (W. Punt, per-sonal communication of unpublished data). Our observations show *Hepatica nobilis* to be very similar except for the observation that distinct microechinae are visible with the LM over the coarse columellae on careful examination of the wall in optical section. Also the columellae are more densely spaced in the apocolpia than in the mesocolpia of *Hepatica*. It is the opinion of W. Punt that *Hepatica* is distinct because of its endocracks, which are much more prominent than in *Anemone* (personal communication of unpublished data). It is noted that some of

Ranunculus type may key out here if the fine columellae are not detected at all. Phase contrast reveals that there is, in fact, some extremely fine patterning between the coarse columellae in *Anemone*. This will be at the limits of resolution of most light microscopes. *Pulsatilla vulgaris* may also key out here — see note at the end of this key]

8b Columellae less coarse and much shorter compared to grain size. Surface tectate-perforate, with small verrucae or scabrae scattered over the tectum (phase contrast may be needed to detect these features). Colpi narrow, acute-ended, often sunken, not granulate, but often with opercula which have the same exine structure as the rest of the grain. Exine the same thickness all over the grain surface: *Saxifraga granulata* type [includes *S. granulata*, *S. cespitosa*, *S. hypnoides* and *S. rosacea*. Verbeek-Reuvers, 1977a, includes *S. exarta*, *S. hartii*, *S. moschata*, *S. rotundifolia*, *S. tridactylites*, *S. cernua*, *S. hirculus* and *S. rivularis*. In our opinion the last four species can be differentiated from the rest if phase contrast is available to the observer. These four species have faint, broad striations underlying the scabrae which can be seen in phase contrast but not in bright field. See also Ferguson and Webb, 1970]

Note on differentiation within 'Ranunculus type'

It may be possible to distinguish the following species if sufficient type material is available for comparison:

Eranthis hyemalis seems distinct in that the coarse columellae are crowded together at the poles with no fine columellae between them; both sizes are present in the mesocolpia, with the coarse columellae widely spaced and numerous finer columellae occurring between them. Microechinae are clearly visible over the coarse columellae. W. Punt (personal communication of unpublished data) reports that the most distinctive feature of *Eranthis* is the broad margo to each colpus in which the sexine is thinner than in the adjacent mesocolpia and the columellae are much smaller than those in the mesocolpia.

Pulsatilla vulgaris is distinct in that it has a large size; an elliptic shape in equatorial view; and in the observation that what looks like a uniform carpet of fine columellae between the coarse columellae, appears to be a layer of denser rods or channels traversing the thick tectum (see optical section of the grain wall ×1000). Only the coarse columellae

reach down to the nexine. Microechinae may be visible on the tectum.

With regard to the remainder of the species included under the 'Ranunculus' type, Andersen (1961) first reported that the presence or absence of the 'ring' of more or less fused, small columellae around each coarse columella can be used to distinguish two major groups within *Ranunculus*. Birks (1973) also indicated such a division. Santisuk (1979) recognized three groups of species based on columellae patterns whilst Andrew (1984) reports two. Our own investigations support the 'two groups' division, but our type material is too limited for definite conclusions. W. Punt and co-workers have used this feature extensively in their distinction of types within the Ranunculaceae (personal communication of unpublished data).

Trizonocolpate, baculate, clavate or gemmate

1a Exine processes all of uniform height and quite densely packed so that some of the capita (heads) appear polygonal in surface view. Processes can be bacula, flat-topped gemmae or truncated clavae (observe an exine optical section ×1000):   2

1b Exine processes usually varying in height, densely or sparsely packed, usually clavae or bacula: 7

2a Exine processes more or less monomorphic (although rather variable in size, there are never two distinct size classes) regularly arranged — apparently in rows. Each process with a number of microechinae or scabrae on the caput in a ring (only visible under ×1000):   3

2b Exine processes dimorphic — in two distinct size classes. Processes with wider capita are circular in surface view and scattered irregularly amongst more numerous polygonal ones with smaller capita. Each process with only one major microechina or scabrum in the centre of each caput (only visible under ×1000):   5

3a Grain small, <25 μm. Less than 10 exine processes across a mesocolpium at the equator, colpus margins not appearing ruptured: *Radiola*

3b Grain larger, >25 μm. More than 15 exine processes across a mesocolpium at the equator: 4

4a Nexine markedly thicker than sexine. Nexine thinning near the colpus edge so that there is a broad band next to the colpus covered only by sexine (i.e. an endocolpus is present, longer and wider than the ectocolpus):

Linum austraicum type, short-styled [includes half of the individuals of: *L. austraicum* ssp. *austraicum*, *L. perenne* ssp. *perenne*, and *L. viscosum*. In addition it includes all individuals of *L. leonii*. Data from Punt and den Breejen, 1981]

4b Nexine not markedly thicker than sexine (it may be slightly thicker than, equal to, or thinner than, sexine). Nexine thins in a narrow band next to the colpus, or not at all:

Linum catharticum type

[includes *L. catharticum*, *L. trigynum*, *L. tenuifolium*, *L. suffruticosum* ssp. *salsoloides* and short styled individuals of *L. flavum*. Data from Punt and den Breejen, 1981]

5a Exine processes very small and short in comparison to grain size. Sexine about $\frac{1}{2}$ of the thickness of the nexine, capita of the processes so close as to be almost touching or overlapping:

Linum bienne type

[includes *L. bienne* and *L. usitatissimum*]

5b Exine processes larger in comparison to grain size, spaces between capita relatively larger: 6

6a Nexine markedly thicker than sexine. Nexine thinning near the colpus edge so that there is a broad band next to the colpus covered only by sexine (i.e. an endocolpus is present, which is longer and wider than the ectocolpus):

Linum austraicum type, long styled

[includes half of the individuals of *L. austraicum* ssp. *austraicum*, *L. perenne* ssp. *perenne* and *L. viscosum*. Data from Punt and den Breejen, 1981]

6b Nexine not markedly thicker than sexine (it may be slightly thicker than, equal to, or thinner than, sexine). Nexine thins in a narrow band next to the colpus, or not at all:

Linum flavum, long styled individuals

[data from Punt and den Breejen, 1981]

7a Grain with densely packed, large clavae which are variable in height and in thickness and which approach pila or bacula: *Ilex* type

[includes *Ilex* and, according to McAndrews *et al.*, 1973, *Nemopanthus**]

7b Grain with sparse, long, large bacula; and between them, many densely packed, slim, short columellae (appearing to be bacula under LM) the heads of which are more or less fused sideways to form a thin tectum layer (see SEM Plate 69). Colpi wide, with granules and bacula on the membrane: *Viscum*

Trizonocolpate, rugulate-striate

1a Grain with predominantly **striate** sculpturing in the mesocolpia (muri long, mostly parallel-sided and usually running parallel to one another, showing only slight anastomosing or interweaving): 2

1b Grain with more **rugulate** sculpturing in the mesocolpia (muri shorter, often not parallel-sided and with much anastomosing and interweaving): 7

2a Muri straight or curving slightly, but always running **meridionally** in the mesocolpia, never travelling in wide swirls or fingerprint patterns. Some of these meridional striae run up to, and meet, the edges of the end thirds of each of the colpi (never running smoothly round or past the ends of the colpi). Striae may meet at right angles at the poles. Columellae in a fine, even carpet. Muri duplicolumellate or pluricolumellate; striae clearly on top of the tectum:

Menyanthes trifoliata

2b Muri may or may not be straight and may or may not run meridionally, but usually curving in swirling or fingerprint patterns which run smoothly round or past the ends of the colpi. Striae running smoothly over the poles. Grain tectate or apparently semitectate: 3

3a Grain small, <25 μm long, narrowly elliptic. Exine thickest, and columellae longest and coarsest, in the apocolpia and in the region just outside them. Striae fine, narrow and faint; often running meridionally adjacent to the colpi but equatorially in the middle of the mesocolpia (i.e. in a fingerprint pattern): *Trollius europaeus*

3b Grains may be larger or smaller than 25 μm long, may or may not be narrowly elliptic. Exine roughly the same thickness all over the grain. Striae faint or very distinct, broad or narrow, directions various, may or may not be in fingerprint patterns: 4

4a Muri narrow, well defined, very long, simplicolumellate, and of similar width. They run predominantly meridionally, parallel to one another, and are sometimes branched. They appear 'beaded' at high focus due to the underlying columellae. Occasionally swirling fingerprint patterns occur instead of the meridional arrangement: *Acer campestre* type

[according to Clarke and Jones, 1978, this includes *A. campestre*, *A. pseudoplatanus*, *A. platanoides*, *A. opalus* and *A. monspessulanum*. For American species of this type see Biesboer, 1975]

4b Muri broader i.e. as wide as, or distinctly wider than, the lumina (grooves). Striae running predominantly equatorially in the middle of the mesocolpium. No obvious 'beading' of the muri although there may be **sparse** granules or scabrae on the tops. Striate pattern faint or clearly defined: 5

5a Without granules or scabrae on top of the muri

(at least under the LM, ×1000). Muri very clear cut, obviously wider than the grooves (lumina) more or less anastomosing and interweaving. They run predominantly in an equatorial direcion in the middle of the mesocolpia, but also meridionally along the margins of the colpi, and thus a fingerprint pattern may be clear:

Saxifraga oppositifolia type

[includes S. oppositifolia, S. aizoides, S. adscendens, S. cotyledon, S. cymbalaria, S. osloensis and S. paniculata. The type is similar to that defined in Verbeek-Reuvers, 1977a; but see also Ferguson & Webb, 1970. Distinctions may be made within this type by consulting these references and type slides. Note that S. oppositifolia is distinct under SEM in having no scabrae on top of the muri; unfortunately these features are not visible under the light microscope]

5b With clear granules or scabrae (view ×1000 and use LO analysis or phase contrast) on top of the muri. Striae clear or faint, may or may not run equatorially: 6

6a Striae clear, even without phase contrast:
                                        Saxifraga hirsuta type
[includes S. hirsuta, S. umbrosa, S. spathularis and hybrids. Type as defined in Verbeek-Reuvers, 1977a]

6b Striae faint, not clear cut (muri best seen in phase contrast). With one row of what appear to be very coarse columellae under each murus (see note on Saxifraga exine structure at the end of this key):   Saxifraga cernua type
[includes S. cernua, S. rivularis and S. tridactylites. The last species may be separated on its smaller size — consult type slides]

7a Colpi very short, little more than elongated pori. Grain large, >50 μm, exine thick and semitectate with coarse, parallel-sided muri which show extensive interweaving: Erodium

7b Colpi longer, grains <50 μm: 8

8a Exine structure complex, apparently two layers of columellae present with the coarsest ones in the outermost layer of the exine (see optical section, ×1000, and note on exine structure at the end of this key). Muri coarse for the grain size, showing much winding and interweaving, but remaining roughly parallel sided and bluntended. Muri simplicolumellate, being only slightly wider than the columellae and with granules on top: Saxifraga androsacea
[data from Verbeek-Reuvers, 1977a]

8b Exine structure without such a three-layered sexine as described above. Muri coarse or fine, simplicolumellate, duplicolumellate or even pluricolumellate in some parts; may or may not

be parallel-sided, no granules visible on top: 9

9a Muri sometimes running parallel, but often characteristically curved and branched with sudden changes in width (knotted effect). Never completely straight and meeting in irregular star-patterns: Dryas octopetala
[both size and sculpturing are rather variable in this species. In sculpture it closely approaches some of the Rosaceae e.g. Crataegus type. See p. 157 and consult type slides]

9b Muri never running parallel. Often with straight muri meeting and anastomosing or interweaving in irregular star-patterns or winding patterns. Width of muri variable: Acer negundo* type
[for included species see Biesboer, 1975. Although this sculpture type is seen in all grains of A. negundo, it has also been observed on some grains on type slides of Acer species which usually have parallel striae]

Note on Saxifraga exine structure

Some species (members of S. hirsuta type, S. oppositifolia type and S. androsacea) would seem to have a thin layer of rather indistinct columellae supporting a tectum which bears coarser columella-like or bacula-like projections which in their turn support the characteristic muri of the rugulate or striate sculpture. This gives the sexine a three-layered appearance in optical section with an apparently semitectate outer layer. See Verbeek-Reuvers, 1977a. The basal layer of 'true columellae' cannot easily be discerned with the light microscope, but LO analysis or phase contrast will sometimes reveal them to an experienced observer. It is much easier to detect the columella-like tectum projections and these are what are mentioned in the key above.

TETRAZONOCOLPATE, PENTAZONOCOLPATE (Plate 37)

Tetrazonocolpate grains are sometimes found in taxa that are usually trizonocolpate e.g. Papaver, Limonium, Convolvulus, Oxalis, Spergula.

1a Grain reticulate, echinate or microechinate: 2
1b Grain psilate or scabrate-verrucate: 7
2a Echinate or microechinate, with 4 very short, narrow colpi and large apocolpia. Each colpus crossed by a short, very **wide**, transverse endocolpus: Lonicera
[includes mostly L. coerulea and some grains of other Lonicera or Linnaea borealis. See p. 152 for further separation]
2b Reticulate, colpi short or long: 3

3a With 4 very short colpi. Grain shape oval to rectangular-obtuse in both polar and equatorial views. Reticulum rather thin-walled and delicate, simplicolumellate: *Impatiens*

3b With 4 or 5 slightly longer colpi. Grain shape circular to elliptic in both polar and equatorial views. Muri simplicolumellate or duplicolumellate: 4

4a With 4 colpi: 5

4b With 5 colpi: 6

5a Lumina size similar all over the grain surface, muri mostly simplicolumellate, but often duplicolumellate at the junctions: *Fraxinus* [see also p. 147]

5b Lumina varying in size, small and large mixed; and showing a slight decrease towards the colpi. Muri always simplicolumellate, columellae more or less angular in cross section: *Gentianella detonsa* [see Punt and Neinhuis, 1976]

6a Microreticulate, greatest grain dimension <20 μm: *Primula veris* type, long-styled individuals [includes *P. veris*, *P. vulgaris* and *P. elatior*. See Punt *et al.*, 1976]

6b Eureticulate, muri thin. Mesocolpium only slightly convex. Greatest dimension of grain >20 μm: *Primula veris* type, short-styled individuals [includes *P. veris*, *P. vulgaris* and *P. elatior*. See ref. under 6a]

7a Colpi wide, obtuse-ended, with edges very ill-defined and diffuse. Granulate colpus membranes often present so that the four colpi are difficult to distinguish from the mesocolpium, except for the difference in exine thickness (slightly thicker in mesocolpium): *Hippuris vulgaris*

7b Colpi with well-defined margins that may or may not be thickened. Colpi so short as to be little more than elongated pori: 8

8a Colpus thickenings markedly protruding from the grain outline. The ends of the short colpi are obtuse: *Myriophyllum spicatum*

8b Colpus margins hardly thickened, slightly protruding from the grain outline (grain tends to be quadrangular-obtuse in polar view). The ends of the short colpi are acute: *Myriophyllum verticillatum*

HEXAZONOCOLPATE, POLYZONOCOLPATE (Plate 40)

1a Grains psilate, non-perforate. Shape in equatorial view elliptic; with very long colpi that are almost syncolpate at the poles. Each mesocolpium is seen as a marked ridge between the valleys where the colpi are: 2

1b Grains psilate-scabrate with perforate tectum; microreticulate; or reticulate. Mesocolpia flat or convex but no marked 'ridge and valley' pattern as above: 3

2a With numerous colpi (usually 8 or more) with smooth straight edges: *Ephedra fragilis* type [see Andrew, 1984. This type may include many species. It seems equivalent to the *E. torreyana** type of McAndrews *et al.*, 1973, which includes *E. aspera*, *E. trifurcata* and others. Erdtman, 1957, illustrated *E. antisyphilitica* which appears to belong to this type]

2b With 8 or less colpi, edges with irregular margins and marked cracks or fissures radiating into the mesocolpia. These fissures may join up to form an irregular network: *Ephedra distachya* type [see Andrew, 1984. This type may also include many species. It seems equivalent to the *E. nevadensis** type of McAndrews *et al.*, 1973, which includes, amongst others, *E. coryi* and *E. viridis*. Erdtman, 1957, illustrates *E. equisetina* which appears to be of this type]

3a Nexine very thin and delicate (see wall optical section, ×1000) thickness <$\frac{1}{4}$ of width of sexine. Grain eureticulate or microreticulate, with simplicolumellate muri: 4

3b Nexine thicker, up to half as thick as sexine, grain tectate-perforate, microreticulate or reticulate: 5

4a Microreticulate, greatest grain dimension <20 μm: *Primula veris* type, long-styled individuals [includes *P. veris*, *P. vulgaris* and *P. elatior*. See Punt *et al.*, 1976]

4b Eureticulate, muri thin. Mesocolpia only slightly convex. Greatest dimension of grain >20 μm or more: *Primula veris* type, short-styled individuals [includes *P. veris*, *P. vulgaris* and *P. elatior*. See ref. under 4a]

5a Surface psilate-scabrate with perforate tectum; or microreticulate. Mesocolpia usually markedly convex, colpi sunken: *Galium* type [includes *Galium*, *Asperula*, *Rubia* and *Sherardia*]

5b Surface eureticulate or suprareticulate (sometimes the reticulum on top of the tectum has perforations in the floors of the lumina). Mesocolpia only slightly convex, colpi not sunken: 6

6a Grains suprareticulate with very large lumina and narrow muri in the centre of the mesocolpia, smaller lumina towards the poles and the colpi. Columellae in an even carpet: *Prunella* type

[includes *Prunella*, *Nepeta*, *Glechoma* and others]

6b Grains eureticulate, or suprareticulate with small lumina and large perforations in the floors of these lumina (examine ×1000 and use phase contrast). Columellae sometimes in a reticulate pattern under the muri: *Mentha* type [includes *Mentha*, *Lycopus*, *Thymus*, *Origanum*, *Clinopodium*, *Acinos* and possibly others. The introduced *Polygonum polystachyum** may key out here as it is hexazonocolpate and coarsely reticulate — surface sculpturing similar to that of *P. persicaria*. See van Leeuwen *et al.*, 1988]

TETRAPANTOCOLPATE, PENTAPANTOCOLPATE, HEXAPANTOCOLPATE, POLYPANTOCOLPATE (Plate 41)

These aperture systems are found in all grains of the taxon keyed out except where indicated. Some species produce a mixture of the 'normal' grains e.g. trizonocolpate, and variously pantocolpate grains. The proportion of each aperture type in the grains on a type slide may vary from one individual to another. Some examples where this has been observed are: *Trientalis europaea*, *Saxifraga granulata*, *Rorippa sylvestris*, *Convolvulus arvensis*, *Oxalis acetosella*, *Anemone*, *Cuscuta*. The pantocolpate condition in species that are normally trizonocolpate is thought to be associated with meiotic irregularities or increasing ploidy levels, sometimes associated with hybridization. Jonsell (1968) reports that the large pantocolpate grains in some individuals of *Rorippa sylvestris* are most likely to represent unreduced pollen mother cells.

If not found in the key below, some pantocolpate grains may be identified by the type of exine structure characteristic of the genus or species. This identification can be achieved by following the trizonocolpate key, although it must be remembered that size criteria will be useless as pantocolpate grains are always bigger than the trizonocolpate versions.

1a Grain reticulate and circular with approximately 30 colpi that divide up the grain surface into regular pentagonal plates. Reticulum with the lumina size uniform and free bacula or clavae visible on the floors:
                          Polygonum amphibium [see van Leeuwen *et al.*, 1988]

1b Grain psilate, scabrate-verrucate or echinate but not reticulate: 2

2a Columellae fine to invisible (the nexine may, however, have endocracks and endosculptures which may be confused with columellae). Sexine thick, with tectum either psilate, undulating or verrucate. Small perforations may be visible under phase contrast. Colpus membranes granulate: 3

2b Columellae visible, tectum psilate, echinate or undulating (irregularly scabrate-verrucate):   5

3a Tectum psilate with clear perforations. Grain mostly 6-pantocolpate, grain shape that of an obtuse cube with a colpus diagonally crossing each of the flat sides. Sexine thickest where the ends of the colpi meet, and in the middle of the mesocolpium. Sexine thins towards the colpus margins. Nexine thin with prominent endocracks: *Corydalis claviculata* [data from Kalis, 1979]

3b Tectum faintly to markedly verrucate-undulate, exine not thinning towards the colpi. Grain 6- to 12-pantocolpate and shape always roughly spherical. No clear endocracks, although endosculptures are present: 4

4a Tectum coarsely verrucate. Colpi parallel-sided and with obtuse ends. Grain never syncolpate:
                            Corydalis solida type [according to Kalis, 1979 this includes all the Northwest European representatives of Section *Bulbocapnos*]

4b Tectum only slightly verrucate. Apocolpia very small or grain syncolpate: *Corydalis lutea* type [according to Kalis, 1979, this type includes *C. lutea* and *C. ochroleuca*]

5a Exine thinning gradually towards each colpus margin. The exine is thickest, and the columellae are longest, in the mesocolpia: 6

5b Exine much the same thickness in the mesocolpia as adjacent to colpus margins: 7

6a Grain psilate or microechinate, tectate-perforate. Perforations smaller in diameter than the columellae. Exine thickest, and columellae longest in the regions where the ends of the colpi approach one another:
                           Adonis aestivalis type [*Adonis aestivalis* type produces a proportion of pantocolpate grains. It is also trizonocolpate — see p. 119]

6b Grain clearly echinate, tectate-perforate; with 12 colpi that are often syncolpate, cutting off square portions of exine. Columellae usually quite coarse, sometimes sitting in a roughly reticulate pattern under the tectum in the centre of the mesocolpia: *Montia fontana* type [includes *Montia fontana* and possibly grains of

M. perfoliata. Most grains of *M. perfoliata* have apertures of irregular shape and spacing rather than neatly pantocolpate like *M. fontana*]

7a Grain clearly echinate, echinae conical: 8

7b Grains without obvious conical echinae, although microechinae may be present. Columellae either uniform in width or coarse and fine mixed: 9

8a Grain <30 μm. Columellae in a fine, dense carpet. Colpi rather short, ill-defined and with irregular margins: *Koenigia islandica* [see van Leeuwen *et al.*, 1988]

8b Grain >40 μm. Columellae coarser and sparser. Colpi well-defined, longer and with straighter edges meeting in a regular pattern which cuts off hexagonal plates of exine. Small perforations visible between the echinae (×1000): *Portulaca oleracea*

9a Columellae of two sorts: sparse, coarse ones in a carpet of more densely spaced, finer ones. Each coarse columella may appear to be surrounded by a light ring (space) and then by a ring of more or less fused small columellae. Tectum often undulating in optical section as the sculpture is irregularly scabrate-verrucate: *Ranunculus* type

[many species in *Ranunculus* type as defined on pp. 127–128 produce a proportion of pantocolpate grains. Type slides of German *Artemisia* material labelled '*A. salina* Willd.', a taxon which is now included within *A. maritima*, have been observed to produce a proportion of irregularly pantocolpate grains which correspond well to the description above. Grains identical to these have been found in the Late Devensian of Scotland; details in Webb, 1977]

9b Columellae more uniform in thickness. Tectum not usually undulating, but rather more regularly scabrate-verrucate: 10

10a Grains with a more or less clear infrareticulum joining the columellae. Microechinae present on top of the tectum (phase contrast helps to see both of these features): *Papaver radicatum* type

[includes *P. radicatum* agg., some grains of *P. argemone* type on p. 106 and some grains of *Meconopsis cambrica*. This last is normally a member of *P. rhoeas* type, p. 127. See also Kalis, 1979]

10b Tectum psilate, perforate, not microechinate. Columellae uniformly coarse with no suggestion of an infrareticulum joining them: *Spergula* type [includes some grains of *Spergula* and *Spergularia*. Odd pantocolpate grains of *Saxifraga*

granulata may key out here, but they may be distinguished by their microechinae and operculate colpi]

TRIZONOCOLPORATE (Plates 41–57, 65–71)

This class covers very many grains and so it is divided into the following sections based on sculpturing type (see p. 00 for details of the sculpturing):

The unique **crested** grain of *Trapa natans* falls into the trizonocolporate aperture class, but because the apertures are obscured, this species is keyed out on p. 91.

Trizonocolporate, psilate to scabrate-verrucate

This is a very difficult group as there are few features available for differentiating the types and many species fall into this category. The first few key divisions are necessarily rather non-specific and it is very easy to end up in the wrong branch of the key. If there is any doubt or ambiguity at a branch point, the best course is to take the grain through alternatives (a) and (b). Some difficult or variable types have been keyed out in several places in order to try and overcome this problem. Confirmation by consulting a type slide is essential in this section.

1a Grain approximately as long as broad or broader than long (P/E ratio up to about 1.2): 2

1b Grain longer than broad (P/E ratio >1.2): 35

2a Grain shape in equatorial view between circular and rhombic-obtuse, or nearly rhombic-obtuse, or rhombic-obtuse with apocolpia truncated (approaches hexangular). Colpi usually straight, travelling through a sharp bend at the equator. **Mesocolpia tending to be flattened or even concave.** Shape in polar view triangular-truncate, triangular-obtuse or tending towards either of these. Pori may or may not project a little at the equator: 3

2b Grain shape in equatorial view circular, elliptic, or rectangular-obtuse; **mesocolpia always convex.** Shape in polar view usually circular or lobed, but may appear more triangular if the pori protrude at the equator: 16

3a Colpi not parallel-sided but **widening appreciably** towards the equator. In polar view optical section, the exine thins slightly towards the colpus margins; and in equatorial view a complex H-shaped endoaperture may be visible in phase contrast (the 'legs' of the H go up the colpus margins beside the costae). Very faint striae may be visible on close examination (×1000):   4 (Crassulaceae)

3b Colpi **parallel-sided**, may or may not be narrow, but not widening appreciably towards the equator. H-endoaperture may or may not be visible: 5

4a Grain less than about 27 μm long. Columellae rarely visible, wall appears structureless:
                            Umbilicus rupestris type
[includes *U. rupestris* and *Crassula*. Distinction may be made between these by reference to type slides]

4b Grain more than about 27 μm long. Columellae clearly visible, at least at poles. Colpus margins irregular towards the ends, colpus membrane slightly granulate: *Jovibarba*

5a Porus represented by a bridge or constriction underlain by a clear circular, elliptic, rectangular or 8-shaped endoporus or by an H-endoaperture (use phase contrast). In the case of the last mentioned, the 'legs' of the H travel up and down roughly parallel to the ectocolpi so they appear under low power like faint extra colpi, either side of the true colpi: 6

5b Porus a mere constriction, bridge or rupture to the colpus which is not underlain by any discernible endoaperture, even in phase contrast (but see note on Cornaceae after exit 14b). Costae to the colpi absent or very small: 12

6a With a more or less clear H-endoaperture to each ectocolpus: 7

6b With circular, elliptic, rectangular or apparently 8-shaped endoaperture without any projecting portions out under the mesocolpia: 8

7a Grain with flattened or concave apocolpia. Sculpture apparently coarsely rugulate-verrucate in the apocolpia and much finer (psilate-scabrate) in the mesocolpia. Grain >35 μm, often crumpled or collapsed, but detectably hexangular in equatorial view, giving an overall three dimensional shape of a triangular prism with the pori projecting from this shape at the equator. Central portion of the H-endoaperture is usually wide:
                            Anthyllis vulneraria
[the coarse sculpture in the apocolpia appears to be due to irregular structural elements **under** the tectum; in this area the nexine appears partially dissolved. N.B. *Ornithopus perpusillus* may key out here as it occasionally has 'horn-like' extensions from its endoporus that curve out under the mesocolpia and allow it to be classed as having an H-endoaperture. See exit 19b]

7b Grain with more rounded apocolpia that have psilate sculpturing. Grain shape thus rhombic-obtuse to circular in equatorial view. Central portion of the H-endoaperture is narrow and like a short transverse colpus. 'Legs' of each H-endoaperture as long as the ectocolpus and may join around its end at the pole:
                            Verbena officinalis
[see Punt and Langewis, 1988. Grains of *Frangula alnus* may also key out here as a faint H-endoaperture is visible in this species if phase contrast is available]

8a Grain with columellae longest and coarsest at poles and well-defined, more or less rectangular, equatorially elongate endopori.
                            Hydrocotyle vulgaris

8b Grain with columellae (if visible) more or less the same length at the poles as elsewhere; each porus either circular, elliptic, figure-8 shape or represented by an acute-ended transverse endocolpus which is never rectangular):   9

9a Colpi deeply sunken or inrolled except at the equator. Each porus represented by a narrow, acute-ended transverse endocolpus which underlies a clear bridge or constriction to the colpus. A small fastigium may be present over the porus. Colpi long and straight, travelling through a sharp bend at the equator, polar area very small: 10

9b Colpi may or may not be inrolled. Pori well-defined, circular, elliptic or transversely elongated but not acute-ended and always rather sunken and thickened. Not fastigiate. Colpi in equatorial view curving more and not travelling through such a sharp bend at the equator: 11

10a Grain small, any diameter is <16 μm and often <14 μm. Columellae invisible, even in phase contrast. Exine very thin, surface completely psilate: *Solanum dulcamara*
[see ref. under 10b]

10b Grain bigger, any diameter >18 μm. Columellae visible in acetolysed grains only. Psilate to scabrate: *Solanum nigrum* type
[includes *S. nigrum*, *S. luteum*, *S. triflorum*, *S. tuberosum**, *Lycopersicon esculentum** and *Capsicum annuum**. Apparently more or less aberrant grains occur in the last three species which are, of course, cultivated plants. Data from Punt and Monna-Brands, 1977. It is possible that those members of the Ericaceae and

Pyrolaceae with single grains i.e. *Erica termi-nalis* and *Orthilia secunda*, may key out here. They are not fastigiate and sometimes have distinctive endocracks in apocolpia and mesocolpia]

11a With large (>3 μm diameter) circular sunken endopori with prominent costae all round them. In polar view optical section, exine thins markedly just outside the thickened porus margin. Tectate-perforate, perforations smaller and more numerous than columellae (×1000, phase contrast helps). Grain circular in equatorial view: *Nyssa**
[e.g. *N. sylvatica*, *N. aquatica*]

11b With pori rather smaller in relation to grain size and more equatorially elliptic or transversely elongated. No perforations. Columella-like structures (black dots in surface view, phase contrast) visible near the pori and along the costae to the colpi, but not detectable else-where. Mesocolpia flat to concave, grain nearly always broader than long, thus elliptic in equatorial view: *Frangula alnus*
[each porus may be seen to be part of a faint, irregular H-endoaperture, if phase contrast is available]

12a Columellae longest and most prominent at the poles, grain neatly rhombic-obtuse with pori projecting at the equator (i.e. grain approaches acuminate-obtuse in shape): *Pleurospermum* type
[includes *Pleurospermum* and *Physospermum*. See Punt, 1984. Grains in these genera are more usually longer than broad]

12b Columellae the same length at poles as else-where on the grain, shape not acuminate-obtuse, colpi may be sunken or inrolled: 13

13a Colpi not inrolled, margins markedly irregular. Colpus membrane visible and with isolated granules and islands of exine. Colpus ends mostly irregular-obtuse. Columellae mono-morphic, in a fine dense carpet under the tectum. No perforations: *Robinia pseudoacacia**

13b Colpi either inrolled or narrow and without isolated granules or islands, margins may or may not be irregular. Columellae monomorphic or dimorphic. Tectum with or without perforations: 14

14a Grain with relatively large apocolpia. Surface tectate-perforate (×1000, use phase contrast if possible) with perforations denser in the meso-colpia and sparse or absent in the apocolpia. Columellae monomorphic, endocracks often radiating from the porus region into the mesocolpium and around the ends of the colpi (phase contrast). Never with regular or irre-

gular scabrae or microverrucae on the tectum: *Medicago sativa*
[includes both subspecies *falcata* and sub-species *sativa*. Grains of this species are more usually longer than broad]

14b Grain with smaller apocolpia, in polar view often truncated triangular with margins of colpi protruding at the equator. Surface may or may not be tectate-perforate (most species without perforations) but always with clear scabrae or microverrucae and/or dimorphic columellae (sparse, coarse ones scattered amongst more numerous slimmer ones visible under ×1000). Tectum may be thrown into small undulations. The black dots that the coarse columellae pre-sent in LO analysis may be mistaken for the scabrae. In fact they seem to underlie the scabrae, but this is visible only in the species with larger grains: 15 (Cornaceae)
[Depending on the preparation technique, a large, transverse, bone-shaped endoporus or H-endoaperture may be apparent, combined with irregular channels in the innermost nexine of the mesocolpium. These features seem visible only in grains acetolysed for short times. In addition, some Rosaceae may key out here, see additional notes on p. 157]

15a Grain <35 μm: *Cornus suecica* type
[according to Lieux, 1983, this includes most grains of *C. suecica*, *C. canadensis* and *C. mas*. Possibly also some grains of *C. florida*]

15b Grain >35 μm: *Cornus sanguinea* type
[according to Lieux, 1983, this includes *C. sanguinea*, *C. amomum*, *C. racemosa*, *C. alternifolia*, *C. rugosa*, *C. stolonifera*, *C. drummondii*, *C. foemina* and *C. florida* possibly others. Some distinction may be made within this type using grain size e.g. *C. sanguinea* and *C. amomum* may be well over 60 μm]

16a Colpi short, whole length of colpus usually visible in an equatorial view of the grain. Apocolpia correspondingly large: 17

16b Colpi longer, curving round the grain and with whole length not usually visible in an equatorial view: 23

17a Pori not fastigiate, therefore not showing pro-jections in a polar view of the grain: 18

17b Pori fastigiate, margins projecting in a polar view. Each fastigium may be small and narrow or large and wide: 20

18a Grain circular in equatorial view. Endopori large; circular or slightly elliptic in shape. Grain sculpture finely rugulate on close inspection: *Fagus*

18b Grain shape quadrangular-obtuse in equatorial view. Endopori circular, elliptic or 8-shaped: 19
19a With circular or meridionally elliptic endopori which are large in comparison with the grain size. Colpi without granules, apocolpia never concave: *Trifolium spadiceum*
19b With equatorially elongated, elliptic or 8-shaped endopori. In phase contrast each endoporus sometimes has curved 'horns' into the mesocolpia. Colpi with granulate membranes, grain with concave apocolpia:
                              Ornithopus perpusillus
[note that *Anthyllis* may key out here if the grain is more rectangular-obtuse than hexangular]
20a With operculate colpi, each porus with a narrow, slightly equatorially elongate fastigium. Grain sculpture psilate to faintly rugulate in the mesocolpia with fine swirling striae in the apocolpia (×1000, phase contrast helps):
                         Sanguisorba minor ssp. *minor*
[this description is of the typical race with chromosome number of 56 according to Erdtman and Nordborg, 1961. These authors state that the diploid 2n = 28 race show no trace of colpi in the pollen grains]
20b Colpi not operculate, always narrow and slit-like: 21
21a With each ectocolpus reaching beyond the polar limits of the equatorially elongated fastigium:
              Elaeagnus multiflorum and possibly *E. argentea**
[see McAndrews *et al.*, 1973]
21b With each ectocolpus very short, such that it only just reaches the polar limits of the large, wide fastigium (grain may appear trizonoporate until examined carefully — short, transverse nexinous ridges crossing under the ends of the ectocolpi may be the only indication of the endoapertures):   22
22a Inner layer of the fastigium walls separating into many lamellae, so that the inside appears ragged and the fastigium itself may appear full of irregular exine fragments: *Ludwigia palustris*
22b Fastigium without such a ragged internal wall:
        some Caprifoliaceae e.g. *Symphoricarpus** and
            Linnaea borealis; plus some members of the
                Elaeagnaceae e.g. *Elaeagnus angustifolia*
23a In equatorial view sexine varying around the grain such that it is thinnest at the poles and thickest in the centre of the mesocolpium. Columellae fine, dense and cylindrical; or coarse, irregularly shaped and spaced in surface view; or a mixture of both sorts. Pori circular, elliptic or represented by a transverse endocolpus. Careful examination will (×1000) nearly always reveal regularly spaced micro-

echinae on top of the tectum: *Artemisia*
[some separation of types may be achieved by examination of type slides. See Webb & Moore, 1982. Some grains of *Nyssa* may key out here]
23b In equatorial view sexine not varying in thickness around the grain; or if it does vary, then it is always slightly thicker at poles than elsewhere: 24
24a Circular-elliptic endopori, no fastigia present, clear costae to the colpi. Grain may appear faintly rugulate-striate or tectate-perforate on close examination: 25
24b Each porus represented merely by a bridge or constriction to the colpus or by an elliptic endoporus or by a narrow transverse endocolpus. Fastigia may or may not be present: 26
25a Exine not thinning from mesocolpium to the colpus edge when seen in polar view, optical section. Shape circular to slightly lobed in polar view. Surface tectate-perforate. Pori may or may not be strongly ringed by costae:
                                          Oxyria type
[includes *Oxyria digyna*, *Rumex crispus*, *R. conglomeratus*, *R. sanguineus*, *R. pulcher*, *R. maritimus* and possibly others — even some grains of *R. acetosella*. Distinction of *Oxyria* may be made by examination of type slides. According to van Leeuwen *et al.*, 1988, *Oxyria* and *R. acetosella* are distinct by their smaller size of <26 μm. The other species are >26 μm in diameter. *Oxyria* in addition has a porus that is more clearly circular and ringed in phase contrast]
25b Exine markedly thinning from the mesocolpium to the colpus edge when seen in polar view, optical section. Shape in polar view tending towards triangular-obtuse. Grain may appear microrugulate. Pori very strongly ringed by costae: *Nyssa**
[e.g *N. sylvatica*, *N. aquatica*]
26a Each fastigium prominent, narrow, beak-like and equatorially elongated. It is underlain by an elliptic endoporus or rarely by a transverse endocolpus. Colpi never inrolled or sunken, being flat on the surface and curving round the grain. Grain always a circular shape in equatorial view: *Hippophaë rhamnoides*
[note that from McAndrews *et al.*, 1973, it seems that *Elaeagnus argentea** may also key out here]
26b Each fastigium either less prominent and not beak-like (just appearing as 'lips' at the equator in polar view) or entirely absent. Colpi either flat on the surface or inrolled, may or may not curve round the grain: 27
27a Fastigia present but not very prominent. Each

may be equatorially elongated and is always underlain by a **transverse endocolpus**. Colpi usually rather sunken, contributing to a lobed appearance in polar view. Colpi straight and travelling through a sharp bend at the equator: 28

27b Not fastigiate, colpi may or may not be sunken, porus usually indicated by a mere bridge or constriction: 30

28a Close examination shows grain surface to be regularly microechinate and the columellae to be aggregated in lines or roughly C-shaped units (×1000, phase contrast helps): *Filipendula*

28b No microechinae or aggregated columellae on close examination. Surface psilate: 29 Solanaceae

29a Grain small, any diameter is <16 µm and often <14 µm. Columellae invisible, even in phase contrast. Exine very thin and surface completely psilate: *Solanum dulcamara* [see ref. under 29b]

29b Grain bigger, any diameter >18 µm. Psilate to scabrate: *Solanum nigrum* type [includes *S. nigrum*, *S. luteum*, *S. triflorum*, *S. tuberosum**, *Lycopersicon esculentum** and *Capsicum annuum**. Apparently more or less aberrant grains occur in the last three species which are, of course, cultivated plants. Data from Punt and Monna-Brands, 1977]

30a Grain syncolpate in both apocolpia; colpi widening in the fusion zone here so that there is a sunken area at each pole. Grain surface scabrate, colpus membranes granulate. Grain size <30 µm and most grains <25 µm: *Soldanella* [data partly from Punt *et al.*, 1974]

30b Grains not syncolpate in the apocolpia: 31

31a With one apocolpium very small, the other somewhat larger. Exine thin, surface completely psilate, no columellae visible (×1000, phase contrast). Grain size usually <about 28 µm: *Gratiola*

31b With apocolpia equal in size. Columellae visible, surface may or may not have perforations (×1000, phase contrast): 32

32a Surface microreticulate-perforate or finely rugulate-striate on close examination. Columellae visible as a dense and even carpet (×1000, phase contrast). This exit leads to the following grains or types: *Lobelia* (e.g. *L. dortmanna*, *L. urens*) and some grains in *Digitalis purpurea* type (e.g. *D. purpurea*, *Sibthorpia europaea*, *Kicksia*) [distinction is possible only if type slides are available. *Lobelia* grains may be distinguished by their minutely rugulate-striate surface and granulate colpus membranes, but these are often difficult to detect. *Digitalis* has rather

short colpi, and *Sibthorpia* has very long colpi which tend to have square ends]

32b Grain not microreticulate-perforate or finely rugulate-striate on close examination. If perforate, then perforations are more widely spaced, at least in the apocolpia. May or may not be scabrate-verrucate: 33

33a Grain with relatively large apocolpia. Surface tectate-perforate (×1000, use phase contrast if possible) with perforations denser in the mesocolpia and sparse or absent in the apocolpia. Columellae monomorphic, endocracks often radiating from the porus region into the mesocolpium and around the ends of the colpi (phase contrast). Never with regular or irregular scabrae or microverrucae on the tectum: *Medicago sativa* [includes subspecies *falcata* and subspecies *sativa*. Grains of this species are more usually longer than broad]

33b Grain with smaller apocolpia, in polar view often truncated-triangular shape. Surface may or may not be tectate-perforate (most species not so) but always with clear scabrae or microverrucae and/or dimorphic columellae (sparse coarse ones scattered amongst more numerous slimmer ones, visible under ×1000). Tectum may be thrown into small undulations. The black dots that the coarse columellae present in LO analysis may be mistaken for the scabrae. In fact they seem to underly the scabrae, but this is visible only in the species with larger grains: 34 (Cornaceae) [depending on the preparation technique, a large, transverse, bone-shaped endoporus or H-endoaperture may be apparent, combined with irregular channels in the innermost nexine of the mesocolpium. These features seem visible only in grains that are acetolysed for very short times]

34a Grain <35 µm: *Cornus suecica* type [according to Lieux, 1983, this includes most grains of *C. suecica*, *C. canadensis* and *C. mas*. Possibly also some grains of *C. florida*]

34b Grain >35 µm: *Cornus sanguinea* type [according to Lieux, 1983, this includes *C. sanguinea*, *C. amomum*, *C. racemosa*, *C. alternifolia*, *C. rugosa*, *C. stolonifera*, *C. drummondii*, *C. foemina* and *C. florida* possibly others. Some distinction may be made within this type using grain size e.g. *C. sanguinea* and *C. amomum* may be well over 60 µm]

35a Grain shape in equatorial view: elliptic to rhombic-obtuse, or rhombic-obtuse, or apiculate at the equator, or hexangular (with flattened apocolpia). Never concave at equator forming a bone shape: 36

35b Grain shape in equatorial view: elliptic to rectangular-obtuse, or rectangular-obtuse, or bone-shaped (concave or constricted at equator). The bone-shape may be restricted to the internal contour of the grain wall. If bone-shaped the grain may occasionally be slightly pointed at the poles: 59

36a Sexine varying in thickness over the surface of the grain; in particular there is always a difference between that at the poles and that elsewhere e.g. in the middle of the mesocolpium or at the grain 'shoulders':   37

36b Sexine relatively uniform in thickness over the grain surface: 44

37a Sexine **thinnest at poles** and becoming gradually thicker towards the centre line of each mesocolpium: 38

37b Sexine **thickest at the poles** and in the centre line of each mesocolpium: 40

38a With short transverse endocolpi that are elliptic-acute or narrow and parallel-sided. Sexine only slightly thicker in the equatorial mesocolpia than at the poles and shoulders. Colpi not sunken, therefore polar view never lobed, but triangular-obtuse with colpi between the angles:

Apium inundatum type and *Trinia glauca* [*A. inundatum* type includes *A. inundatum*, *A. nodiflorum* and *A. graveolens* and may be distinguished by the short, elliptic-acute transverse endocolpi and faint rugulate sculpturing upon close examination. *Trinia glauca* has longer, narrower, transverse endocolpi which have diffuse or rounded ends. Data partly from Punt, 1984]

38b With clear circular or elliptic endopori, or endopori represented merely by a rupture. Sexine markedly thicker in the middle of the mesocolpia than at the poles or the colpus margins. Polar view lobed or slightly triangular-obtuse: 39

39a Close examination will nearly always reveal regularly spaced microechinae on the tectum (×1000). Columellae can be any of the following combinations: fine, dense and cylindrical; coarse, irregularly shaped and spaced in surface view; a mixture of both sorts. Pori circular to equatorially elliptic. Grains usually lobed in polar view: *Artemisia* [some separation of types may be achieved by examination of type slides. See Webb and Moore, 1982]

39b Close examination never reveals microechinae, but faint striae may be seen or the grain may

appear psilate (×1000, phase contrast helps). Porus most commonly represented by an irregular, meridionally elongated rupture:

Alchemilla type [includes some grains of *Alchemilla* and *Aphanes*. Some *Alchemilla* species would not key out here as they produce only deformed grains, other grains in this type appear without pori. Some *Artemisia* species are nearly psilate e.g. some subspecies of *A. maritima*, and care must be taken not to identify these with *Alchemilla* type]

40a Pori fused laterally to form a complete equatorial girdle or circular endoaperture. Sexine thickest at poles and in a ridge down the centre of each mesocolpium. Thick nexinous bands (costae) are on the margins of this girdle on either side of the equator. Grain triangular-obtuse in polar view with pori between the angles: *Fallopia convolvulus* type [includes *F. convolvulus* and *F. dumetorum*]

40b Pori not forming a complete equatorial girdle; being either very indistinct, or elliptic to rectangular, or represented by transverse endocolpi that do not fuse laterally. If grain triangular-obtuse in polar view, then pori between the angles or on the angles: 41

41a In equatorial view, grain with pori projecting or pouting giving it an acuminate-obtuse shape. Mesocolpia sunken (concave) and colpi long. Endopori indistinct or, if seen, then roughly rectangular or quadrangular. Columellae longest at poles, costae indistinct: 42

41b In equatorial view, pori not pouting; if grain rhombic-obtuse in shape then with straight or convex sides. Mesocolpia can be flat, concave or convex. Colpi long or short. Endopori distinct, clearly broadly rectangular or represented by short transverse endocolpi: 43

42a P/E ratio <1.6:   *Pleurospermum* type [according to Punt, 1984, this includes *Pleurospermum* and *Physospermum*. Cf. also type slides of *Bupleurum rotundifolium*]

42b P/E ratio >1.6:   *Smyrnium olusatrum* [see ref. under 42a]

43a Polar view triangular with apertures **on** the angles. Shape neatly rhombic-obtuse. Rectangular, transversely elongated endopori or short transverse endocolpi present: *Bupleurum* [see also Punt, 1984. Rare grains of *Hydrocotyle* that are longer than broad may key out here. They may be distinguished from *Bupleurum* by their much stronger costae which project inwards at the equator and by their narrower transverse endocolpi. In addition some grains of *Hypericum androsaemum* type may key out here, however the transverse endocolpi in this are always acute-ended]

43b Polar view triangular with apertures **between** the angles. Shape more elliptic, or if rhombic external contour, then elliptic internal contour. Columellae usually longer in the equatorial region of each mesocolpium. Endopori or transverse endocolpi rectangular, diffuse-ended or acute-ended. This exit leads to the following grains or types which may be distinguished by consulting type slides:

Anthriscus caucalis, Petroselinum segetum and *Sison amomum*

[*Sison* and *Petroselinum segetum* have rather narrow, parallel-sided transverse endocolpi. *Anthriscus caucalis* is distinct in its short and not sunken colpi together with its slightly undulating tectum in the equatorial region. See information in Punt, 1984]

44a Each porus represented by a mere bridge, constriction or rupture to the colpus: 45

44b Each porus a clearly defined circle, ellipse or well-marked transverse endocolpus or H-endoaperture. If endocolpus present, it may be wide or narrow and equipped with 'horns' or not: 52

45a Colpi narrow and parallel-sided, often crack or slit-like, with colpus membranes not usually visible. Margins may be inrolled: 46

45b Colpi widening from their acute ends to the equator so that colpus membrane is visible. If membrane not visible it is because the colpus edges are inrolled and the membrane sunken and covered (thus only fully expanded grains will correctly key out from this point on). Surface may or may not be rugulate-striate on close examination: 49

46a Colpus margins often with undulations along their length. Grain tectate, close examination (×1000, phase contrast) may reveal that it is faintly rugulate-striate. Mesocolpia convex, never flattened:

some Rosaceae e.g *Sorbus aria, Rubus, Rosa* or *Crataegus*. See additional notes on p. 157 [*Viola palustris* type may key out here if the wide, granulate membranes to the colpi are not detected, however this type never shows any sign of being rugulate-striate]

46b Colpus margins never undulating. Grain surface either minutely scabrate or tectate-perforate on close examination. Mesocolpia commonly flattened: 47

47a Grain with relatively large apocolpia. Surface tectate-perforate (×1000, use phase contrast if

possible) with perforations denser in the mesocolpia and sparse or absent in the apocolpia. Columellae monomorphic, endocracks often radiating from the porus region into the mesocolpium and around the ends of the colpi (phase contrast). Never with regular or irregular scabrae or microverrucae on the tectum:

Medicago sativa

[includes subspecies *falcata* and subspecies *sativa*. Note that *Anthyllis* may key out here if the endopori are not clear]

47b Grain with smaller apocolpia, in polar view often truncated-triangular or with margins of colpi protruding at the equator. Surface may or may not be tectate-perforate (most species not so) but always with clear scabrae or microverrucae and/or dimorphic columellae (sparse coarse ones scattered amongst more numerous slimmer ones − visible under ×1000). Tectum may be thrown into small undulations. The black dots that the coarse columellae present in LO analysis may be mistaken for the scabrae. In fact they seem to underly the scabrae, but this is visible only in the species with larger grains: 48 (Cornaceae)

[depending on the preparation technique, a large, transverse, bone-shaped endoporus or H-endoaperture may be apparent, combined with irregular channels in the innermost nexine of the mesocolpium. These features seem visible only in grains that are acetolysed for short times]

48a Grain <35 μm: *Cornus suecica* type [according to Lieux, 1983, this includes most grains of *C. suecica, C. canadensis* and *C. mas.* Possibly also some grains of *C. florida*]

48b Grain >35 μm: *Cornus sanguinea* type [according to Lieux, 1983 this includes *C. sanguinea, C. amomum, C. racemosa, C. alternifolia, C. rugosa, C. stolonifera, C. drummondii, C. foemina C. florida* and possibly others. Some distinction may be made within this type using grain size e.g. *C. sanguinea* and *C. amomum* may be well over 60 μm]

49a In polar view, optical section, exine thins slightly towards the colpus margins at the equator. Close examination (×1000, phase contrast) of these margins and the bridge in equatorial view may reveal a complex H-shaped endoaperture (phase contrast helps). Slight costae may be detectable on the colpi. Colpus membranes psilate, never granulate: 50

49b Exine not thinning towards the colpi in polar view, optical section. Never any indication of an H-endoaperture on each colpus under phase contrast. Costae may or may not be present: 51

50a Grain less than about 27 μm long. Columellae rarely visible, wall appears structureless:

Umbilicus rupestris type

[includes *U. rupestris* and *Crassula*. Distinction may be made between these two by reference to type slides]

50b Grain more than about 27 μm long. Columellae clearly visible, at least at the poles. Colpus margins irregular towards the ends, colpus membrane slightly granulate: *Jovibarba*

51a Grain completely psilate, only a dense carpet of columellae with no perforations visible under phase contrast. Membranes of colpi granular. Colpi very long and apocolpium correspondingly small. No costae to the colpi:

Viola palustris type

[includes *V. palustris*, *V. hirta*, *V. odorata* and some grains of *V. riviniana* and *V. canina*. The last two have also tetrazonocolporate grains]

51b Surface microreticulate-perforate or finely rugulate-striate-perforate on close examination (×1000, phase contrast):

this exit leads to *Lobelia* (e.g. *L. dortmanna*, *L. urens*); some grains in *Digitalis purpurea* type (e.g. *D. purpurea*, *Sibthorpia europaea*, *Kicksia* — see also p. 146) and *Hypericum androsaemum* type (see also p. 150)

[distinction is possible only if type slides are available. *Lobelia* grains may be distinguished by their minutely rugulate-striate surface and granulate colpus membranes, but these features are often difficult to detect. *Digitalis* may be distinct in its rather short colpi, and *Sibthorpia* in its very long colpi which tend to have square ends]

52a Pori represented either by H-endoapertures or wide transverse endocolpi with varying numbers of 'horns' into the mesocolpia: 53

52b Endopori or transverse endocolpi present, but never with horns or extended into H-endoapertures: 54

53a Apocolpia always convex. On close examination (×1000) grain tectate-perforate with thick costae to the ectocolpi. Crossing the ectocolpi at the equator are transverse endocolpi with obtuse ends and with or without a varying number of 'horns' travelling in a curving fashion out under each mesocolpium (phase contrast helps to see these). Nexine may be very thick compared to sexine, but this is a variable feature between grains: *Glaux maritima*

53b With apocolpia flat or concave and with apparently coarse rugulate-verrucate sculpturing.

Mesocolpia flat and with finer sculpture (psilate-scabrate). Grain usually rather crumpled or collapsed, but shape can be seen to be basically that of a triangular prism, with the pori projecting from this shape at the equator. H-endoaperture present in which each transversely elongate, elliptic or figure of 8-shaped endoporus has extensions which travel up and down, roughly parallel to the ectocolpi. These extensions appear under low power magnification like faint extra colpi, one either side of each true colpus: *Anthyllis vulneraria*

[the coarse sculpture in the apocolpia appears to be due to irregular structural elements **under** the tectum; in this area the nexine appears partially dissolved]

54a Close examination shows grain covered in regularly spaced microechinae that are less densely spaced than the columellae. A small fastigium is present over each narrow, transverse endocolpus at the equator (fastigium best seen in polar view as small 'lips' projecting from grain outline). Columellae aggregated in groups, lines or roughly C-shaped units (best seen in phase contrast): *Filipendula*

54b Grain psilate or faintly rugulate, never microechinate. Columellae may or may not be aggregated but mostly present as a fine even carpet. Fastigia may or may not be present: 55

55a Grains neatly rhombic-obtuse (both internal and external contours) always with transverse endocolpi which are covered by fastigia. Grain shape in polar view circular to triangular-obtuse with apertures **on the angles**: 56

55b Grains more narrowly elliptic in shape. Each porus represented by a transverse endocolpus. Polar view circular to triangular-obtuse with apertures **between the angles**: 58

56a Porus represented by a wide rectangular endoporus or short wide transverse endocolpus:

Bupleurum

[see also Punt, 1984. Grains where the exine thickness variations are not visible key out here]

56b Porus represented by a narrow, acute-ended transverse endocolpus which underlies a clear bridge or constriction to the colpus. Equatorially elongate fastigia may be visible: 57

57a Grain small, any diameter is <16 μm and often <14 μm. Columellae invisible, even in phase contrast. Exine very thin and surface completely psilate: *Solanum dulcamara*

[see ref. under 57b]

57b Grain bigger, any diameter >18 μm. Psilate to scabrate: *Solanum nigrum* type

[according to Punt and Monna-Brands, 1977;

this type includes *S. nigrum*, *S. luteum*, *S. triflorum*, *S. tuberosum**, *Lycopersicon esculentum** and *Capsicum annuum**. Apparently more or less aberrant grains occur in the last three species *Hypericum androsaemum* type may also key out here, see p. 150. It is possible that the members of the Ericaceae and Pyrolaceae with single grains i.e. *Erica terminalis* and *Orthilia secunda*, may key out here. They have no fastigia and often characteristic endocracks in the apocolpia and mesocolpia]

58a Grain <20 µm. No fastigium, but small extensions like lips projecting above the endoporus at the equator in polar view. Transverse endocolpi with diffuse or acute ends, rarely with small 'horns'. Shape in polar view circular or triangular-obtuse with convex mesocolpia:

Castanea type

[includes *Castanea* and probably *Chrysolepis*]

58b Grain >20 µm, no fastigia or small exine extensions over the endocolpus at the equator. Endocolpus with acute or diffuse ends. Polar view usually triangular-obtuse with the apertures between the angles. This exit leads to the following grains or types which may be distinguished by consulting type slides:

Apium inundatum type (includes *A. nodiflorum*, *A. inundatum* and *A. graveolens*) *Anthriscus caucalis*, *Trinia glauca*, *Sium latifolium* type (including *S. latifolium* and *Berula erecta*)

[*Apium inundatum* type may be identified by the presence of broad, elliptic-acute transverse endocolpi and a faint rugulate sculpturing. *Trinia* has rather narrow, parallel-sided transverse endocolpi, *Sium latifolium* type is distinct in its large, wide, slightly elliptic transverse endocolpi. *Anthriscus caucalis* is distinct in its short and not sunken colpi. See further information in Punt, 1984]

59a Exine varying in thickness over the surface of the grain (excluding the region of the costae). Most of the thickness differences due to sexine variations. Grain may or may not be bone-shaped (the bone-shape may be restricted to the inner contours): 60

59b Exine of roughly the same thickness all over the surface of the grain (excluding the region of the costae). Grain never bone-shaped: 63

60a Grain with pori represented by transverse endocolpi which fuse laterally to form a complete equatorial girdle. This girdle is bordered by thick band-like costae which project inwards (see equatorial view, optical section; also Fig 5.6, p. 72). Sexine thinnest and columellae shortest near the poles, thickest and columellae

longest in a broad strip down the centre of each mesocolpium. Columellae coarse and branched in the centre of the mesocolpia. Grain shape that of an obtuse, triangular prism:

Centaurea cyanus type

[includes *C. cyanus* and *C. montana*. Distinction may be made by reference to type slides. *C. montana* has less heavy costae bordering the equatorial girdle. Note that rare grains of *Polygonum aviculare* type show a complete equatorial girdle]

60b Grain with pori or transverse endocolpi that are separate, never fused laterally to form a complete girdle, although there may be a complete ring of costae around the equator. Columellae may be any of the following combinations: short in the polar area and longer in the mesocolpia; long at the poles and short in the mesocolpia; or longest at the poles **and** in the mesocolpia. Endopori circular, elliptic, rectangular, or represented by equatorially elongated transverse endocolpi. Columellae may or may not be coarse and branched: 61

61a Grain obviously bone-shaped, at least in the internal contour, P/E usually >1.5. Columellae may or may not be longer at poles, often long in other areas as well. Tectum may undulate down the centre of each mesocolpium. Pori never circular, but may be elliptic, rectangular or represented by wide or narrow transverse endocolpi: Umbelliferae

[this key exit covers many genera in this family. Distinctions may be made by reference to type slides. Otherwise see Punt, 1984, for descriptions, keys and photographs and a full list of references to other works on this family. Some distinctions are discussed in the brief account on p. 143]

61b Grain never bone-shaped, even in the internal contour; shape always elliptic to rectangular-obtuse, P/E usually <1.5. Columellae longer and sexine thicker at **poles only**. Pori circular, or equatorially elongated ellipses, never either rectangular or very narrow endocolpi: 62

62a Columellae all fine, never branched; endopori transversely elliptic-acute with clearly defined ends, very rarely fusing to a complete equatorial girdle. Surface tectate, never perforate:

Polygonum aviculare type

[according to van Leeuwen *et al.*, 1988; this includes *P. aviculare*, *P. arenastrum*, *P. boreale*, *P. maritimum*, *P. oxyspermum*, *P. patulum*, and *P. rurivagum*. Several of these species also produce a significant proportion of pantocolporate grains. According to the above authors, *P. oxyspermum* and *P. maritimum* may be distinguished by their relatively thin sexines and short columellae]

62b Columellae either all fine except at the poles, or mostly rather coarse and branched (irregular in outline in exine surface view) branching being most apparent in the long columellae at the poles. Each endoporus is circular to transversely elliptic; if elliptic then with ends obtuse or diffuse but not acute. Surface may be tectate-perforate, if so then each perforation is directly over a branched columella (×1000, phase contrast helps). This feature is most clear in the apocolpia: *Polygonum bistorta* type [includes *P. bistorta* and *P. viviparum*. The morphology of grains of these species shows much variation on type slides and it is worth noting that (i) grains vary greatly in size (33−76 µm according to van Leeuwen *et al.*, 1988), (ii) the smaller ones have the exine thick, condensed and apparently structureless, whilst the larger have clear columellate structure (iii) the pori may not be very clear or not at all apparent, (iv) the perforations may be clear or invisible with LM]

63a Colpi widening from acute ends towards the equator. Each colpus membrane covered by coarse angular echinae or bacula or verrucae. Each porus well-defined, circular or elliptic (if elliptic, then meridionally elongate). Careful examination under ×1000 reveals fine rugulate-striate sculpture on the mesocolpia: *Aesculus*

63b Colpi narrow and crack-like, no angular projections on the colpus membranes. Pori may or may not be well-defined. Each porus: circular, rectangular, a transverse endocolpus, an H-endoaperture or a mere bridge or constriction to the ectocolpus: 64

64a With apocolpia flat or concave and apparently with coarsely rugulate-verrucate sculpturing. Mesocolpia flat and with finer sculpture (psilate-scabrate). Grain usually rather crumpled or collapsed, but shape can be seen to be basically that of a triangular prism, with the pori projecting from this shape at the equator. H-endoaperture present in which each transversely elongate, elliptic or figure of 8-shaped endoporus has extensions which travel up and down, roughly parallel to the ectocolpi. These extensions appear under low power like faint extra colpi, one either side of each true colpus. Grain >35 µm: *Anthyllis vulneraria* [the coarse sculpture in the apocolpia appears to be due to irregular structural elements **under**

the tectum; in this area the nexine appears partially dissolved]

64b Never the exact combination of rugulate at poles and psilate in the mesocolpia (may be rugulate in the latter area). Shapes and apertures various but no such apparent extra colpi (H-endopertures). Grain more or less than 35 µm: 65

65a Ectocolpi short (whole length visible in an equatorial view of the grain). Polar area consequently larger: 66

65b Ectocolpi longer (whole length not visible in an equatorial view of grain). Polar area consequently smaller: 69

66a Grains with very thick costae to the colpi (see equatorial view with mesocolpium facing and colpi in optical section). Usually >27 µm long: 67

66b Grains with smaller costae to the colpi, size usually <25 µm: 68

67a Grain with transverse endocolpi that have indistinct, obtuse ends. Surface psilate or faintly granular, but with slight indications of perforations or reticulate sculpturing in a broad equatorial girdle over the mesocolpia and transverse endocolpi. Ectocolpi with irregular margins and often forked or branched at their ends: *Anchusa arvensis*

67b Grains with pori very well-defined, circular or elliptic. If elliptic, then either equatorially or meridionally elongated with clear, **thick costae** around them: *Vicia cracca* type [includes *V. cracca* and some grains of the following species that are more commonly suprareticulate: *V. hirsuta*, *V. tetrasperma*, *V. orobus*, *V. sepium*, *Lathyrus palustris*, *L. niger* and *L. montanus*. Some distinction between these may be made by reference to type slides]

68a Colpi with granulate membranes, grain concave at the poles giving it a truncated shape. Sculpture sometimes irregularly rugulate-verrucate or pitted over the equator. Endoporus an equatorially elongated ellipse or 8-shape, sometimes with curved 'horns' into the mesocolpia (phase contrast): *Ornithopus perpusillus*

68b Colpi without granulate membranes, grains convex at the poles. Endoporus an equatorially elongated ellipse or 8-shape. Grain surface completely psilate even in phase contrast: *Lotus* type [includes *Lotus* and *Tetragonolobus*. The latter may be distinguished on type slides by the larger grain size and the acute ends to the transversely elongated endopori]

69a Grains small (<20 µm) and usually neatly

rectangular-obtuse in shape. Columellae not more obvious at the poles. Pori or transverse endocolpi always fairly large and wide in comparison to grain size. Exine thin, sexine/nexine boundary difficult to see:

Androsaceae elongata type [according to Punt, 1974, this includes *A. elongata, A. carnea, A. septentrionalis, A. lactea* and *A. villosa. Castanea* may key out here if its shape is not clearly rhombic-obtuse. It may be distinguished from this type by its thicker exine and slightly wider colpi, also by the small exine extensions over the endoaperture]

69b Slightly larger grains (>20 μm) columellae often more obvious at the poles. Sexine distinct from nexine. Sculpturing may be faintly rugulate-striate. This exit leads to the following species or types in the Umbelliferae which are not really all that similar, but which arrive together because they have few positive distinguishing features:

Crithmum maritimum, Cicuta virosa, Carum verticillatum, Trinia glauca and *Sium latifolium* type (which includes *S. latifolium* and *Berula erecta*)

[there are surface, size and aperture differences which are revealed by type slides and data in Punt, 1984]

UMBELLIFERAE – EXTRA NOTES ON DISTINCTIVE FEATURES OF SOME GENERA OR SPECIES

These notes are not presented in the form of a key as they are not, on their own, meant to lead to identification. Their purpose is to encourage readers to consult the relevant type slides and more detailed texts, in order to carry out identification as far as is possible. Data from Punt, 1984, and our own observations. The groups 1–12 below refer to distinctive morphological characteristics and it should be noted that any one grain may fall into more than one group if it has several distinctive features.

1 Grains neatly bone-shaped in external contour with columellae shortest at the poles: a large number of grains fall within this category — it is the most common and characteristic 'Umbellifer' that will be seen. A few examples are:

Peucedanum, Foeniculum, Bunium, Daucus, Conopodium, Aegopodium, Pimpinella

2 Grains more elliptic to straight-sided in external contour but slightly bone-shaped in internal contour:

Oenanthe, Conium, Cicuta, Heracleum, Pastinaca, Carum verticillatum

3 Grains bone-shaped, with endoapertures (pori or transverse colpi) noticeably protruding (pouting) at the equator:

Eryngium, Ligusticum scoticum, Aethusa cynapium, Chaerophyllum tementulum, C. hirsutum, Meum amanthicum, Smyrnium olusatrum, Bifora radians, Silaum silaus, Torilis japonica, Coriandrum sativum, Laserpitium

4 Grains with extremely long colpi (reaching up to the poles in an equatorial view):

Oenanthe, Astrantia, Sanicula, some *Eryngium* species, *Cicuta, Carum verticillatum*

5 Grains with coarse, long, **branched** columellae at the poles: *Heracleum*

6 Grains with subacute apocolpia:

Astrantia, Sanicula, Cicuta, Eryngium

7 Grains with short to very short colpi (about $\frac{1}{2}$ distance between the poles, or less):

Myrrhis odorata, some *Eryngium* species (colpus length seems variable within *Eryngium*), *Anthriscus caucalis, Pimpinella anisum, Caucalis platycarpos, Scandix pecten-veneris, Bifora radians, Coriandrum sativum*

8 Grains with tectum undulating markedly in the equatorial area:

Torilis japonica, T. nodosa, Myrrhis odorata, Anthriscus cerefolium, Daucus carota

9 Grains with sculpture noticeably rugulate-striate on close examination (×1000, phase contrast):

Astrantia, Apium

10 Grains with columellae varying in length such that they are markedly longest at various regions down the centre of the mesocolpium:

Pimpinella anisum, Myrrhis odorata, Cuminum cyminum, Anethum graveolens, Seseli libanotis, Falcaria vulgaris, Torilis arvensis, Petroselinum segetum

11 Grains with shape markedly asymmetric in equatorial side view (one colpus facing observer) due to a slight flattening of each mesocolpium at the equator: *Eryngium*

12 Grains with columellae rather fine and short compared to the grain size:

Eryngium, Astrantia, Sanicula

Trizonocolporate, eureticulate

1a Grain shaped like a tetrahedron, with pori at three of the four apices. One polar area small and psilate, the other large and reticulate. Mesocolpia reticulate, with lumina largest in the centre of each: *Thesium humifusum*

1b Grain not shaped like a tetrahedron, being circular, elliptic or rhombic-obtuse in equatorial view: 2

2a Sculpture microreticulate (lumina of reticulum <1 μm in width):   3

2b Sculpture reticulate (at least some lumina of reticulum e.g. those in the centre of the mesocolpium) >1 μm width: 21

3a Columellae indistinct to invisible in optical section of grain wall ×1000, very fine or indistinct in surface view: 4

3b Columellae visible in optical section of grain wall ×1000 and in surface view: 11

4a Pori represented by equatorially elongated, elliptic endopori or transverse endocolpi: 5

4b Each porus represented by: an indistinct endoporus, a rupture, a constriction or bridge to the colpus: 7

5a Faint, **narrow**, small transverse endocolpi (may need phase contrast to detect these). Grain narrowly elliptic to rhombic-obtuse with exine slightly thicker at the poles:

Hypericum perforatum type and *Hypericum androsaemum* type

[*H. perforatum* type includes *H. perforatum*, *H. tetrapterum*, *H. maculatum*, *H. hirsutum*, *H. pulchrum* and others. *H. androsaemum* type includes *H. androsaemum*, *H. calycinum*, *H. hircinum* and *H. inodorum*. Most data from Clarke, 1976. Distinction may be achieved by consultation of type slides. *H. perforatum* type is more lobed in polar view and *H. androsaemum* type is more triangular-obtuse-convex in polar view with the colpi on the angles. This difference is not, however, true for every grain. *H. androsaemum* type is actually suprareticulate with grouped perforations of the tectum in the floor of each shallow lumen, but this feature is very difficult to detect with the light microscope. Note that some grains in *Saxifraga stellaris* type may key out here if narrow, small transverse endocolpi are detected. However in this type there are usually 2–6 of these present in the equatorial region of each colpus. See exit 10b]

5b Clear, **wider** elliptic endopori or transverse endocolpi: 6

6a With elliptic endopori or transverse endocolpi without 'horns'. Grains small, 18 μm or less, elliptic to rectangular-obtuse:

Androsace elongata type

[according to Punt *et al.*, 1974; this includes *A. elongata*, *A. septentrionalis*, *A. villosa*, *A. lactea* and *A. carnea*]

6b With wide, obtuse-ended transverse endocolpi;

each with or without a varying number of 'horns' or extensions travelling in a curving fashion out under the mesocolpial sexine (phase contrast helps to see these). Costae to the colpi prominent, nexine may be very thick compared to sexine, but this is a variable feature — not seen on all grains. Size 18 μm or more:

Glaux maritima

[see ref. under 6a]

7a Grain with small irregular verrucae or gemmae projecting inwards from innermost side of nexine. Grain shape circular to rhombic-obtuse in equatorial view, triangular-obtuse in polar view (with flattened mesocolpia):

Trientalis europaea

[this species also produces large, pantoaperturate grains. See ref. under 6a]

7b Without such verrucae or gemmae, grains elliptic to rhombic-obtuse in equatorial view. Mesocolpia rounded: 8

8a Exine slightly thicker at the poles, grain narrowly elliptic to rhombic-obtuse:

Hypericum perforatum type and *H. androsaemum* type

[*H. perforatum* type includes *H. perforatum*, *H. tetrapterum*, *H. maculatum*, *H. hirsutum*, *H. pulchrum* and others. *H. androsaemum* type includes *H. androsaemum*, *H. calycinum*, *H. hircinum* and *H. inodorum*. Most data from Clarke, 1976. Distinction may be achieved by consultation of type slides. *H. perforatum* type is more lobed in polar view and *H. androsaemum* type is more triangular-obtuse-convex in polar view with the colpi on the angles. This difference is not, however, true for every grain. *H. androsaemum* type is actually suprareticulate with grouped perforations of the tectum in the floor of each shallow lumen, but this feature is very difficult to detect with the light microscope]

8b Exine same thickness at the poles as elsewhere on the grain surface: 9

9a Apocolpium larger (full length of colpi usually visible in equatorial view). Quite pronounced costae to the colpi. Equatorial bridge projecting or pouting: *Elatine*

[*E. hexandra* up to 29 μm, *E. hydropiper* considerably smaller — usually <20 μm]

9b Apocolpium very small (full length of colpi not visible in equatorial view). No costae to the colpi. Close examination may reveal that each porus is represented by 2–6 narrow, parallel, transverse endocracks or very small transverse endocolpi: 10

10a Grains usually <18 μm in length. Apocolpia

tectate-perforate with only a few perforations: *Chrysosplenium*
[this grain is very difficult to distinguish and type slides should always be consulted. See ref. under 10b]

10b Grains usually >18 µm in length, apocolpia microreticulate. Sometimes syncolpate at the poles: *Saxifraga stellaris* type
[this type is difficult to distinguish from *Chrysosplenium* and type slides should always be consulted. According to Verbeek-Reuvers, 1977a, it includes *S. stellaris*, *S. nivalis*, *S. clusii*, *S. tenuis*, *S. hieracifolia* and *S. foliolosa*. *Primula hirsuta* may also key out here according to the description in Punt *et al.*, 1974]

11a Colpi narrow, slit-like and parallel-sided. Caution — wide colpi may appear narrow if they are inrolled: 12

11b Colpi widening towards the equator from the acute ends (may be narrowed or nipped-in at the equator): 18

12a With each porus represented by a transverse endocolpus that may end in branches or 'horns' into the mesocolpium (careful LO analysis or phase contrast is needed to detect this. Sometimes a complete equatorial girdle is present: 13

12b Endopori circular, elliptic, elliptic-acute or represented by a mere bridge or rupture, but no such transverse endocolpi present: 14

13a Grain length 21 µm or more: *Anagallis arvensis* type
[includes *A. arvensis* and *Lysimachia nemorum*. Some grains may be as large as 35 µm, but in this case the sculpture may be bigger than microreticulate. See ref. under 13b]

13b Grain length 22 µm or less: *Anagallis tenella* type
[includes *A. tenella* and according to Punt *et al.*, 1974, also *A. minima*, *A. crassifolia* and *Asterolinum linum-stellatum*]

14a A bridge or rupture to each colpus: 15

14b A circular, elliptic or elliptic-acute endoporus to each colpus: 16

15a Grain triangular-obtuse in polar view, small verrucae or gemmae projecting inwards from the innermost side of the nexine. Usually >18 µm in size: *Trientalis europaea*

15b Grain circular to lobed in polar view. No verrucae projecting from the nexine inwards, size greater or smaller than 18 µm. This decription leads to the following taxa which may be distinguished only by very careful examination of type slides using phase contrast:
Samolus valerandi, *Hottonia palustris* (long-styled individuals) *Anagallis tenella* type where endocolpus not apparent (see also exit 13b) *Chrysosplenium*,

Saxifraga stellaris type (see also exit 10b) and probably also *Primula hirsuta* according to the description in Punt *et al.*, 1974.

16a Columellae coarse, often irregularly shaped in cross section. If columellae long, then nexine thin — up to $\frac{1}{4}$ of total exine thickness. Sexine and nexine always **markedly thinner** in a band around each colpus margin and sometimes projecting or pouting in the equatorial region of each colpus. Reticulum tends towards foveolae i.e. the muri are sometimes wider than the lumina: *Euphorbia*
[some distinction between the species is possible if type slides are available. There is considerable variation in grain size, in size of the reticulum lumina or foveolae, and in relative thickness of the nexine. Branching of the columellae may be detected in some species — see also the suprareticulate key, p. 150]

16b Columellae shorter and finer, not irregularly shaped in cross section. Nexine = or >$\frac{1}{4}$ total exine thickness. Exine never thinning towards the colpus margins or pouting at the equator: 17

17a Colpi relatively short (full length easily visible in equatorial view). Endoporus and costae to the colpi rather faint. Grain size 28 µm or less. Consult type slides: *Rumex acetosa* type
[according to data in van Leeuwen *et al.*, 1988, this type includes *R. acetosa*, *R. thyrsiflorus* and possibly *R. alpestris*]

17b Colpi longer (full length not usually visible in equatorial view). Endoporus and costae to the colpi clearly defined. Grain size greater or smaller than 28 µm. Consult type slides: *Oxyria* type
[includes *Oxyria digyna*, *Rumex maritimus*, *R. crispus*, *R. conglomeratus*, *R. sanguineus*, *R. pulcher*, some grains of *R. acetosella* type and possibly others. It includes occasional grains of species which are normally pantocolporate cf. *R. obtusifolius* type, p. 162. *Oxyria* may be distinguished from the rest on comparison with type slides by its endopori which are circular and very clearly thickened, combined with its very prominent columellae. Van Leeuwen *et al.*, 1988 state that *Oxyria* and *R. acetosella* are distinct from the rest of the grains in this type by their small size of <26 µm]

18a Exine in optical section looks pilate i.e. with long columellae which have swollen heads — only faintly united into a reticulum in surface view. Columellae very prominent, quite coarse and inordinately (irregularly) arranged, not in a reticulate pattern under the muri. Colpus margins may or may not be well defined: 19

18b Exine in optical section not looking pilate but with shorter, finer, parallel-sided elements, united into a reticulum in surface view. Colpus membrane usually psilate, margins of colpi well-defined and commonly inrolled: 20

19a Edge of each colpus ill-defined, colpus membrane granulate. A long transverse equatorial bridge apparent on each colpus at the equator. Grain shape rhombic-obtuse in equatorial view: *Mercurialis*

19b Edge of each colpus well defined, colpus membrane psilate. Equatorial bridge not so elongated. Consult a type slide:
Gentianella campestris type
[according to Punt and Neinhuis, 1976, this includes *G. campestris*, *G. germanica*, *G. anglica* and *G. uliginosa*. It may be possible to distinguish *G. uliginosa* by its very short columellae]

20a Exine showing a **gradual** slight decrease in thickness from the middle of the mesocolpia to each apocolpium. The thinning at the poles is achieved by shortening the columellae, which also become slimmer and more crowded together. Lumina decrease gradually in size towards the apocolpium. Consult a type slide:
Sambucus nigra type
[includes *S. nigra* and possibly *S. racemosa*, although the decrease in exine thickness towards the poles seems less marked in the last species. See ref. under 24a]

20b Exine not gradually decreasing in thickness from mid-mesocolpia to each apocolpium. Surface with extremely small reticulations (sometimes tending towards tectate with numerous small perforations). Columellae in a regular or irregular carpet. Consult type slides:
Digitalis purpurea type
[includes *Digitalis purpurea*, *Sibthorpia europaea*, *Kickxia*, *Limosella aquatica*, *Cymbalaria muralis*, *Chaenorhinum minus* and grains of some *Linaria* species that appear microreticulate. This type is not uniform and some distinction is possible within it if type slides are available. *Cymbalaria* has the smallest grains and *Sibthorpia* the largest. *Digitalis* has the shortest colpi and *Sibthorpia* the longest colpi — these latter often with square ends]

21a Each porus represented by a mere bridge, constriction or rupture to the colpus (caution — very careful examination of the equatorial region is required, even so, some well defined endopori may be missed if phase contrast is not available): 22

21b Each porus a well defined circle or ellipse or elongated equatorially into a transverse endo-colpus (often nexine features only, more easily detectable with phase contrast): 35

22a With largest lumina in the middle of each mesocolpium; **lumina size decreasing towards the poles**. Lumina also gradually decrease in size towards the colpi: 23

22b Lumina in the middle of each mesocolpium not much larger than those at the poles i.e. lumina size **not decreasing towards the poles**, it may or may not decrease towards the colpi: 25

23a Sexine and nexine of approximately the same thickness, sexine does not vary in thickness over the grain surface: *Adoxa moschatellina* [see Reitsma and Reuvers, 1975]

23b Sexine thicker than nexine, at least in the centre of each mesocolpium, sexine may or may not vary in thickness over the surface: 24

24a Sexine markedly decreases in thickness from mid-mesocolpia to each apocolpium. Muri with coarse columellae arranged in a reticulate pattern. Relatively large lumina in the mesocolpia (up to 3 μm). Granules or short bacula visible on the floor of each lumen: *Sambucus ebulus* [See Punt *et al.*, 1974. Rare grains of *Euonymus* may key out here, but these lack granules on the lumina floors]

24b Sexine shows only a slight decrease in thickness towards the apocolpia. Columellae finer and lumina relatively small (about 1 μm or just over this) in the centre of each mesocolpium. Columellae rather irregularly arranged in a carpet under the muri. No granules or bacula in the lumina: *Sambucus nigra* type [includes *S. nigra* and *S. racemosa*. A distinction may be possible if sufficient type slides are available. See ref. under 24a]

25a Reticulum coarse with granules or short bacula visible on the floors of the lumina. These granules or bacula about the same thickness as the columellae, but appear fainter: 26

25b Reticulum coarse or finer, but no such granules or bacula on the lumina floors: 27

26a Sexine and nexine of similar thickness (at least in the centre of each mesocolpium) columellae rather short and apparently incompletely fused sideways to form muri i.e. they appear to be pila, especially in the apocolpia: *Viburnum lantana* [see ref. under 24a]

26b Sexine thicker than nexine, columellae longer and more clearly fused: *Viburnum opulus* type [includes *Viburnum opulus* and *V. tinus*. See ref. under 24a]

27a Colpi rather short, full length easily visible in equatorial view: 28

27b Colpi longer, full length not usually visible in equatorial view: 30

28a Narrow slit-like colpi, sexine and nexine of similar thickness. Sexine not thinning towards the colpi and reticulum meshes not decreasing in size towards the colpi: *Fraxinus* [includes *Fraxinus excelsior* and *F. ornus*. The latter may be distinct because of its larger size, coarser reticulum and thicker sexine relative to nexine. Type slides should be consulted if it is likely to occur]

28b Colpi wider, with granulate or psilate colpus membranes. Sexine much thicker than nexine, Columellae and reticulum very coarse. Sexine thins from mid-mesocolpium to colpus margins (see polar view, optical section): 29

29a Lumina decreasing markedly in size towards the colpus margins so that the edge is actually microreticulate. Each porus may appear as a large ruptured area: *Euonymus europaeus* [grains where the pori are not well-defined]

29b Lumina not decreasing so much in size towards the colpus margins. Edge with lumina >1 μm wide. Porus always represented by a bridge: *Ligustrum vulgare* [the commonly cultivated *Syringa* may appear similar]

30a Colpi widening towards the equator, with psilate membranes. Lumina decrease slightly and gradually towards a tectate or microreticulate margin to the inrolled colpi (margin may be difficult to see because of the inrolling). Exine thickness also decreases towards margins of colpi: 31

30b Colpi narrow, membranes not visible. Exine thickness not decreasing towards colpi. Lumina not, or only very slightly decreasing towards colpi: 32

31a Exine remaining roughly the same thickness from mesocolpia to poles, lumina may or may not decrease in size towards the poles. Consult type slides: *Scrophularia* type [includes *Scropularia*, *Verbascum*, *Misopates orontium* and those grains of *Linaria* that are not microreticulate. *Linaria* and *Misopates* are smaller than the other two taxa, but unfortunately there is a size overlap with the bigger *Scrophularia* and *Verbascum* grains. If type slides are available it is possible to make a fairly reliable distinction between a *Scrophularia* type (which includes *Verbascum*) and a *Linaria* type (which includes *Misopates*). Note that occasional grains of *Salix* may key out here if they have a bridge to each colpus]

31b Exine **gradually** decreasing in thickness from middle of the mesocolpia to the poles. The thinning is achieved by shortening the columellae, which also become slimmer and more crowded.

Lumina decrease gradually in size towards the apocolpia, which may be microreticulate. Consult a type slide: *Sambucus nigra* [see ref. under 24a]

32a Columellae clearly visible in surface view, ×1000, as black dots under the muri of the reticulum. Grain usually >25 μm: *Diapensia lapponica*

32b Columellae not clearly visible in surface view, ×1000. Grains usually <25 μm: 33

33a Lumina size not decreasing from mid-mesocolpium to colpus. No tectate margo — reticulum continues to edge of inrolled colpus (consult a type slide): *Hottonia palustris* [the larger grains that are produced by short-styled individuals — see Punt *et al.*, 1974]

33b Lumina gradually decrease in size from mid-mesocolpium to a narrow tectate margo to each colpus. Consult a type slide: *Saxifraga stellaris* type [this exit covers the members of the type that have the largest lumina, which may be slightly over 1 μm in diameter e.g. *S. stellaris*, *S. foliolosa*. See Verbeek-Reuvers, 1977a, and exit 10b. No subdivision of the type appears possible as lumina size is variable within the above species]

34a Pori represented by transverse endocolpi. Lumina size decreases towards the colpi, so that a relatively wide tectate margo borders each colpus: 35

34b Circular or elliptic endopori, Lumina may or may not decrease towards the colpi: 38

35a Grain large (>35 μm in length) with long colpi that widen gradually towards the equator. Lumina rather rounded and even sized in the middle of the mesocolpium but decreasing in size markedly towards the **poles**. Muri appearing narrow at the top and thicker on focusing down. Muri always duplicolumellate or pluricolumellate: *Hypericum elodes* [this grain is not actually eureticulate, but is suprareticulate with a small perforation in the centre of the funnel-shaped roof to each lumen. The roof to each lumen is difficult to detect with the light microscope. See ref. under 8a]

35b Grain larger or smaller than 35 μm. Colpi narrow and slit-like, not appreciably widening towards the equator. Lumina size either not decreasing, or only slightly decreasing towards the poles. Colpi may be sunken or inrolled. Muri may appear simplicolumellate or duplicolumellate: 36

36a Each transverse endocolpus with diffuse ends. Mesocolpia flattened (grain triangular-obtuse in polar view, circular to rhombic-obtuse in equatorial view) and colpi rather short. Muri sinuous and lumina large and of irregular shape in the middle of the mesocolpia and also in the apocolpia. Grains >34 μm in length: *Hedera helix* [considerable variation in size is seen in collections of this species and smaller grains may be seen. The muri also show variation, being pluricolumellate, duplicolumellate, or occasionally simplicolumellate. See van Helvoort and Punt, 1984]

36b Each transverse endocolpus tapering, with distinct, acute ends (sometimes with 2 or more 'horns' into the mesocolpia or occasionally fused sideways with other endocolpi to form a complete equatorial girdle). Mesocolpia more commonly convex, grains circular to elliptic in equatorial view. Colpi long or short. Grains <36 μm in length: 37

37a Colpi relatively short (whole length usually visible in an equatorial view). Lumina in centre of mesocolpia large compared to those bordering the colpus margins, muri simplicolumellate or duplicolumellate. Fastigia may or may not be detectable: *Lysimachia vulgaris* type [includes *L. vulgaris*, *L. thyrsiflora*, *L. nummularia*, and *L. punctata*. See ref. under 6a]

37b Colpi always long (full length not visible in equatorial view). Lumina in centre of mesocolpia smaller, muri always simplicolumellate. Small fastigia present: *Anagallis arvensis* type [includes *Anagallis arvensis* and *Lysimachia nemorum*. See ref. under 6a. Some grains of this type may also be classed as microreticulate]

38a Margins of colpi and apocolpium striate, exine thickest in the middle of the mesocolpia and thinning markedly towards the colpi:
                                        Blackstonia perfoliata

38b Margins of colpi reticulate or tectate, exine may or may not thin towards the colpi: 39

39a Mesocolpia markedly flattened or concave, thus grain shape triangular-obtuse to hexangular in polar view. Shape elliptic to rhombic-obtuse in equatorial view. Colpi very long, narrow and sunken; with marked costae and small neatly circular pori: *Vitis vinifera*

39b Mesocolpia always convex, thus grain circular, lobed or triangular-convex in polar view. Shape circular to elliptic in equatorial view. Colpi may or may not be sunken: 40

40a Colpi narrow and slit-like, margins not inrolled although the colpi might be slightly sunken in a polar view: 41

40b Colpi widening towards the equator, often sunken with inrolled margins: 43

41a Muri rather sinuous or variable in width and

usually **duplicolumellate** (best seen in phase contrast). Narrow costae along colpus margins. Grain circular in equatorial view and circular to slightly lobed in polar view due to slightly sunken colpi. Consult a type slide:
                              Rumex acetosella type [includes *R. acetosella* and *R. tenuifolius*. See van Leeuwen *et al.*, 1988. These authors state that *R. acetosella* type is not sufficiently distinct to be separable from the grains in *Oxyria* type on the basis of sculpture alone, as not all grains have the surface as described. However we feel that the detection of sinuous or winding muri is still a useful pointer to this very ecologically important type. This should be combined with their observation that *R. acetosella* is <26 μm in diameter. It is noted that *Euphorbia peplus* also conforms to the description above except for the fact that it lacks costae along the colpi]

41b Muri not sinuous, usually simplicolumellate. Columellae coarse compared to grain size: 42

42a Columellae irregularly (inordinately) arranged, not in a reticulate pattern. If columellae are long, then nexine is thin, up to $\frac{1}{4}$ of the thickness of the exine. Sexine and nexine always **markedly thinner** in a band around each colpus margin and sometimes projecting or pouting in the equatorial region of each colpus. Reticulum lumina tend towards foveolae i.e. the muri are sometimes wider than the lumina: *Euphorbia* [some distinction between the species is possible if type slides are available. There is considerable variation in grain size, in size of the reticulum lumina or foveolae, and in relative thickness of the nexine. See note under 41a above. Branching of the columellae may be detected in some species — see also the suprareticulate key, p. 150]

42b Columellae in a reticulate arrangement under rather wide muri. Muri may be microechinate. Exine does not decrease in thickness towards the colpi: *Fumana* type [includes *Fumana* and some species of *Cistus*]

43a Lumina of similar size in the apocolpia and in the mesocolpia. Lumina may or may not decrease in size towards the colpi: 44

43b Lumina largest in the mesocolpia and decreasing in size towards the **poles** (compare a polar view with a mid-mesocolpial view). Lumina always decrease in size towards colpi. Colpi long, broad and sunken: 45

44a Columellae very coarse and inordinately (irregularly) arranged. Lumina not decreasing in size towards the colpi: *Gentianella amarella* type [includes *G. amarella* and *G. tenella*. *G. tenella* may be distinct because of its smaller size — <35 μm according to Punt and Neinhuis, 1976]

44b Columellae finer and arranged in a reticulate pattern under the muri. Lumina decreasing markedly towards the colpi (margins are micro-reticulate). Muri narrow, acute in cross section and rather sinuous in the mesocolpia:
Bryonia dioica
[see also *Euonymus*, which may key out here]

45a Columellae at low focus (level of their bases) inordinately (rather irregularly) arranged, sometimes crowded together. Lumina not much over 1 μm: *Gentianella campestris* type [according to Punt and Neinhuis, 1976, this includes *G. campestris*, *G. germanica*, *G. anglica*, *G. uliginosa*. They state it may be possible to distinguish *G. uliginosa* by its very short columellae]

45b Columellae at low focus (level of their bases) arranged in a reticulate pattern under the muri: 46

46a With apocolpium **very small**, 4 μm or less (sometimes syncolpate). Simplicolumellate muri. Endopori rather small in diameter, about 4 μm. Grain usually circular in equatorial view, <30 μm long. Consult type slides:
Parnassia palustris
[see Verbeek-Reuvers, 1977c]

46b Grains with apocolpia larger, and simplico-lumellate or duplicolumellate muri. Endopori usually >4 μm in diameter. Grains circular or elliptic in equatorial view, usually >30 μm: 47

47a Sexine 1 (columellae layer) about as thick as sexine 2 (muri) in mid-mesocolpia. Muri sim-plicolumellate or duplicolumellate. Apocolpia more or less than 7 μm in diameter. Consult type slides: *Gentiana verna* type [according to Punt and Neinhuis, 1976, this includes *Gentiana verna*, *G. nivalis* and *Gentianella ciliata*. They state that only *G. ciliata* appears small enough to have a slight size over-lap with *Parnassia*. N.B. Some species of *Cistus* e.g *C. ladaniferus*, may key out here or under 47b]

47b Sexine 1 (columellae layer) always thicker than sexine 2 (muri) in mid-mesocolpia. Muri always simplicolumellate. Apocolpia usually 7 μm or more in diameter. Consult a type slide:
Euonymus europaeus

Trizonocolporate, suprareticulate to foveolate

The detection of a suprareticulum is not always an easy task. In an optical cross section of the grain wall, ×1000, the walls of the muri appear very shallow. In surface view the columellae are often arranged without any relationship to the muri i.e. in an even carpet under the muri and lumina. Small perforations may be present on the floors of the shallow lumina.

1a Grain triangular-obtuse or triangular-truncated (hexangular) in polar view, often with P/E ratio less than 1.0, i.e. broader than long: 2

1b Grain usually circular to lobed in polar view (rarely triangular-truncated). P/E ratio usually 1.0 or more: 3

2a Grain large (>32 μm in polar view) with very short colpi, coarse columellae and pori situated between the angles in polar view. The pori have large nexinous thickenings. Each lumen of the reticulum is funnel shaped and directly over a coarse columella (i.e the columella forms a solid 'stalk' to the funnel): *Tilia* [*T. platyphyllos* is larger and has coarser exine structure than *T. cordata*. The complex exine structure in this genus is more accurately de-scribed as pertectate — see explanation in Christensen and Blackmore, 1988. See this text also for more details on the distinction between *T. cordata* and *T. platyphyllos*]

2b Grain smaller (<30 μm in polar view) with long colpi and fine columellae that are almost invisible under ×1000 and phase contrast. In polar view the pori are situated on the truncated angles of the triangular shape and thus the mesocolpia are flattened. Lumina of the supra-reticulum widest in the centre of the meso-colpium, decreasing in size towards the colpi. The long colpi have marked costae along their length: *Rhamnus* type [includes *Rhamnus catharticus* and some grains of *Frangula alnus* — the latter is most commonly psilate, but occasional grains show faint lumina in the mesocolpia in phase contrast. In both species the aperture seems to have a rather complex structure with a small short endocolpus connecting to thin areas of exine that travel up and down parallel to the colpus, resembling an 'H' shaped endoaperture. Note that occasional legumes with rather flattened mesocolpia may key out here but they are usually different in shape and porus struc-ture — see *Astragalus danicus* type and *Trifolium* type on p. 151]

3a Exine markedly thickest at the poles: 4

3b Exine not thickest at the poles, either thinner here or same thickness as in the middle of the mesocolpium: 6

4a Columellae longest and coarsest at the poles, being clearly branched here (see optical section of wall ×1000). Foveolae most clearly present at the poles where each hole is located directly over a branched columella. Each colpus narrow and slit-like, margins not inrolled. Grain elliptic in shape: *Polygonum bistorta* type [includes *P. bistorta* and *P. viviparum*. See van Leeuwen *et al.*, 1988; who present information on possible distinction between these two species. This pollen type is very variable and it is worth noting that (i) grains vary greatly in size — the above authors quote 33−76 μm, (ii) the smaller grains have the exine thick, condensed and apparently structureless whilst the larger have clear columellate structure, (iii) the pori may be either unclear or not at all apparent, (iv) the perforations may be abundant and clear or undetectable]

4b Columellae short and unbranched (may be so fine as to be almost undetectable) most of extra thickness at the poles achieved by a thicker nexine and tectum (see optical section of wall ×1000). Surface suprareticulate rather than foveolate. Colpi wider, with membrane exposed and margins inrolled except at the equator where projecting portions of sexine meet over the endoporus. Grain shape elliptic to rhombic-obtuse: 5

5a Size >34 μm, lumina large in the mesocolpium and decreasing markedly in size towards the poles and colpus margins. Muri triangular in cross section, becoming wider as one focuses down. Each lumen is actually a conical shaped depression and small perforations may be visible on its floor. With quite large costae to the colpi and porus clear: *Hypericum elodes* [see Clarke, 1976]

5b Size <30 μm, adjacent lumina may vary in size but there is no marked decrease towards the poles and colpus margins. Shape can be very narrowly elliptic with the long sides straight. A small, narrow transverse endocolpus may be detectable, but commonly the only indication is the bridge formed by the projecting portions of sexine over the colpi at the equator: *Hypericum perforatum* type and   *H. androsaemum* type [according to Clarke, 1976; *H. perforatum* type includes *H. perforatum*, *H. tetrapterum*, *H. hirsutum*, *H. maculatum*, *H. pulchrum* and others. *H. androsaemum* type includes *H. androsaemum*, *H. calycinum*, *H. hircinum* and *H. inodorum*. Distinction may be achieved by consultation of type slides. *H. perforatum* type is more lobed in polar view and *H. andro-saemum* type is more triangular-obtuse-convex

in polar view with the colpi on the angles. This difference is not, however, true for every grain]

6a Grain pear-shaped in equatorial view, i.e. broadest at one pole and narrowest at the other. Pori not in the equatorial region but situated nearer to the broader pole than to the narrower one: *Echium vulgare*

6b Grain elliptic, rectangular-obtuse or rhombic-obtuse. Pori always in the equatorial region: 7

7a With coarse, sometimes branched, columellae (length $\frac{1}{2}$ to $\frac{3}{4}$ of the thickness of the exine). Colpi narrow and slit-like, never sunken or inrolled. Careful examination of the wall in optical section ×1000 and surface view may reveal that the muri of the reticulum rise up **between** the columellae rather than **over** them, i.e. each lumina may be seen to be directly over a coarse columella. Nexine may appear very thin: 8

7b With finer, shorter columellae (up to $\frac{1}{2}$ exine thickness) which are sometimes visible only at the poles. Colpi may or may not be narrow, with inrolled margins: 9

8a With very coarse, sparsely spaced columellae, each of which obviously branches at $\frac{1}{2}$ to $\frac{2}{3}$ of its length to meet the tectum as a group of fine rods. Exine does not thin towards the colpus margins (see polar view, optical section):   *Fagopyrum esculentum* [grains on type slides show wide variation in size and grains over 60 μm can be found]

8b With coarse columellae that are closely packed together and rather irregular in cross section. Sometimes branching may be observed. Exine always **markedly thinner** in a band around each colpus margin, and sometimes projecting or pouting in the equatorial region of each colpus: *Euphorbia* [some distinction between the species is possible if type slides are available. There is considerable variation in grain size, in size of the reticulum lumina or foveolae, and in relative thickness of the nexine]

9a Endopori well-defined, neatly circular or elliptic or transverse endocolpi present: 10

9b Endopori less well-defined, often indicated merely by the presence of a rupture or constriction to the colpus: 15

10a Transverse endocolpi present: 11

10b Neat circular or elliptic endoporus: 12

11a Grain <30 μm, transverse endocolpi narrow: *Hypericum perforatum* type and   *H. androsaemum* type [according to Clarke, 1976; *H. perforatum* type includes *H. perforatum*, *H. tetrapterum*,

H. hirsutum, *H. maculatum*, *H. pulchrum* and others. *H. androsaemum* type includes *H. androsaemum*, *H. calycinum*, *H. hircinum* and *H. inodorum*. Distinction may be achieved by consultation of type slides. *H. perforatum* type is more lobed in polar view and *H. androsaemum* type is more triangular-obtuse-convex in polar view with the colpi on the angles. This difference is not, however, true for every grain]

11b Grain larger or smaller than 30 μm, transverse endocolpus wide, often branched or forked. Costae to the colpi usually prominent, nexine may be very thick compared to sexine (although this feature appears variable): *Glaux maritima*

12a Reticulum continues over the poles, muri thin so that reticulum appears as a delicate lace-like network in phase contrast. Usually >15 lumina across a mesocolpium, sometimes microreticulate: 13

12b Reticulum disappears at poles, or is restricted to a few widely spaced lumina only. Muri wider, usually <15 lumina across a mesocolpium: 14

13a Grain usually larger than 27 μm. Columellae detectable under ×1000 but may need phase contrast. Consult type slides: *Ononis* type [includes *Ononis* and *Melilotus*. Note that some grains of *Astragalus danicus* type may key out here if they have a noticeable reticulum at the poles]

13b Grain smaller than about 27 μm, columellae invisible under ×1000. Consult type slides: *Astragalus alpinus* [the introduced species *Galega officinalis* may key out here but it can be distinguished by its almost quadrangular-obtuse grains and its large, meridionally elongated endopori]

14a Each circular or elliptic endoporus very well-defined because it is surrounded by a ring-like costa. Very thick costae present along the margins of the colpi (exine adjacent to colpus just above equator 2−3 times as thick as exine elsewhere e.g. at poles): *Vicia* type [includes *V. sylvatica*, *V. hirsuta*, *V. sativa*, *V. faba*, *V. sepium*, *V. lutea*, *V. orobus*, *V. tetrasperma*, *Lathyrus sylvestris*, *L. japonicus*, *L. pratensis*, *L. montanus*, *L. palustris*, *L. nissiola* and possibly introduced *Lupinus* and *Lathyrus* species. Some distinction within this type is possible on the size of the reticulum meshes if type slides are consulted. The biggest meshes compared to grain size are found in *V. faba*, *V. sativa*, *V. sepium* and *V. sylvatica*; the smallest compared to grain size in *Lathyrus pratensis*, *L. palustris* and *L. japonicus*. The

smallest grain size is in *V. tetrasperma*. *V. cracca* occasionally appears reticulate, and may key out here, but it is more commonly seen with irregularly rugulate-verrucate sculpturing]

14b Each circular or elliptic endoporus less well defined — costae present at the meridional margins, but these do not extend around the equatorial margins. Costae to the colpi may be prominent but rarely as thick — up to 2 times exine thickness at poles. Reticulum meshes sometimes smaller in an equatorial band: *Astragalus danicus* type [includes *Astragalus danicus*, *A. frigidus*, *A. arenarius*, *A. glycyphyllus*, *Oxytropis campestris*, *O. halleri*. Most grains of this type have the reticulum reduced or absent at the poles, but some may have the reticulum detectable here and thus key out elsewhere. Grains of some *Trifolium* species may also key out here. Type slides should be consulted]

15a Muri and lumina usually wide; reticulum usually different at poles i.e. reticulum most commonly reduced to a few, sparse lumina here (lumina may even be absent at poles). Columellae thicker and more prominent at poles. Apocolpium relatively large (whole length of colpi visible in equatorial view) grain shape rectangular-obtuse to rhombic-obtuse in equatorial view, truncated-triangular in polar view: *Trifolium* type [includes most species of *Trifolium* and some of *Medicago* i.e. *T. pratense*, *T. repens*, *T. scabrum*, *T. squamosum*, *T. incarnatum*, *T. medium*, *T. fragiferum*, *T. striatum*, *T. subterraneum*, *T. campestre*, *T. arvense*, *T. dubium*, *T. micranthum*, *T. badium*, *T. montanum*, *T. hybridum*, *T. ornithopodioides*, *Medicago lupulina*, *M. minima* and possibly *M. polymorpha* and *M. arabica* (these last have rather irregular sculpturing). Some differentiation is possible if type slides are available e.g. *T. campestre*, *T. arvense*, *T. badium*, *T. micranthum* and *T. dubium* form a group distinguishable by their short colpi, large apocolpia and shape which may approach quadrangular-obtuse in equatorial view. The size of the reticulum meshes seems quite variable; within the same species (e.g. *T. repens*, *T. campestre*, *T. ornithopodioides*) we have collections that have much less than 15 lumina across a mesocolpium as well as collections with equal to or just over this number]

15b Small delicate reticulum with thin muri (microreticulate); reticulum continues unchanged across the poles. Colpi long, grain shape never rectangular-obtuse: 16

16a Grain narrowly elliptic to rhombic-obtuse. Colpus margins inrolled except for the 'bridge' portion at the equator. Colpi not parallel-sided, but slightly gaping, so that the psilate colpus membrane is visible (see polar view) along the length except for the bridge portion at the equator. Surface microreticulate, grain <30 μm: *Hypericum perforatum* type and *H. androsaemum* type
[according to Clarke, 1976; *H. perforatum* type includes *H. perforatum*, *H. tetrapterum*, *H. hirsutum*, *H. maculatum*, *H. pulchrum* and others. *H. androsaemum* type includes *H. androsaemum*, *H. calycinum*, *H. hircinum* and *H. inodorum*. Distinction may be achieved by consultation of type slides. *H. perforatum* type is more lobed in polar view and *H. andro-saemum* type is more triangular-obtuse-convex in polar view with the colpi on the angles. This difference is not, however, true for every grain]

16b Grain broadly elliptic or rhombic-obtuse. Colpi narrow and parallel sided, margins not inrolled, grain larger or smaller than 30 μm: 17

17a Large portions of exine project to form a bridge to the equator of each colpus. Costae present on the colpi. Consult a type slide: *Elatine*
[*E. hexandra* up to 29 μm, *E. hydropiper* considerably smaller — usually <20 μm]

17b No projecting portions of exine at the equator. Muri sometimes of irregular widths separating lumina of a regular size (gives distinctive black 'holly leaf' pattern under phase contrast). No costae to colpi: *Ulex* type
[includes *Ulex*, *Genista* and *Cytisus*]

Trizonocolporate, echinate or echinate-verrucate

1a Each porus represented by a large, **very wide** transverse endocolpus that is indicated by its margins — two curved nexinous ridges that cross near the ends of each ectocolpus. Grain triangular-convex (with colpi on the angles) to circular in polar view. Colpi short and narrow, apocolpia consequently very large: 2

1b Each porus circular or represented by a **narrow** transverse endocolpus. Grain shape circular or lobed in polar view. Colpi long or short, wide or narrow: 4

2a Echinae <or = 1.5 μm (mostly microechinae) and densely clothing the surface. Margins of transverse endocolpus very short and situated right at the ends of each ectocolpus, near to the poles of the grain. Grain size 50 μm or less:
                                    Linnaea borealis

2b Largest echinae >or = 1.5 μm, densely or sparsely clothing the grain surface. Echinae may be dimorphic (long and short mixed). Grains mostly >50 μm: 3

3a With long echinae, >1.5 μm and up to 3 μm, widely spaced on the surface:
                          Lonicera xylosteum type
[includes *L. xylosteum*, *L. alpigena*, *L. nigra* and some grains of *L. coerulea* that are not 4-aperturate. See Punt *et al.*, 1974. Some separation of types is apparently possible e.g. *L. coerulea* has detectably dimorphic columellae]

3b With shorter echinae, up to 1.5 μm long, more densely spaced on the exine:
                          Lonicera periclymenum type
[includes *L. periclymenum* and *L. caprifolium*. See ref. under 3a. Separation is apparently possible depending on whether each transverse endocolpus has complete or open ends]

4a Tectum processes <1.5 μm in height, all echinae, microechinae or low verrucae: 5

4b Tectum processes >1.5 μm in height, echinae or verrucae: 9

5a Columellae distinct and clearly longest in the centre of each mesocolpium, gradually decreasing in length towards the colpi and the poles. Tectum covered in rather regularly spaced microechinae. Pori small, not equatorially elongated: *Artemisia*
[some separation is possible between the species of this genus depending on the prominence and density of the microechinae, see Webb and Moore, 1982. *A. norvegica* has the longest, sparsest microechinae; some forms of *A. maritima* have a verrucate or undulate tectum and columellae more or less fused into irregular islands]

5b Columellae (if present) of similar length all over the grain surface. Pori represented by transverse endocolpi or equatorially elongated endopori: 6

6a Small, slim true columellae present joining nexine to tectum. Porus narrowly rectangular in shape, edges of mesocolpia protrude over each porus at the equator. Consult a type slide:
                                    Filipendula

6b Tectum thick, traversed by many slim rods (infratectal rods) that give the appearance of columellae in surface view. True columellae (i.e. connections between the echinate tectum and the nexine) rudimentary or absent resulting in a clear **gap** (study optical section of grain wall, ×1000) in this area. Transverse endocolpi present. These may or may not fuse to a complete equatorial girdle: 7

7a Exine thickest at poles and in middle of mesocolpia. Grain shape thus tends towards rhombic-obtuse in equatorial view, triangular-obtuse in polar view (with colpi between the angles). Microechinae only about 0.5 μm and sometimes indistinct; grain outline thus may or may not be undulating from ×400:

Centaurea scabiosa

7b Exine roughly the same thickness all over the grain surface. Microechinae larger, about 1.0 μm; wide-based, giving a verrucate or undulating outline to the grain from ×400: 8

8a Grains always elliptic in equatorial view, ectocolpi always long and transverse endocolpi fused to a complete equatorial girdle:

Centaurea nigra type

[includes *C. nigra* and *C. nemoralis*]

8b Grains usually circular in equatorial view. Ectocolpi long, or short and resembling elliptic pori. Transverse endocolpi always short and never fused to an equatorial girdle:

*Ambrosia** type

[includes *Ambrosia*, *Iva* and *Xanthium*. Distinction may be made within this type depending on the length of the colpi — see McAndrews *et al.*, 1973. Note that the smallest members of *Aster* type, e.g. *Gnaphalium* or *Filago*, may key out here if the echinae are borderline in length]

9a With cylindrical echinae which are always solid and constricted at the point of attachment to the tectum. They are often seen leaning over at various angles: *Rubus chamaemorus*

9b With verrucae or conical echinae. If the latter, then never constricted at the base, and either solid or traversed by numerous slim rods (infratectal bacula): 10

10a Echinate, with no true columellae visible underneath the tectum, only infratectal bacula visible in optical section and surface view:

Aster type

[there is considerable variation in size in this large type. Unfortunately the range of sizes in this type shows such a degree of overlap that further subdivision within the type is very difficult. The type includes *Bidens*, *Galinosoga*, *Filago*, *Gnaphalium*, *Senecio*, *Tussilago*, *Petasites*, *Antennaria*, *Anaphalis*, *Inula*, *Conyza*, *Pulicaria*, *Solidago*, *Bellis*, *Aster*, *Arnica*, *Erigeron*, *Eupatorium*, *Carduus* and possibly more. Some distinction may be possible if sufficient type slides are available]

10b Echinate or verrucate. True columellae visible underneath the tectum (the true columellae may be distinguished from any infratectal bacula by their position in an optical section and by the observation that they are always coarser than the infratectal bacula): 11

11a Columellae longer under each echina and here arranged slantingly, so that a radiating or 'star' pattern may be visible on LO analysis of a single echina. Colpi fairly short, grain circular in equatorial and polar views: *Cirsium*

11b Columellae always perpendicular under the tectum beneath an echina or verruca (no 'star' pattern on LO analysis). Colpi longer, grain circular to elliptic in equatorial view: 12

12a Grains large, usually >40 μm in size, echinate or verrucate: *Serratula* type

[includes *Serratula*, *Saussurea*, *Arctium* and *Carlina*. Further separation is possible by consultation of type slides. In *Saussurea* the echinae are so low and rounded that they may appear to be verrucae]

12b Grains smaller, usually <40 μm; always echinate: *Anthemis* type

[includes the following: *Anthemis*; *Achillea*; *Leucanthemum*; *Chrysanthemum*; *Tanacetum*; *Chamaemelum*; *Otanthus*; *Cotula*; and *Matricaria*]

Trizonocolporate, rugulate-striate

1a With operculate colpi (membrane covering the colpus is thickened, except for a thin area all round which joins the operculum to the colpus margin). Operculum often bears the same sculpture as the rest of the exine: Rosaceae

[see special key on p. 156]

1b No opercula visible on the membranes of the colpi. Membrane surface psilate or with granules, bacula or echinae: 2

2a In the mesocolpia there are rugulae which are **straight** and long. These rugulae **randomly intersect** or cross one another or fuse into **star** shapes (×1000, phase contrast helps): 3

2b In the mesocolpia there are either parallel striae or irregular, sinuous rugulae. The muri sometimes anastomose, taper-off, or dive under one another, but never fuse into star shapes (×1000, phase contrast helps): 6

3a Colpi widening (gaping) appreciably towards the equator. Porus represented by an equatorial bridge or rupture underlain by a complex H-shaped endoaperture, the 'legs' of which travel up the margins of the colpi (this is best seen in phase contrast of an equatorial view, with one colpus facing; in polar view, optical section, the legs can be detected by the thinning of the exine adjacent to the colpus):

4 (Crassulaceae)

3b Colpi narrow, not widening towards the equator. Each endoporus elliptic or 8-shaped, with or without 'horns' which may make it rather H-shaped. Apocolpia may be flattened: 5

4a Grain >20 µm, rugulae clearly visible:

Sedum type

[includes *Sedum* and *Sempervivum*. Some distinction may be made within this type by reference to type slides. The smallest grains in this type belong to *Sedum annuum* and *S. rosea*]

4b Grain <21 µm. Rugulae very fine — only faintly visible under phase contrast: *Crassula*

5a Apocolpia flattened; rugulate pattern may disintegrate into small reticulations. Each elliptic or 8-shaped endoporus without horns:

Hippocrepis comosa

[consult a type slide]

5b Apocolpia flattened or rounded; rugulate pattern is found all over the grain. Each elliptic or 8-shaped endoporus commonly with 'horns' into the mesocolpium (best seen in phase contrast):

Coronilla varia

[consult a type slide]

6a With clear transverse endocolpi or marked equatorially elongated endopori (these endoapertures may be difficult to detect without phase contrast, ×1000). Grain shapes rhombic-obtuse to elliptic: 7

[NB Some *Rosa* species may occasionally present rather clear transverse endocolpi and follow this branch of the key. Their identification is rather difficult — see discussion on p. 157]

6b With circular or meridionally elongate endopori, or endopori unclear, or pori represented by mere constrictions or ruptures to the colpi. Grain shapes rhombic-obtuse, elliptic or rectangular-obtuse: 13

7a Mesocolpium flattened (grain triangular-obtuse in polar view). Grain may or may not be longer than broad: 8

7b Mesocolpium convex (grain circular or lobed in polar view). Grain longer than broad, P/E ratio >1.2:   10

8a Colpi widening appreciably towards the equator, edges may or may not be sunken or inrolled. Grain surface strongly striate with short, parallel muri in the mesocolpia with rows of perforations between the muri; sculpture gradually becoming euriticulate at the poles. Grain as long as broad, P/E ratio 1.2 or less:

Atropa bella-donna

8b Colpi narrow, not widening towards the equator, edges slightly inrolled. Surface striate or

rugulate, may or may not be perforate. Grain usually longer than broad, P/E ratio >1.2:   9

9a Grain small, <25 µm, rugulate in apocolpia and rugulate to microreticulate in the mesocolpia; apparently without perforations between the muri. Endoaperture with clear margins and ends, rectangular obtuse or 8-shaped. Portions of sexine clearly project over the endoaperture at the equator:

Androsace maxima

9b Grain larger, >30 µm. Strongly striate-perforate in the mesocolpia with quite long parallel muri travelling meridionally or in swirling patterns. The perforations occur in single rows between the muri; this 'ladder' effect gradually becoming tectate-perforate in the small apocolpia. Endoaperture with diffuse or flaring ends (best seen under phase contrast):

*Rhus typhina**

10a Grain large, >35 µm, surface striate or rugulate-striate: 11

10b Grain smaller <30 µm, surface rugulate or rugulate-striae: 12

11a With fine, predominantly transverse (equatorial) striae. Portions of exine clearly projecting over transverse endocolpus at the equator. Exine thickest and columellae longest and coarsest at the poles. Grain never slightly equatorially constricted, size larger or smaller than 45 µm: *Agrimonia eupatoria*

11b Sculpture more meridionally striate or irregularly rugulate-striate. No projecting exine at the equator, grain slightly equatorially constricted, at least in the inner contour. Colpi very long, ends not visible in equatorial view. Apocolpia subacute; grain usually >45 µm:

Astrantia major

12a Each porus represented by a transverse endocolpus which is neatly elliptic-acute. Ectocolpi longer, >⅔ of distance from pole to pole:

Apium inundatum type

[includes *A. inundatum*, *A. nodiflorum* and *A. graveolens*. See Punt, 1984]

12b Each porus represented by a transverse endocolpus with obtuse or diffuse ends, never acute. Ectocolpi rather shorter (approximately ⅔ of distance from pole to pole) not at all sunken or inrolled. Columellae may be slightly longer at poles: *Anthriscus caucalis*

13a With elongate zones of thin, granulate exine, one down the centre of each mesocolpium. These are bordered on either side by bands of thicker striate exine. Each porus neatly circular or meridionally elongated, strongly ringed:

Lythrum salicaria type

[includes *L. salicaria* and *L. hyssopifolia*. Grain

size may vary widely in the first species in association with tristyly in the flowers. Three size classes exist, but there is some overlap. Any one plant of *L. salicaria* has only two of the three size classes]

13b Without such elongate, granulate zones in the mesocolpia. Pori various: 14

14a Striae long and coarse, more or less straight and unbranched, running meridionally and stretching from pole to pole in the mesocolpia. Grains tectate, not perforate: 15

14b Striae (or rugulae) shorter, more numerous and finer, running meridionally or transversely across the mesocolpia, or in curves and swirls, or irregularly. Muri may or may not have perforations between them: 16

15a Striae simplicolumellate, unbranched; with straight muri running past the margins and ends of the colpi, and smoothly over the poles. Colpus with a very prominent, wide, bridge to it at the equator: *Geum*
[it may be possible to distinguish *Geum rivale* by its larger size, but there is some degree of overlap with *G. urbanum* on type slides]

15b Striae duplicolumellate to pluricolumellate with a carpet of fine columellae. Muri occasionally tapering-off, anastomosing or interweaving, but most commonly so in the apocolpia where they may be very disorganised and approach a rugulate pattern with straight, randomly crossing muri. Muri in the mesocolpia run up to, and meet, the edges of the end thirds of each colpus: *Menyanthes trifoliata*

16a Muri simplicolumellate, surface striate-perforate with perforations often neatly in rows between the muri of the striations giving a 'ladder' effect (observe ×1000, phase contrast helps): 17

16b Muri simplicolumellate, duplicolumellate or pluricolumellate. Grain surface tectate and rugulate-striate with either no perforations or small perforations but no clear 'ladder' effect: 22
[see however *Rubus*, *Malus*, *Sorbus*, p. 157]

17a Grain elliptic and slightly pointed at the poles (subacute apocolpia) apocolpia small due to long colpi. Striations always meridional. Perforations narrower than muri: 18

17b Grains circular to elliptic with more rounded poles. Striations may or may not be meridional, perforations may or may not be narrower than muri: 19

18a Pori usually clearly circular with no acute extensions, smaller in relation to grain size. Colpi very narrow, grain commonly >40 μm in size: *Helianthemum*
[some grains have insignificant pori. Small

grains may look very much like *Gentiana pneumonanthe* and type slides should always be consulted for these]

18b Pori more or less circular; each with two small, acute, lateral equatorial extensions into the colpus margin which reach just to the point where the margin starts to inroll. Pori larger relative to grain size, colpi slightly wider and rather more sunken or inrolled. Grain <40 μm in size: *Gentiana pneumonanthe* type
[includes *G. pneumonanthe*, *G. cruciata*, *G. lutea*, *G. purpurea* and possibly others, see Punt and Nienhuis, 1976]

19a Grain >50 μm in size, coarsely striate with perforations as wide as the muri. Colpi narrow: *Tuberaria guttata*

19b Grain <50 μm in size, perforations wider or narrower than muri, colpi narrow or wide (gaping): 20

20a With straight, parallel striae which are rather variable in direction within and between grains. Striae most commonly running in sets which abruptly change direction, thus the individual striae often meet at sharp angles (V-shapes). However, predominantly meridional and transverse striae are also seen. Small perforations between the muri. Each colpus rather wide (gaping) and granular; with a more or less clear H-shaped endoporus, (detectable in polar view, optical section by a marked thinning of the exine towards the apertures). Also sometimes with a granular frill and costae around the central, circular part of the endoporus: *Centaurium*
[includes *Centaurium* and *Cicendia filiformis*. According to Punt and Nienhuis, 1976; the latter may be distinguished by its smaller size and slightly different sculpturing; cf. also *Helianthemum*]

20b Striae running mainly meridionally, at least in the mesocolpium (may or may not be more irregular at the poles). Perforations larger, at least equal to the width of the muri. No H-shaped endoaperture, having an endoporus which is circular or an equatorially elongated ellipse: 21

21a Colpi narrow with indistinct margins. Endoaperture a large, elliptic, equatorially elongated endoporus, which is covered by distinctive, thin flaps of sexine which protrude from the grain outline: *Lomatogonium rotatum*
[consult a type slide]

21b Colpi wider, slightly granulate. Endoaperture may or may not be circular, often rather obscured, but with no projecting sexine portions covering it: *Swertia perennis*

[*Gentiana pneumonanthe* type may key out here if the apocolpia are not detectably subacute]

22a With large (5 µm or more in diameter) neatly circular or meridionally elongate endopori, each surrounded by a clear costa. Fine transverse (equatorial) striae in the mesocolpia. Colpi acute-ended but widening towards the equator, each colpus membrane covered by prominent echinae, bacula or verrucae: *Aesculus*

22b Without neatly circular pori, but with a mere bridge, constriction or rupture to each colpus. Colpi wide or narrow, but never with such echinae, bacula or verrucae: 23

23a With long, coarse, sharply-cut, parallel-sided muri which are duplicolumellate or pluricolumellate. Columellae fine and densely spaced in an even carpet. Muri show some anastomosing and interweaving, especially in the polar areas. Muri run up to, and meet, the edges of the end thirds of each colpus: *Menyanthes trifoliata*
[*Prunus spinosa* may approach this sculpturing type, but is distinct because of its coarser columellae and the small undulations of the colpus margins seen in equatorial view, one colpus facing]

23b Without muri so long and coarse, usually being much shorter and finer and simplicolumellate or duplicolumellate: 24

24a Exine clearly thickest in a zone down the centre of the mesocolpium, thinner at the poles and towards the colpi. Shape in polar view markedly triangular-obtuse. Thick tectum with extremely faint meridional striae (×1000, phase contrast):   *Alchemilla* type
[includes *Alchemilla* and *Aphanes*. The striae are very difficult to detect, so it is most likely that *Alchemilla* type will key out in a psilate sculpturing section]

24b With exine approximately the same thickness all over the grain, striae clear or faint: 25

25a With finely granulate membranes to each colpus, a bridge or constriction that is not ruptured and no costae to the colpi. Very fine rugulate-striate sculpturing (only clearly visible in phase contrast). Columellae fine but clearly visible ×1000. Colpus edge does not appear to have small undulations where it starts to inroll. Mesocolpia always convex: *Lobelia*
[examination of type slides is really necessary to separate, e.g. *L. dortmanna* and *L. urens* from some members of the Rosaceae like *Rubus saxatilis* and *Sorbus aucuparia*. Descriptions do not really suffice although the difference is clear to the eye — see Rosaceae notes on p. 157]

25b Membranes to colpi not usually granulate (except *Rubus arcticus*) and usually invisible due to inrolling of colpus margins. Costae may or may not be present on colpi. Each colpus edge may show small undulations where it starts to inroll (see equatorial view with one colpus facing). Sculpture varies from clear, well-defined striae or rugulae to very fine, faint markings. Mesocolpium flattened or convex: non-operculate Rosaceae
[see p. 157 for some additional notes on distinction within this group which includes *Rubus, Rosa, Crataegus, Dryas, Malus, Pyrus, Mespilus, Sorbus, Amelanchier**]

SPECIAL KEY TO OPERCULATE ROSACEAE — FROM EXIT 1a

For use only with a wide range of type slides. Consultation of the works listed below under the notes on the non-operculate Rosaceae will also be useful. Some information from these works has been used in the key

1a With short narrow colpi, grain circular in both polar and equatorial view. Margins of colpi protrude from the grain outline at the equator. Sculpture consisting of fine striae in the apocolpia and with or without coarser, undulating rugulae at the equator of the mesocolpia: *Sanguisorba minor* ssp. *minor*

1b With longer colpi, grain elliptic or rhombic-obtuse in equatorial view. Margins of colpi may or may not protrude at the equator, sculpture of the same type in apocolpia and mesocolpia: 2

2a Grain >35 µm long, markedly longer than broad with very fine, equatorially running striae in the mesocolpia. Columellae coarser at the poles. Operculum very narrow, transverse endocolpus sometimes visible underneath the equatorial bridge and operculum: *Agrimonia eupatoria*

2b Grain <35 µm long, only slightly longer than broad, striae running equatorially or meridionally or in various directions (more like rugulae):   3

3a With long, strong, straight, parallel, meridional striae. Operculum long and prominent, covering all the colpus membrane and often protruding from the colpus: *Potentilla* type
[includes *Potentilla, Sibbaldia* and *Fragaria. Fragaria vesca* may possibly be distinct by its larger, flattened apocolpia and more hexangular shape in polar view. *P. fruticosa* seems to be the smallest in length and may be distinct because of this.]

3b With striae shorter and finer, more variable in distribution and length. Operculum sometimes shorter and narrower, less often protruding. Colpus margins frequently with small undulations just at the point they start to inroll (see equatorial view with one colpus facing). Opercula mostly narrow and short *Rosa*
[includes grains of e.g. *R. rubiginosa*, *R. canina*, *R. pimpinellifolia*, *R. gallica* and possibly others. Grains of some *Rosa* species are more commonly without any opercula e.g. *R. canina*. See the additional notes on non-operculate Rosaceae]

NON-OPERCULATE ROSACEAE – EXTRA NOTES ON DISTINCTIVE FEATURES OF SOME GENERA AND SPECIES FROM KEY EXIT 25b

It is possible to differentiate some types within the the grains included in this group but written descriptions or study of photographs will not be sufficient to permit accurate identification. The features sometimes used show considerable variation between individual grains on type slides (e.g. the stellate endoaperture in *Malus*, surface sculpture in *Prunus padus*) and between different collections of the same species (e.g. the sculpture and size in *Dryas octopetala*). Thus what follows is not a key but a list of certain distinctive features which, if seen, should prompt a worker to compare the grain with a good selection of type slides and permit tentative identification to a species or group of species. This should not be attempted if there is no access to a comprehensive type collection. Useful works to consult are: Reitsma (1966), Eide (1981), Boyd and Dickson (1987a), Hebda *et al.*, (1988), Faegri and Iversen (1989). Information from some of these works has been used in the account which follows; however the exact size measurements of most species are not quoted as our own investigations indicate that size is a very unreliable character in this group. Possibly this is due to the frequency of hybrids and polyploid races in some of the species mentioned below.

Note that the groups given below may not be mutually exclusive.

1 Grains with large, stellate (star-shaped) endoapertures (especially clear in phase contrast) which are sometimes seen as large irregular ruptures to the colpi in the equatorial region. Sculpturing very finely rugulate-striate: equatorially, meridionally or more irregularly. Grain often rather narrowly elliptic or rhombic-obtuse:
Malus sylvestris, *Pyrus pyraster*, *Sorbus aucuparia*, *S.*

rupicola, *S. torminalis* and possibly other *Sorbus* subspecies. *Pyrus pyraster* and *S. aucuparia* are the smallest grains in this group.

2 Grains large, usually over 25 μm and some over 40 μm; only slightly longer than broad and with either extremely fine, faint, rugulate-striate sculpturing, or rather coarser sculpturing featuring shorter, wider, curved and branched striae or rugulae which appear to be amalgamations of groups of finer striae. Sometimes with stellate endoporus or rupture:
Crataegus monogyna, *C. laevigata*, *Sorbus aria*, some *Sorbus* polyploids
[*Sorbus aria* shows only the finer sculpture. Some polyploid *Sorbus* species have grains at the upper end of the size range, e.g. *S. intermedia*. Whilst all our British *Crataegus* type material shows such sculpturing variations, it is noted that Eide, 1981, shows photographs of *C. monogyna* and *C. laevigata* which have only the coarser sculpturing which approaches the pattern on the surface of 'typical' *Dryas octopetala* – see group 3 below. The European *Cotoneaster* species may have sculpture which puts them in group 2]

3 Grains with distinctive (especially in phase contrast) **curved and branched striae** or **knotted rugulae** (muri varying in width considerably – thickest at junctions). The thinner parts of the muri sometimes appearing 'beaded' at high focus (i.e. simplicolumellate). Grains small, circular or rhombic-obtuse, with only very slight constriction to the colpus but no bridge (never ruptured):
Dryas octopetala
[some type slides show this 'typical' morphology whilst others show much larger grains without a trace of the characteristic sculpturing, having very fine parallel striae travelling in wide swirls and showing much less anastomosing. These latter are difficult to distinguish from e.g. some *Rosa*, *Sorbus* or *Crataegus* species. Reitsma, 1966, noted such size variations in *Dryas* and postulated that the larger grains may belong to polyploid races; quoting the findings of Bocher and Larsen, 1955, that diploid and tetraploid plants have been found side by side in the Alps]

4 Grain with very coarse, conspicuously curved or S-shaped, wide striae; which are often short and abruptly tapering-off to their ends: *Prunus padus*
[there seems to be rather a lot of variation between grains on type slides as to the length of the muri; some grains having muri so short as to be almost verrucate, while other grains have longer muri approaching other *Prunus* species]

5 Striae fairly coarse, being clear and running mostly parallel and meridionally in the mesocolpia, also

showing some degree of branching and anastomosing. Muri appear sharply cut in optical cross section of the exine and are mostly duplicolumellate and sometimes simplicolumellate. Undulations along colpus edge quite pronounced:

some *Prunus* species e.g. *P. avium* and *P. spinosa* [other species such as *P. cerasifera*, *P. laurocerasus*, *P. lusitanica*, *P. domestica*, are less coarsely rugulate-striate and commonly simplicolumellate; so are thus difficult to distinguish from e.g. *Rosa* or *Sorbus* species. *P. spinosa* may be distinct because of its tectum, which is much thicker than in *P. avium*. *P. lusitanica* has circular grains with much shorter, broader colpi with more granulate membranes than the other species]

6 Colpi with very irregular margins and markedly **granulate membranes**. They are also broad, obtuse-ended, and not inrolled. Relatively large apocolpia and an insignificant constriction to each colpus at the equator (never any sign of endoporus). Surface very finely rugulate-striate:

Rubus arcticus

[other *Rubus* species e.g. *R. fruticosus*, have longer, more inrolled colpi and no such markedly granular colpus membranes. They are rather similar to *Sorbus* and/or *Rosa*. Slides of *R. fruticosus* agg. show a large proportion of pantocolporate or rather irregular grains. *Prunus lusitanica* has colpi that conform to the above description, but it is a much larger grain, with stronger, striate sculpturing]

7 With fine, small, **irregular rugulae**; having very short muri showing much interweaving or interlocking. Grain usually >31 μm long:

Mespilus germanica

8 Grain surface covered in regular, small perforations between very fine rugulate-striate sculpture (×1000, phase contrast):

most *Rubus* species e.g. *R. arcticus*, *R. idaeus*, *R. saxatilis*; *Malus sylvestris*; some *Rosa* species; some of the *Sorbus* polyploids; such perforations may even be detectable on *Dryas octopetala*.

Trizonocolporate, baculate or pilate

1a Grains rhombic-obtuse to elliptic in equatorial view, with a transverse endocolpus crossing each colpus at the equator. Sexine composed of pila with capita (heads) joined laterally to form a faint reticulum which may or may not be detectable with the light microscope (phase contrast helps, but see SEM): *Mercurialis*

1b Grains circular to elliptic in equatorial view with no transverse endocolpi. Sexine composed of sparse, large, long bacula between which are many densely packed, slim, short columellae (appearing to be bacula under LM) the heads of which are more or less fused sideways to form a thin tectum layer (see SEM): *Viscum*

TETRAZONOCOLPORATE (Plates 46, 57, 58)

1a Grain echinate or microechinate, with very short, narrow colpi and large apocolpia. Each porus actually a very large, **wide** transverse endocolpus with diffuse ends: *Lonicera* [includes most grains of *L. coerulea* and occasional grains of other *Lonicera* species or *Linnaea borealis*. See p. 152 for further identification]

1b Grain not echinate; colpi longer, apocolpia therefore not so large. Pori or endocolpi various shapes, but not very large and wide: 2

2a Grains with clear endoapertures which are usually equatorially elongated into transverse endocolpi or a complete equatorial girdle. Usually prominent costae to the colpi. Grains rectangular-elliptic, may or may not be constricted at the equator: 3

2b Grains either with clear circular endopori or porus represented by a mere constriction or rupture to the colpus, but never an equatorially elongated endocolpus. Grain never equatorially constricted: 9

3a Grain heteropolar, endoapertures situated along the colpi closer to one pole than to the other: 4

3b Grain not heteropolar, all endoapertures situated at the equator. Colpi short, narrow, margins may or may not be irregular: 5

4a Grain <12 μm long, with long colpi and endopori very close to one pole. Shape narrowly rectangular, constricted between the equator and the pori; slightly concave at the poles:

Lithospermum officinale

[see Clarke, 1977]

4b Grain >14 μm long, endopori less close to one pole. Grain only slightly constricted between the equator and the pori, convex at the poles:

Lithospermum purpurocaeruleum

[see Clarke, 1977]

5a Grain <22 μm long. Elliptic endopori present, which may or may not be fused to an equatorial girdle: 6

5b Grain >22 μm long. Transversely elongate endocolpi present, which may or may not be fused to a girdle. Ectocolpi short and narrow, with very irregular margins: 7

6a Grain markedly equatorially constricted and

therefore bone-shaped; endopori distinct, separate and slightly elliptic. Colpi short, margins granular, costae very prominent:
Lithospermum arvense type
[rare grains with low aperture number. Includes *L. arvense* and *Lithodora diffusa*. See Clarke, 1977]

6b Grain only slightly constricted at the equator. Endoapertures linked to form a complete equatorial girdle. Sculpture faintly perforate to microreticulate with lumina largest in the equatorial region. Colpi longer, grain up to 15 μm:
Pentaglottis sempervirens
[see Clarke, 1977]

7a Grain with a clear 'waist' caused by an equatorial constriction. Endocolpi fused to form a complete girdle. Surface tectate-perforate, perforations enlarging to reticulations in the 'waist' region over the endocolpi and near to the margins of the ectocolpi. Sexine thinner at poles than in mesocolpia. Costae present but not very thick. Grain <34 μm: *Nonea versicolor*
[see Clarke, 1977]

7b Grains not markedly 'waisted', being rectangular-obtuse in shape. Endocolpi may or may not be fused to a girdle, surface with or without perforations or reticulations but always with very large costae to the colpi. 8

8a Grain size >44 μm. Surface psilate or faintly granular. Very slight indications of perforations, reticulations or irregular sculpturing in a broad equatorial band over the mesocolpia and endocolpi. Ectocolpi short, often with very irregular margins and forked or branched ends:
Anchusa arvensis
[rare grains of this species. Most grains are trizonocolporate. See Clarke, 1977]

8b Grain size <46 μm. Surface reticulate-perforate along colpus margins and in an equatorial girdle across the mesocolpia and over the endocolpi:
Pulmonaria obscura type
[according to Clarke, 1977, this includes *P. obscura*, *P. angustifolia*, *P. longifolia*, *P. montana*, *P. officinalis*, and *Anchusa officinalis*]

9a Each colpus with equatorial bridge, constriction or rupture. Grain elliptic to rhombic-obtuse in equatorial view, apocolpia flat or concave. Surface psilate-scabrate, sometimes coarsely rugulate-verrucate in the apocolpia: 10

9b Colpi with circular to elliptic endopori. Grains always circular in equatorial view, apocolpia never flat or concave. Colpi narrow, crack-like, often meeting in pairs (strictly this means the grains are pantocolpate). Surface psilate-scabrate, tectate-perforate or eureticulate: 11

10a Grains quadrangular in polar view, with exine projecting in the porus region at the equator. Apocolpium large, commonly flat or concave and coarsely rugulate-verrucate. Colpi with obtuse, irregular ends. Grain large >48 μm in length: Viola arvensis type
[includes *V. arvensis*, *V. tricolor*, *V. lutea*, *V. culcurata* and possibly others in section *Melanium* not examined at the time of writing]

10b Grains more circular in polar view; apocolpia smaller, never rugulate-verrucate. Colpi not so obtuse ended, grains smaller, <50 μm long:
Viola canina, V. riviniana
[occasional grains only, these species more commonly are trizonocolpate or trizonocolporate. It is noted that the commonly naturalized *Buddleja davidii** has tetrazonocolporate grains which may key out here]

11a Grains with relatively short colpi (full length visible in equatorial view) which are not sunken. Sculpture tectate-perforate to microreticulate. Endopori and costae to the colpi rather faint. Usually <28 μm in diameter: Rumex acetosa type
[according to data in van Leeuwen *et al.*, (1988); this type includes *R. acetosa*, *R. thyrsiflorus* and possibly *R. alpestris*]

11b Grains with longer colpi (full length may or may not be visible in equatorial view). Sculpture tectate-perforate, microreticulate or reticulate. Endopori and costae to the colpi clearly defined: 12

12a Grains >28 μm. Sculpture with or without winding muri, sometimes with sexine surface very irregular in optical section:
Rumex obtusifolius type
[according to data in van Leeuwen *et al.*, 1988; this type includes *R. obtusifolius*, *R. longifolius*, *R. palustris*, *R. aquaticus*, *R. hydrolapathum* and some grains of other species that are more commonly trizonocolporate. Apparently it may be possible to separate *R. aquaticus* and *R. hydrolapathum* on their large size of >40 μm]

12b Grains smaller, <28 μm. Sculpture sometimes with muri slightly winding and duplicolumellate or pluricolumellate in phase contrast:
Rumex acetosella type
[includes *R. acetosella* and *R. tenuifolius*. See ref. under 12a].

PENTAZONOCOLPORATE (Plates 57, 58)

1a Grains with two or more pori per colpus. Colpi broad, obtuse-ended, granulate and sunken.

Mesocolpia psilate, grain elliptic in shape:
Ribes uva-crispa
[see Verbeek-Reuvers, 1977b]

1b Grains with only one porus to each colpus, colpi wide or narrow, more acute ended: 2

2a Each colpus with an equatorial bridge, constriction or rupture. Colpi wide, with granulate membranes (if the colpi are deeply inrolled the membrane cannot be seen). Grain pentangular in polar view, with exine projecting in the porus region at the equator. Surface psilate-scabrate, sometimes coarsely rugulate-verrucate in the apocolpia. Apocolpia may be flat or concave. Grain large, >48 μm:
Viola arvensis type
[includes some grains of *V. arvensis*, *V. tricolor* and possibly others in section *Melanium*. See exit 10a in Tetrazonocolporate]

2b Colpi with elliptic endopori or transverse endocolpi or a complete equatorial girdle. Grain often with prominent costae to the colpi: 3

3a Grains markedly constricted at the equator (with a clear 'waist'): 4

3b Grains not markedly constricted at the equator, they may be slightly constricted or rectangular-obtuse: 5

4a Length >22 μm. Endocolpi fused to form a complete equatorial girdle. Surface tectate-perforate, perforations enlarging to reticulations in the 'waist' region and near to the margins of the colpi. Sexine thinner at the poles than in the mesocolpia:
Nonea versicolor
[see Clarke, 1977]

4b Length <22 μm. Endoapertures elliptic pori which are completely separate. Surface psilate, colpi short, with granular margins and prominent costae:
Lithospermum arvense type
[includes *L. arvense* and *Lithodora diffusa*. See Clarke, 1977]

5a Length >20 μm, can be up to 45 μm. Surface reticulate-perforate along colpus margins and in an equatorial girdle across the mesocolpia and over the endocolpi. Transverse endocolpi may or may not fuse to form a complete girdle:
Pulmonaria type
[includes *Pulmonaria* and *Anchusa officinalis*. See Clarke, 1977]

5b Length 12–15 μm. Surface faintly perforate to microreticulate in the equatorial region. Grain slightly constricted at the equator, endoapertures fused to form a complete girdle:
Pentaglottis sempervirens
[see Clarke, 1977]

HEXAZONOCOLPORATE (Plates 57, 58)

1a Grains with two or more pori per colpus. Colpi broad, obtuse-ended, granulate and sunken; mesocolpium psilate. Grain shape elliptic:
Ribes uva-crispa
[see Verbeek-Reuvers, 1977b]

1b Grains with only one porus per colpus, colpi narrower, acute ended: 2

2a Grain markedly constricted at the equator; costae prominent and each endoporus clear — a neat equatorially elongated ellipse. Colpi short, with granular margins: *Lithospermum arvense* type [includes *L. arvense* and *Lithodora diffusa*. See Clarke, 1977]

2b Grain not markedly constricted at the equator, being elliptic to rhombic in equatorial view. Each porus an elongated transverse endocolpus or represented only as a bridge or constriction of the ectocolpus: 3

3a Grain with a thick, smooth, non-perforate tectum (see optical section of grain wall). Each porus represented by an elongate transverse endocolpus: *Sanguisorba officinalis*
[this description is of the typical 2n = 56 race according to Erdtman and Nordborg, 1961. Apparently the diploid 2n = 28 race has smaller grains which tend to have the colpi arranged in three pairs — resembling a trizonocolporate grain with wide opercula to the colpi. Reitsma, 1966, regards *S. officinalis* as trizonocolporate with very wide opercula to the colpi]

3b Grain eureticulate to tectate-perforate, each ectocolpus showing a bridge or constriction at the equator: *Pinguicula*

POLYZONOCOLPORATE (Plate 58)

1a Grains eureticulate to tectate-perforate. No well defined endopori, each ectocolpus has a mere constriction or bridge at the equator: *Pinguicula*

1b Grains psilate to scabrate with clear endopori or an equatorial girdle of fused transverse endocolpi: 2

2a Apocolpia with large circular or irregular shaped depressions. Colpi wide, parallel-sided and obtuse-ended, with an equatorial girdle of fused transverse endocolpi: *Polygala*

2b Apocolpia without such depressions, colpi narrow and slit-like: 3

3a With 8 colpi and 8 neat elliptic endopori. Grain shape rectangular-obtuse in equatorial view:
Symphytum

3b With 10 or more colpi, some of which are syncolpate across the poles. Indistinct endopori or complete equatorial girdle. Grain shape rhombic-obtuse in equatorial view: *Utricularia*

HETEROCOLPATE (Plates 41, 58)

1a With 3 meridional porate colpi and 6 parallel non porate colpi. The non porate colpi are situated one on either side of a porate colpus and are actually crack-like or horn-like extensions of the equatorially elongated endoporus or endocolpus (i.e. part of a large H-endoaperture, best seen in phase contrast): 2

1b With 3 or 4 or 5 porate colpi, and 3 or 4 or 5 nonporate colpi. Non-porate colpi situated exactly between the porate colpi: 3

2a Grain apparently coarsely rugulate-verrucate in the apocolpia and with much finer sculpturing (psilate-scabrate) in the mesocolpia. The coarse sculpture in the apocolpia is actually due to irregular structural elements **under** the tectum; in this area the nexine appears very thin or possibly absent. Mesocolpia flattened, apocolpia flat or concave (grains thus triangular prism-shaped). Pori may or may not project at the equator: *Anthyllis vulneraria*
[this species is included here, but it is not truly heterocolpate as the non-porate colpi are nexinous features only, and are actually part of a complex H-endoaperture]

2b Grain psilate-scabrate or rugulate in both the apocolpia and in the mesocolpia. Porus represented by a transverse endocolpus covered by a fastigium. Apocolpia always convex, grain not triangular prism-shaped, being circular to elliptic in equatorial view. Pori projecting at the equator: *Verbena officinalis*
[this species is included here, but it also is not truly heterocolpate, for the same reason as *Anthyllis*. See Punt and Langewis, 1988]

3a Pori clearly defined, circular and uncovered (not endopori only). Grains never equatorially constricted, colpus membrane clearly granulate, some areas of the exine striate (×1000): 4

3b Pori always covered nexine features (i.e. endopori) which may or may not be circular, sometimes equatorially or meridionally elongated. Grains may be bone-shaped (equatorially constricted) Sculpture never striate: 5

4a Non-porate colpi actually wide areas of intectate, granulate exine. These areas are separated from the porate colpi by bands of thick, tectate-striate exine. Aperture arrangement often irregular and grain asymmetric: *Lythrum portula*

4b Non-porate colpi not occupying wide areas, being narrower and shorter than porate colpi.

Large areas of exine thus have meridional, thick, tectate-striate bands which also cover the apocolpium. Grain symmetrical, apertures clearly zonally arranged: *Lythrum salicaria* type
[includes *L. salicaria* and *L. hyssopifolia*. Grain size may vary widely in the first species in association with heterostyly in the flowers. Three size classes exist but there is some size overlap between the classes. Any one plant has only two of the size classes in its anthers]

5a Grains heteropolar, colpi with endopori offset from the equator; one porus closer to one pole, two pori closer to the other. Size <18 μm, shape clearly equatorially constricted: *Lappula deflexa* type
[according to Clarke, 1977, this includes *L. deflexa* and *L. squarrosa*]

5b Grains not heteropolar, all pori equatorial: 6

6a Grains <14 μm, endoaperture circular or an equatorially elongated ellipse or rectangle. Grain psilate, slightly equatorially constricted: *Myosotis arvensis* type
[according to Clarke, 1977, this includes most species of *Myosotis*, *Asperugo*, *Omphalodes* and *Cynoglossum*. Some separation of types may be made by reference to this work and to type slides]

6b Grains >14 μm: 7

7a Colpi long, narrow and sunken so that the mesocolpia are markedly convex. Endoporus difficult to see as it is meridionally elongated and largely hidden under sexine. Grain psilate. Shape in equatorial view not equatorially constricted but elliptic to rectangular-obtuse. Colpi often S-shaped, grain 33–40 μm: *Heliotropium europaeum*
[data from Clarke, 1977]

7b Endoaperture clear. Grain 34 μm or less, colpus margins granular: 8

8a Grain small, 14–19 μm, distinctly equatorially constricted and bone-shaped. All colpi of the same length: *Mertensia maritima*
[see Clarke, 1977]

8b Grain larger, >17 μm, not, or only slightly equatorially constricted. Colpi not all of the same length: 9

9a With 10 or 12 colpi, the porate ones short, wide at the equator and narrowing sharply to the poles. Non-porate colpi long, narrow, parallel-sided and tending to join at the poles. Endoporus equatorially elongated into an ellipse: *Myosotis discolor*
[data from Clarke, 1977. This is apparently the only European species of *Myosotis* with different morphology]

9b With 6 colpi. Porate colpi long, narrow and

parallel sided. Endoporus large and circular or meridionally elongated. Grain rectangular-elliptic with convex poles. Exine tectate-perforate, granulate: *Amsinkia*
[data from Clarke, 1977]

TETRAPANTOCOLPORATE, PENTAPANTOCOLPORATE AND HEXAPANTOCOLPORATE (Plate 57)

Note that pantocolporate grains may occasionally appear in groups which are normally trizonocolporate e.g. Rosaceae

1a Each ectocolpus crossed by a transverse endocolpus. Grain with 4–12 ectocolpi:
                       Polygonum aviculare type
[according to van Leeuwen *et al.*, 1988, pantocolporate grains occur quite frequently in some members of this type. The highest frequency of pantocolporate grains was observed in *Polygonum oxyspermum* ssp. *raii*]

1b Each ectocolpus with a more or less circular endoporus in the middle. 4 to 9 ectocolpi: 2

2a Each colpus bordered by 2 broad tectate-psilate bands. Colpus membranes and mesocolpia appear granulate: *Lythrum portula*

2b Colpi not bordered by such tectate-psilate bands. Mesocolpia tectate-perforate to eureticulate: 3

3a Ectocolpi rather short, not or only slightly sunken. Grains usually $<28\,\mu m$, with rather indistinct pori. Tectate-perforate to microreticulate. Consult a type slide: *Rumex acetosa* type
[according to van Leeuwen *et al.*, 1988, this includes *Rumex acetosa* and *Rumex thyrsiflorus* and possibly *Rumex alpestris*. Only a proportion of the grains in these species are pantocolporate]

3b Ectocolpi longer, often nearly or actually syncolpate. Pori distinct, sculpture tectate-perforate, microreticulate or eureticulate. Size smaller or larger than $28\,\mu m$:   4

4a Grains small, usually $<28\,\mu m$, sculpture sometimes in phase contrast with rather winding or sinuous duplicolumellate muri. Consult a type slide: *Rumex acetosella* type
[includes *R. acetosella* and *R. tenuifolius*. See ref under 3a]

4b Grains larger, $>28\,\mu m$, sculpture with or without winding muri, sometimes with sexine surface very irregular in optical section:
                       Rumex obtusifolius type
[includes *R. obtusifolius*, *R. longifolius*, *R. palustris*, *R. aquaticus*, *R. hydrolapathum* and some grains of other species that are more commonly trizonocolporate. See ref under 3a. According to these authors it may be possible to separate *R. aquaticus* and *R. hydrolapathum* on their large size of $>40\,\mu m$]

Glossary

Acuminate-obtuse: Shape of a grain in equatorial view where the poles are extended but obtuse. See Fig. 5.7.

Annulus: Border to a porus which is produced by either a thickening or thinning of the sexine. For an example of the latter case see Caryophyllaceae. See Fig. 5.6.

Apocolpium: The area at a pole of a zonocolpate grain delimited by the ends of the colpi. See Fig. 5.5.

Baculate: With bacula.

Baculum: (pl. **Bacula**): Pillar or rod-like element always longer than broad and higher than 1 μm. Also used as a synonym for columella by other authors, e.g. Erdtman *et al.*, (1961) See Fig. 5.8.

Bifurcated: Branched into two in the manner of a 'Y'.

Caput: (pl. **Capita**): The apical swollen part or 'head' of a pilum.

Clava (pl. **Clavae**): A projecting element which is higher than broad and tapering towards the base, i.e. club-shaped. Height greater than 1 μm.

Clavate: With clavae.

Colpate: With one or more colpi.

Colporate: With a colpus and porus combined in the same aperture, e.g. *Fagus*.

Colpus (pl. **Colpi**): An elliptic aperture resembling a groove or furrow and with a length/breadth ratio greater than 2, e.g. *Quercus*.

Colpus membrane: Thin, usually structureless layer of exine which covers a colpus in the living pollen grain and is the area through which water is lost or absorbed. This membrane is often lost during fossilization. See also Operculum.

Columellae: General term for small, rod-like elements, radially directed and forming the inner layer of the sexine. They are attached at their bases to the nexine and at their heads to the tectum (when present). On focusing down through the exine they appear as small black dots. See p. 63 and Fig. 5.9.

Columellate: With columellae.

Costa: (pl. **Costae**): A thickening of the nexine near an aperture. See Fig. 5.6; see also Annulus and Margo.

Cylindrical-conical: Projection parallel-sided for most of its length, with a conical apex.

Dimorphic: Existing in two forms.

Distal face: That part of a spore which faces away from the centre of the tetrad during meiosis. See Fig. 5.5 and Proximal face.

Distal pole: That pole of a zonoaperturate grain which faces away from the centre of the tetrad during meiosis.

Dizonocolpate: With two colpi arranged in an equatorial zone.

Duplicolumellate: With columellae in two rows under each murus.

Dyad: Two grains united. See Fig. 5.4.

Echinae: Any sharply pointed sculpturing elements. They may vary from cylindrical to cone-shaped. See Fig. 5.8a.

Echinate: With echinae.

Echinolophate: With a pattern of high echinate ridges separating large 'gaps' which are arranged in a geometrical pattern.

Ectoaperture: An aperture in the sexine.

Ectocolpus (pl. **Ectocolpi**): An elliptic ectoaperture with a length/breadth ratio higher than 2.

Ectoporus (pl. **Ectopori**): Circular or faintly elliptic ectoaperture with a length/breadth ratio smaller than 2.

Ektexine: Outer layer of exine which comprises tectum (if present), columellae and foot layer (nexine 1). It is equivalent to sexine plus nexine 1.

Endexine: Inner, sometimes laminated layer of the exine, synonymous with nexine 2 and underlying nexine 1 (foot layer).

Endoaperture: An aperture in the nexine.

Endocolpus (pl. **Endocolpi**): A colpus, i.e. an elliptic endoaperture with a length/breadth ratio which is greater than 2. The endocolpus may be covered by a layer of sexine. Endocolpi often cross ectocolpi at 90°, and adjacent endocolpi may fuse to form a girdle in zonoaperturate grains, e.g. *Centaurea cyanus*, Plate 45. See Fig. 5.6.

Endocracks: Elongated fissures or gaps in the nexine, sometimes branched.

Endoporus (pl. **Endopori**): Circular or faintly elliptic endoaperture with a length/breadth ratio smaller than 2.

Equatorial axis: A straight line equivalent in length to the diameter of the grain at its equator.

Equatorial bridge: An interruption to a colpus at the equator, formed by two portions of the adjacent mesocolpia meeting over it.

Equatorial girdle: A complete circular endoaperture around and under the equator of a zonocolporate grain.

Equatorial view: View of a zonoaperturate grain where the equatorial plane is directed towards the observer. See Fig. 5.5.

Eureticulate: With partially dissolved tectum, i.e. heads of columellae joined in only one or two directions to form the muri of the reticulum. Distribution of columellae often corresponds to that of the muri although there may be free columellae in the lumina, e.g. *Salix, Ligustrum*. See Fig. 5.8a and Suprareticulate.

Fastigium: A small chamber formed by the sexine and nexine splitting apart from one another in the vicinity of each porus in a zonocolporate grain.

Fastigiate: With a fastigium to each porus.

Fossulate: With perforations of the tectum that are elongated sideways to form straight or sinuous channels.

Foveolae: Large perforations in a tectum (i.e. >1 μm in diameter) Perforations always narrower than the area between the perforations. See Fig. 5.8a.

Foveolate: With foveolae.

Gemma (pl. **Gemmae**): Sculpturing element with width approximately the same as height and constricted at its base. Height always greater than 1 μm.

Gemmate: With gemmae.

Granules: General term for very small (<1 μm high structures on a grain surface that are not assignable to gemmae, bacula, echinae or clavae, e.g. *Populus*.

H-endoaperture: An endoaperture in a zonocolporate grain where the 'porus' is seen to occupy a position similar to the central horizontal bar on a letter H. Endocracks or nexine channels extend out from this bar, running up and down parallel to the colpus in a manner resembling the 'legs' (vertical parts) of the letter H.

Heterocolpate: With some apertures colpi and some apertures colpi + pori.

Heteropolar: With apocolpia differing in size i.e. one smaller, one larger.

Hexangular: Grain the shape of a six-sided polygon.

Hexapantocolpate: With six colpi scattered all over the surface of the grain.

Hexapantocolporate: With six colpi scattered all over the surface of the grain, each colpus with a porus in its centre.

Hexapantoporate: With six pori scattered all over the surface of the grain.

Hexazonocolpate: With six colpi arranged in an equatorial zone.

Hexazonocolporate: With six colpi arranged in an equatorial zone, each colpus having a porus in its centre.

Hexazonoporate: With six pori arranged in an equatorial zone.

Inaperturate: Without apertures (pori or colpi).

Infrareticulate: With a reticulate pattern produced by the distribution of the columellae underneath a complete or partially dissolved tectum, e.g. *Papaver, Glaucium*.

Infratectal bacula: Small, slim rods which traverse the thickness of the tectum and sometimes also the bases of the echinae which may be on the tectum.

Inordinate: An arrangement of structures where there is no discernible pattern, more or less irregularly distributed.

Intectate: Sculpturing type in which the tectum is absent, i.e. the heads of the structural rods are free. Intectate grains can be baculate, echinate, clavate, gemmate, verrucate or granulate. See Fig. 5.8b and p. 73.

Intine: The cellulosic innermost layer of the pollen grain wall which underlies the exine. It is the cell wall of the living pollen grain.

Lacuna (pl. **Lacunae**): Large gap in the sexine of echinolophate grains. The lacunae are separated by high sexinous ridges arranged in a fixed pattern. See page 92.

Laesura: The single slit or furrow which forms the aperture in monolete pteridophytes.

Lamellae: Tangential layers of exine material which are usually seen in the nexine.

LO analysis: Analysis of exine structure by careful focusing down through the structures in a surface view under ×1000. 'LO' is the term given by Erdtman (1956) to the sequence of diffraction images which produces the pattern: bright islands and dark channels at high focus followed by dark islands and light channels at low focus. If the reverse sequence occurs, i.e. dark islands and light channels followed by light islands and dark channels, it was given the term 'OL'. See Fig. 5.9.

Lumen (pl. **Lumina**): A gap or space between the walls of a reticulate, striate or rugulate sculpture.

Margo: Zone around a colpus formed by a sudden thinning or thickening of the sexine or by any other sexine structure different from the remaining sexine. See Fig. 5.6.

Meridional: Term applied to grain features (e.g. colpi) which run along lines joining the distal pole to the proximal pole over the grain surface.

Mesocolpium (pl. **Mesocolpia**): The area of grain surface between two adjacent colpi. In a zonocolpate grain it is usually delimited by transverse lines drawn through the ends of the colpi. See Fig. 5.5.

Mesoporium (pl. **Mesoporia**): The area of grain surface between two adjacent pori. In a zonoporate grain it is delimited by transverse lines drawn through the polar edges of the pori. See Fig. 5.5.

Microechinate: With minute echinae, less than 1 μm in height.

Microgemmate: With minute gemmae, less than 1 μm in height. See also Granules.

Microreticulate: With lumina of reticulum less than or equal to 1 μm in diameter.

Microrugulate: With minute rugulae, length less than or equal to 1 μm.

Microverrucate: With minute verrucate, less than 1 μm in height.

Monocolpate: With one colpus.

Monolete: Term given to pteridophyte spores with one aperture that resembles a groove or furrow.

Monoporate: With one porus only.

Murus (pl. Muri): A ridge or wall separating two lumina of reticulate, striate or rugulate sculpture.

Nexine: Inner unsculptured part of the exine which appears as a solid or occasionally laminated layer. See Fig. 5.2 and page 63.

Octangular: Grain the shape of an eight-sided polygon.

Operculate: With an operculum on some or all of the apertures.

Operculum: Thick membrane covering a porus or colpus. Generally there is a very thin area just inside the immediate margin of the aperture, so that the operculum is easily lost.

Optical section: View where focal plane is half-way through grain, i.e. thickness and structure of grain wall is visible.

Papillae: Hollow finger-like projections which are always longer than broad.

Pentapantocolpate: With five colpi scattered all over the surface of the grain.

Pentapantocolporate: With five colpi scattered all over the surface of the grain, each colpus having a porus in its centre.

Pentapantoporate: With five pori scattered all over the surface of the grain.

Pentazonocolpate: With five colpi arranged in an equatorial zone.

Pentazonocolporate: With five colpi arranged in an equatorial zone, each colpus having a porus in its centre.

Pentazonoporate: With five pori arranged in an equatorial zone.

P/E ratio: The number produced by dividing the length of the polar axis of the grain by the length of the equatorial axis.

Perforate: With tectum pierced by small holes. Diameter of holes less than 1 μm. See also Foveolate.

Perine: The outermost part of the exine in bryophytes and pteridophytes which may be more or less detached from the inner part.

Perisporium: See Perine.

Pilate: With pila.

Pilum (pl. Pila): Element consisting of a rod-like part (columella) and an apical swollen part (caput). See also Clava.

Pluricolumellate: With columellae in several rows beneath each murus.

Polar axis: A straight line connecting the distal and proximal pole of a pollen grain.

Polar view: View of a zonoaperturate grain where the polar axis is directed straight towards the observer. See Fig. 5.5.

Polyad: More than four grains united in a group. See Fig. 5.4.

Polypantocolpate: With more than six colpi scattered all over the surface of the grain.

Polypantoporate: With more than six pori scattered all over the surface of the grain.

Polyplicate: With more than six plicae (ridges) arranged in an equatorial zone. In practice this condition is almost indistinguishable from polyzonocolpate, and for this reason we have included *Ephedra* in the latter class.

Polyzonocolpate: With more than six colpi arranged in an equatorial zone.

Polyzonocolporate: With more than six colpi arranged in an equatorial zone, each colpus having a porus in its centre.

Porate: With one or more pori.

Porus (pl. Pori): A circular or slightly elliptic aperture with a length/breadth ratio smaller than 2.

Porus membrane: Thin, usually structureless layer of exine which covers a porus in the living pollen grain and is the area through which water is lost or absorbed. This membrane may be lost during fossilization. See also Operculum.

Proximal face: That part of a spore which faces towards the centre of the tetrad during meiosis. See Fig. 5.5. In trilete spores the three-slit aperture is on the proximal face.

Proximal pole: That pole of a zonoaperturate grain which faces towards the centre of the tetrad during meiosis.

Psilate: With completely smooth, sculptureless surface.

Rectangular-obtuse: Shape of a grain in equatorial view where the grain resembles a rectangle with the corners rounded. See Fig. 5.7.

Reticulate: With a reticulum.

Reticulum: A network or mesh-like pattern.

Rhombic-obtuse: Shape of a grain in equatorial view where the grain resembles a rhomboid with the corners rounded. See Fig. 5.7.

Rugulate: With sculpturing elements elongated sideways, length greater than twice breadth (i.e. muri). Elements very irregularly distributed. See Fig. 5.8a.

Saccate: With two sacci.

Sacci (sing. Saccus): Large hollow projections from the

main body of the grain or spore. In some conifers the sacci are thought to be formed by sexine and nexine being pushed apart at the lower end of the columella-like structures, and the sexine then being very much expanded and stretched e.g. *Pinus*.

Scabrae: Very small (<1 μm high), isodiametric sculpturing elements, (e.g. *Cornus sanguinea*. See Fig. 5.8a.

Scabrate: With scabrae.

Semitectate: Sculpturing type in which the tectum is partially absent, i.e. the perforations in the tectum are wider than or equal to the areas between the holes. Semitectate grains can be striate, reticulate or rugulate. See Fig. 5.8a and p. 73.

Sexine: Outer sculptured part of the exine which in Angiosperms takes the form of a set of radially directed rods (columellae) supporting a roof (tectum). The tectum may be complete, partially dissolved or completely lacking. See Fig. 5.2 and p. 63.

Sexine 1: The innermost layer of the sexine, composed of columellae.

Sexine 2: Any complete or partially complete tectum (e.g. muri of a reticulum or of striae).

Sexine 3: Sculptural elements on top of the tectum e.g. echinae, granules, suprareticulum.

Simplicolumellate: With columellae in one row under each murus.

Striate: With a pattern consisting of approximately parallel muri and lumina. See Fig. 5.8a.

Suprareticulate: With a reticulum on top of the tectum. Pattern is thus independent of the distribution of the columellae, e.g. *Galeopsis*. Columellae are not always visible, e.g. in some members of the Leguminosae. See Fig. 5.8a.

Syncolpate: With two or more colpi fused at their ends. See p. 69.

Tectate: With a complete roof layer joining the heads of the columellae. See Fig. 5.8a.

Tectum: 'Roof' layer which joins the heads of the columellae and thus forms the outer layer of the sexine. It may be more or less dissolved. See Semitectate and Fig. 5.8a.

Tetrad: Four grains united in a group. See Fig. 5.4.

Tetrahedral: Arrangement of four grains such that their centres are on the four apices of a regular tetrahedron.

Tetrapantocolpate: With four colpi scattered all over the surface of the grain.

Tetrazonocolpate: With four colpi arranged in an equatorial zone.

Tetrazonocolporate: With four colpi arranged in an equatorial zone, each colpus having 1 porus in its centre.

Tetrazonoporate: With four pori arranged in an equatorial zone.

Trilete: A three-slit aperture, appearing like a Y. This aperture is on the proximal face of a spore.

Trimorphism: Existing in three forms.

Triradiate: Three lines or ridges running out straight from a central join, along radii of a circle.

Trizonocolpate: With three colpi arranged in an equatorial zone.

Trizonocolporate: With three colpi arranged in an equatorial zone, each colpus having a porus in its centre (porus may be well defined and circular or represented merely by a bridge or constriction in the middle of the colpus).

Trizonoporate: With three pori arranged in an equatorial zone.

Verrucae: Wart-like processes, usually broader than high and never constricted at the base (i.e. nearly hemispherical), e.g. *Plantago major*. See Fig. 5.8a.

Verrucate: Possessing verrucae on the surface of the grain.

Vestibulate: With a vestibulum to each porus.

Vestibulum: Small chamber formed by the sexine and nexine splitting apart from one another in the vicinity of a porus. The chamber thus communicates with the inside and outside of the grain, e.g. *Betula, Alnus*. See page 71 and Fig. 5.6.

Zonoaperturate: With apertures (pori or colpi) arranged in a band running round the equator of the grain.

Zonoporate: With pori arranged in a band-running around the equator of the grain.

Zonocolpate: With colpi arranged in a band running around the equator of the grain, the long axes of the colpi at right angles to the equator.

Zonocolporate: With colpi arranged in a band running around the equator of the grain, each colpus having a porus in its centre.

7

ASSEMBLING POLLEN DATA

COUNTING TECHNIQUE

The identification and counting of pollen grains is the basis of pollen analysis. Once the analyst is familiar with the major pollen types, it is usually possible and convenient to identify and count samples from prepared slides by scanning at a magnification of ×400. This provides sufficient magnification for routine identifications, and has the advantage that the field of view resulting from the use of a ×40 objective lens is usually appropriate for slides with a reasonable pollen density. Oil immersion microscopy at ×1000 will be necessary for dubious grains. It is not normally very useful for routine scanning because of the very small field of view it offers. Only where pollen density is very high does it become possible to count under oil immersion.

Some workers prefer to count using phase contrast microscopy. This is largely a matter of taste, though it does have the advantage of providing more information about sculpturing than does bright field illumination.

The most useful eyepiece lenses to use are ×8 or ×10, preferably with wide angle adaptation. Having chosen appropriate lenses for counting, it is important to ensure that the area of the slide scanned is covered only once and that it offers a representative coverage of the total area available. In particular it is important to avoid counting only in the centre of the coverslip or only at the edges as there is some evidence to suggest that the migration of pollen grains under the weight of the coverslip is inversely related to the size of the grain. So larger grains may be more frequent in the centre and smaller grains at the edges (Brooks & Thomas 1967). This pattern is best overcome by sampling the slide by means of linear traverses passing from one edge to another.

The operation of traverses demands the use of a mechanical stage on the microscope and it is essential to have this graduated so that coordinates can be recorded for any grain that needs to be inspected again on a later occasion. It is a good idea to move the slide in one direction only rather than reversing direction each time a new traverse begins. Not only is this easier on the optic nerves, but if the analyst is interrupted there is no problem concerning the direction in which the traverse was proceeding.

SEQUENTIAL SAMPLING

The process leading up to the preparation of slides containing concentrations of pollen is essentially one of sequential sampling and subsampling. Each sampling stage involves a successive approximation and hence the introduction of greater errors. These stages in successive estimates were first discussed by Woodhead and Hodgson (1935) and they can be recognized as follows:

1 A single borehole is taken to be representative of a body of sediment. When considered in terms of area, this clearly represents a minute proportion of the total contemporaneous surface of sediment at any given point in history. Even if several borings are taken and a number of pollen diagrams constructed for a site, the area sampled

is still proportionally very small (see Edwards 1983).

2 The diameter of the core is usually more than 1 cm, which means that the 1 cm sample taken for pollen analysis is only a subsample of the cross-sectional area available.

3 In all except certain types of absolute pollen analysis only a proportion of the final suspension is actually mounted on slides for counting. This subsample may be replicated by preparing a number of slides for each level.

4 A sample of the pollen present on a given slide is identified and counted. The proportion of pollen actually counted on a slide will vary with the density and with the degree of accuracy required, particularly with respect to minor components of the pollen assemblage.

In the final outcome, one takes a relatively small sample of pollen grains (very few analysts count more than a total of 1000 pollen grains/level) to represent the pollen fallout over the contemporaneous sediment surface. In situations where several pollen diagrams are available from a single site, however, there is usually a considerable degree of correspondence between them, particularly for the major components. This encourages one to believe that, despite the sampling problems, the overall technique does yield reproducible results.

Sampling errors can be kept to a minimum by replicating samples wherever possible and keeping the final pollen sum as large as is feasible. To some extent the use of a small vertical sampling interval can compensate for smaller pollen counts where the major components are of greatest interest (Green 1983). Obviously, such replication and enhanced counting is time consuming and it is important to ensure that the effort does supply the extra confidence required. In the case of determining the total pollen count to be used, it may be useful to check the evenness of pollen dispersion on the slide (which is the major contributor to error at this stage). One simple way to determine the most efficient pollen count is to count one sample in blocks of 100 pollen grains (Fig. 7.1) and find the minimum count needed to

provide a representative and acceptable estimate of the pollen proportions. What is evident from Fig. 7.1 is that this selected count may not be optimum for all pollen types. The proportion of commoner pollen types will generally be adequately estimated by relatively small total counts, such as 600 or less, whereas minor components require much higher total sums counted if the estimate of their proportional contribution to the assemblage is to be relied upon.

Often the nature of the question asked in the research project determines the size of the pollen count. If the main interest is in gross forest history and tree pollen comprises most of the pollen assemblage, then a sum of about 600 grains is likely to be adequate for these purposes. If, however, the interest centres upon the minor fluctuations of a rarer component species, whether a tree or, perhaps, a weed type indicative of agriculture, then it is probable that a larger sum must be counted in order to produce reliable data.

It is possible to combine these two requirements by counting all pollen taxa to a low sum, say 600, noting the number of slide traverses necessary to achieve this sum, then counting on further but neglecting selected common taxa. When adequate counts of the rarer types have been achieved the final number of traverses can be noted and, if the common types form part of the pollen sum used in the expression of the rarer taxa (see below), estimates of the total sum can be made.

POLLEN DETERIORATION

Pollen is easily damaged during the extraction process, especially if subjected to prolonged acetolysis or oxidation. However, poor quality pollen may reflect the conditions of fossilization. The quality of pollen grain preservation in sediments is determined to some extent by the conditions under which the sediments were accumulated. For this reason it is often informative to record the degree and type of grain deterioration found in any given sample. Some

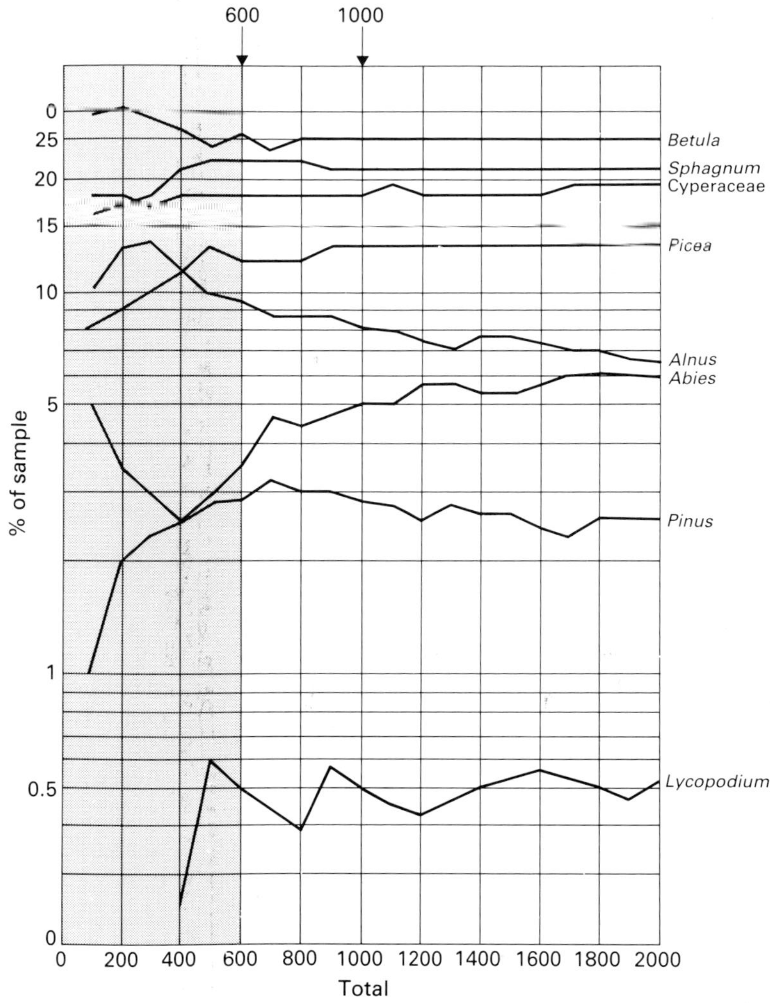

FIG. 7.1. The relationship between the percentages recorded for a series of pollen types and the size of the sample counted, expressed on a logarithmic scale. For major components (over 10% representation) a pollen count of 600 total grains provides an adequate sample, but for minor components (below 5% representation) a count of 1000 grains is preferable. Redrawn with permission from Birks and Birks (1980).

research workers have recorded the types of deterioration only in the unidentified fraction of the pollen, but it can be useful, and certainly provides a larger sample, if deterioration type is noted even in those grains sufficiently well preserved to be identifiable. At Lake Lamminjarvi, Tolonen (1980a) found that up to 30% of the total grains after Iron Age times were deteriorated and these were interpreted as being rebedded as a result of forest clearance and soil erosion. H.J.B. Birks (1970) also used pollen deterioration as a means of identifying inwashed pollen assemblages.

Various authors have described a range of deterioration types (e.g. Cushing 1967a; Tolonen 1980a — including some sketches), but probably the most useful classification, accompanied by a discussion of the causes of deterioration types, is that of Delcourt and Delcourt (1980), reproduced in summary form in Berglund and Ralska-Jasiewiczowa (1986). The types of deterioration they describe are as follows:

1 *Corrosion*. The exine may have complete perforations penetrating it, often quite neat and circular, or its surface may merely be scored, etched or pitted. The cause of such deterioration

is usually microbial activity in which fungi and bacteria are responsible for local oxidation of the exine. Since the bulk of fungal and bacterial activity occurs under conditions of at least periodic aeration, the implication of this type of deterioration is that oxygen was in contact with the grain, either after its deposition on the sediment surface, or at some prior period in its history, such as in soil humus. Microbial attack, particularly by anaerobic bacteria, will continue in waterlogged sediments (Clymo 1965), but at a very much reduced rate.

2 *Degradation*. This is characterized by a general thinning of the exine rather than local perforation. In its extreme form it can result in the exine elements becoming obscure or even apparently fused together in a structureless, waxy mass. Such deterioration provides evidence of exposure of the grain to air with its resultant chemical oxidation. In a peat, the grain may have experienced this by periodic drying of the surface materials.

3 *Mechanical damage*. The exine of the grain may be bent, crumpled or ruptured, but does not necessarily show thinning or perforation. The cause of such deterioration is generally the physical stress to which it has been exposed in the course of its depositional history, e.g. stream transport into a lake. Strongly crumpled grains are also to be found in the faecal pellets of invertebrates, such as Collembola. Compaction, crushing and folding of grains can take place after deposition.

4 *Obscured grains*. A common cause of failure to identify pollen grains and spores is because they are obscured by detritus on the slide. This is usually due to certain components of the sediment matrix being resistant to the chemical processing employed. Delcourt and Delcourt (1980) also point out that pollen grains may become infilled with minerals crystallizing *in situ* in the sediment, which can render identification difficult. Scoring obscured grains during routine counting can provide an indication of the abundance of resistant materials in the sediment.

FLUORESCENCE MICROSCOPY

Estimation of pollen deterioration provides a means of assessing the proportion of secondarily deposited grains in a sediment, but it does not indicate the age of the derived grains. One approach to this problem is the use of fluorescence microscopy. When excited by ultraviolet light, pollen exines fluoresce, and the wavelength of the emitted radiation varies with the geochemical modification of the grain and therefore with its age. In general, the grains tend towards the orange and red end of the spectrum as they age, but different pollen types behave in somewhat different fashion. Details of equipment and techniques are given by Phillips (1972). Colour illustrations are given by Traverse (1988).

THE POLLEN SUM

Where absolute techniques of analysis have been employed, it is possible to express pollen and spore data in terms of numbers accumulating/ unit area of sediment per unit time. This means that the value for any given taxon is independent of all others in its expression, and one can follow trends in the abundance of the pollen rain of that taxon in time. In those forms of analysis where absolute techniques are not employed, pollen taxa are expressed as a percentage of a given sum, which leads to certain numerical problems.

The most obvious difficulty arising from the use of percentage expression of pollen data is the influence of each type upon all others within the sum. Figure 7.2 shows a hypothetical situation in which the total pollen assemblage is made up of three types, **a**, **b** and **c**. At the lowest depth these three types are shown with equal proportions, each constituting 33% of the total pollen. Consider a situation in which the total pollen input of type **a** into the sediment varies while the absolute quantities of **b** and **c** remain unaltered. If, for example, type **a** is reduced in abundance to such an extent that it now makes up only 10% of the pollen sum, the effect on the other two types is to produce

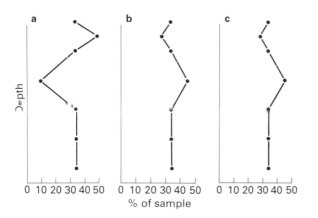

FIG. 7.2. Hypothetical representation of a pollen assemblage consisting of three types: a, b and c. The proportional representation of all pollen types is affected by real fluctuations in the abundance of any of the others. See text.

compensatory rises, thus maintaining the required total of 100%. Despite the fact that the absolute input of pollen of types **b** and **c** is unchanged, their proportional representation has risen by almost 12% each. Similarly, a real increase in **a** in the upper part of the diagram causes apparent decreases in **b** and **c**.

In a real pollen diagram the situation is more complicated because one can never be sure that only one component of the pollen assemblage is varying; very often several types will move in concert, others may genuinely vary in the opposite direction because of competitive and community interactions among the plants, and yet others will vary only because they are members of the pollen sum which must always maintain itself at 100%.

This problem becomes particularly severe if certain local pollen taxa with a high pollen input into the assemblage undergo large fluctuations during the course of time represented by a pollen diagram. A few samples containing large quantities of a local aquatic or mire surface species, for example, if included in the pollen sum must produce a corresponding decrease in all other components of the sum. One answer to this difficulty is the removal of such local species

from the pollen sum. This demands the recognition of which plants are indeed local and need to be removed (Rybnickova & Rybnicek 1971), and it also creates difficulties in the comparison of pollen diagrams in which different pollen sums have been used for expression.

Several methods have been used in pollen sum construction:

1 *Arboreal pollen.* In the early days of temperate zone pollen analysis the emphasis was very strongly on the pollen of trees, often to the complete exclusion of herbaceous pollen and spores. In part this was due to lack of precision in the identification of herbaceous pollen types, but also to the initial over-riding interest of many palynologists in the reconstruction of broad patterns of environmental (including climatic) history in the temperate latitudes. The expression of data on the basis of tree pollen has the advantage that aquatic taxa and those herbaceous taxa often associated with wetlands, such as Gramineae and Cyperaceae, are automatically excluded from the sum and therefore do not contribute their fluctuations to the expression of tree percentages.

The method works reasonably well as long as (i) the tree pollen input is regional rather than local, and (ii) the influx of total tree pollen remains reasonably steady. If some trees are locally very abundant, perhaps as members of the wetland communities, for part or all of the profile, it can obscure fluctuations in other tree types. Janssen (1959) found this to be the case with *Alnus* and consequently excluded it from his pollen sum. If the total input of trees varies during the course of a diagram, this is not evident from the tree pollen curves, which can be misleading. The extinction of forest at the commencement of a cold, glacial period, for example, is not documented if a tree pollen sum is used. Neither is it apparent if forests are cleared by human activity, though the proportional representation of various trees may change.

Those pollen taxa that do not fall within the tree pollen sum have often been expressed on a tree pollen sum basis. This has the effect of producing exaggerated values for herbaceous and other taxa, especially when tree pollen values are low. This can be visually impressive in situations where human activity has resulted in a decrease in trees and an increase in weed types, for the latter, when expressed in terms of the former, will become inflated in value. The expression of pollen data in this way has been used to emphasize periods of intense human activity, and has proved particularly attractive to those engaged in studies of past land use (e.g. some diagrams of Pott 1986).

2 *Pollen sum in groups*. The inflation of non-tree pollen types referred to above can be overcome by expressing each group of taxa (shrubs, herbs, aquatics, spores, etc.) as a percentage of its own group plus the trees. The outcome of this system is that values of more than 100% are avoided, whereas such values are entirely possible when using a tree pollen basis.

The main disadvantage of this form of expression is that comparability between workers may not be maintained. Differences of opinion are inevitable concerning the definition of groups (such as what constitutes a tree, a shrub and a dwarf shrub). The method has, however, been widely used, e.g. H.H. Birks (1970) (Fig. 7.3).

3 *Total pollen and spores (excluding aquatics)*. This is perhaps the most frequently used basis for modern percentage pollen diagrams and permits the greatest degree of comparability between sites and between workers. Its major problem is the definition of 'aquatic' pollen. This is relatively easy when dealing with lake sites, as the term is usually confined to submerged and floating macrophytes. In peat sites it can include many other taxa, some of which, like Gramineae, may occur both as local species (such as *Phragmites*) and as dry land contributors to the pollen sum. In situations of this kind, where a pollen taxon contains both local wetland and regional components, it is conventional to include it within the sum.

A further problem of definition may occur when the sediments under investigation are of a varied nature, as when a succession from open water to peat is represented. Again it is conventional to exclude only the obligate aquatics of open water from the sum. The outcome is that there may be points in the profile where there is clear local dominance of one taxon, but the causes are normally evident on inspecting the stratigraphic profile. A good example of a diagram constructed using this basis is that of Boyd and Dickson (1987b).

OTHER PROPORTIONAL PROBLEMS

There are certain other difficulties caused by the use of proportional data which need to be considered. Fagerlind (1952) and subsequently Faegri and Iversen (1989) have pointed out that as a pollen taxon approaches 100% or 0% it becomes less efficient in its reflection of absolute changes in the abundance of that type. This can be illustrated by a simple hypothetical example. Consider a site which receives only two pollen types, A and B, and which receives 90 grains of taxon A and 10 grains of taxon B/unit area per unit time. The proportional representation would obviously be 90% A and 10% B. If the pollen input of taxon A doubled because of some change in vegetation, then the situation changes to:

A 180 grains = 94.7% (formerly 90 grains = 90%)
B 10 grains = 5.3% (formerly 10 grains = 10%).

A doubling of the pollen input of A has thus raised its representation by only 4.7% of the total pollen.

If, on the other hand, B were to double its original pollen input, the final representation would be:

A 90 grains = 81.8%
B 20 grains = 18.2%.

The second hypothetical change would have involved less modification to the original vegetation, but it had greater effect on the proportional representation than the first. Faegri

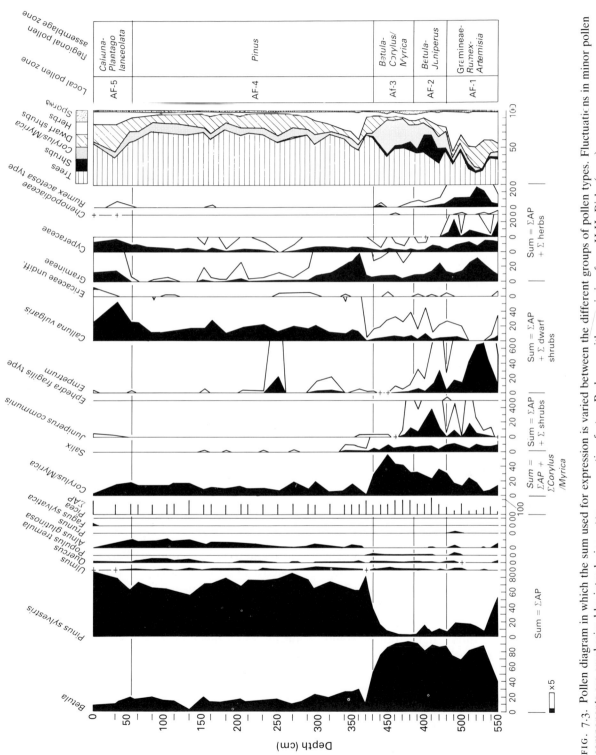

FIG. 7.3. Pollen diagram in which the sum used for expression is varied between the different groups of pollen types. Fluctuations in minor pollen components are emphasized by introducing a ×5 exaggeration factor. Redrawn with permission from H.H. Birks (1970).

and Iversen refer to this statistical effect as an illustration of the 'law of diminishing returns', for as a species approaches 100% a large absolute change in pollen input will cause very little change in its representation.

The practical outcome of these statistical effects is that caution must be applied to the interpretation of pollen types at very high values. Small changes in proportions may reflect very considerable alterations in the influx of pollen. Webb *et al.* (1981) and Prentice and Webb (1986) have used regressions to correct for this 'Fagerlind' effect in proportional pollen diagrams.

THE POLLEN DIAGRAM

Pollen data from stratified sediments are generally displayed graphically in the form of a pollen diagram. These appear in a range of different forms from different research workers, but there are some conventions which allow ease of comparison of diagrams (see Berglund & Ralska-Jasiewiczowa 1986):

1 *Arrangement*. Vertical axis represents depth and horizontal axis the abundance of the pollen taxon in either proportional or absolute terms. The alignment of diagrams in this way is logical given that the sediment core is usually a vertical one, but since it also relates to time (greater depth equals greater age), there is a case for a left to right alignment of a time axis with the pollen curve above it. This can be a very expressive way of representing what are essentially successive expansions and contractions of pollen (and therefore plant) populations in time. This type of diagram has been used by Bennett (1986a,b; 1987) and Bennett and Lamb (1988) for absolute pollen influx data, but it could also be used for proportional expressions (see Fig 7.4).

2 *Stratigraphy*. The stratigraphic sequence of sediment types is represented diagrammatically by a vertical column on the left hand side. This is an important part of the pollen diagram and should not be omitted, for it provides informa-tion that may be essential for an adequate inter-pretation of the fluctuating pollen curves.

The symbols used in representing sediment types have varied greatly in the past and there is a pressing need to establish conventions that will permit easy comprehension. The most widely used system is that of Troels-Smith (1955), which has been simplified and somewhat modified by Aaby and Berglund (1986). This forms a sound basis for establishing an internationally acceptable set of symbols (see Fig. 3.1).

3 *Pollen curves*. The proportions of the various pollen types is indicated at each level of sampling either by a bar histogram (Fig. 7.5) or by a point on a continuous curve (see Fig. 7.3). Both methods are frequently encountered and often the curves are made more distinctive by shading the area beneath them. There are two reasons why the bar graph is preferable, first because it makes no assumptions about the values between sampling levels (joining points in a curve is effectively assuming that intermediate levels have intermediate values), and second because the thickness of the bar indicates the thickness of the sample taken. This is particularly helpful where sample thickness varies at different sam-pling levels to permit higher degrees of resolution (see, for example, the diagram of Sturludottir & Turner 1985). It must be added, however, that the use of a curve, particularly if shaded, can provide a more immediate visual clarity.

4 *Pollen abundance scales*. The horizontal scales of pollen proportions should be consistent throughout the diagram or, where minor com-ponents need emphasis, any variation in hori-zontal scale should be made clearly apparent. An alternative approach to the problem of representing both major and minor components on the same diagram is to exaggerate the scale by an appropriate amount. A ×5 exaggeration is often employed (e.g. H.H. Birks 1970 — see Fig. 7.3) and this can be made more conspicuous by omitting the shading beneath the expanded scale.

5 *Order of taxa*. Conventionally the pollen dia-gram is arranged into groups of taxa, with

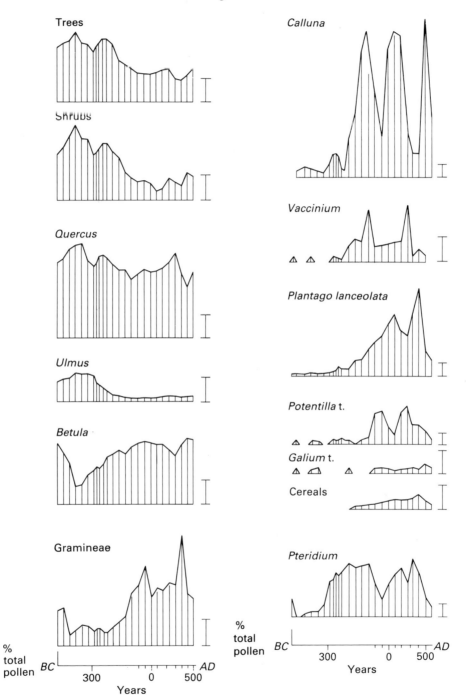

FIG. 7.4. Pollen diagram covering the Iron Age and Roman periods from a site in South West England (The Chains, Exmoor). The horizontal axis represents time, proceeding from left to right. The time axis is non-linear and is based on a series of radiocarbon dates. The vertical bar represents 10% of the total pollen.

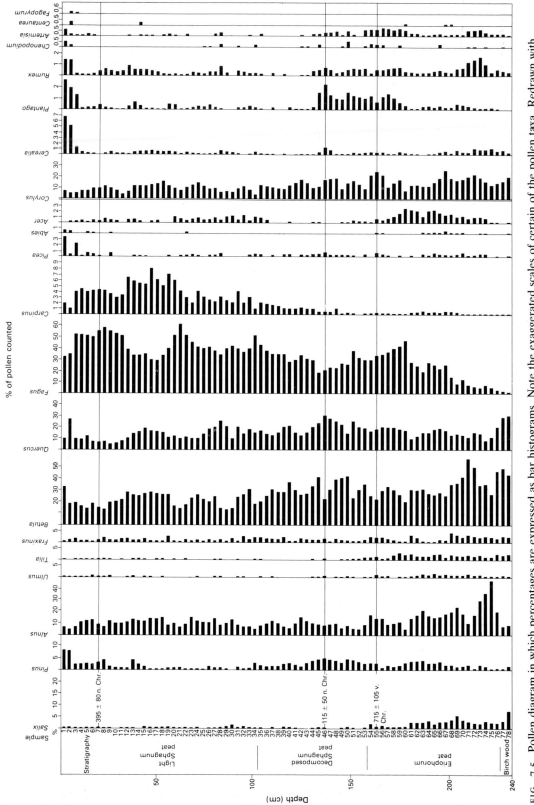

FIG. 7.5. Pollen diagram in which percentages are expressed as bar histograms. Note the exaggerated scales of certain of the pollen taxa. Redrawn with permission from Pott (1986). The diagram is from a bog in northern Germany.

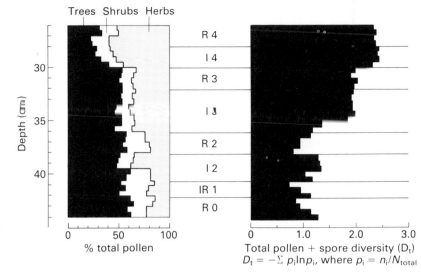

FIG, 7.6. Summary pollen diagram and an index of pollen diversity at each level, using the Shannon function. I and R phases refer to episodes of 'interference' and 'regeneration' respectively, the zones being based upon pollen indicator types and other evidence. It can be seen that the projected interference phases are marked by rises in the diversity of pollen. Redrawn with permission from Moore (1973).

arboreal types first, followed by shrubs, then dwarf shrubs, dry land herbs, aquatics and finally spores. The precise arrangement within any group, such as the herbs, varies from one worker to another. Some use taxonomic arrangements, others use order of appearance in the core. Although the use of a conventional arrangement assists the reader in locating specific taxa on the diagram, it may not be the most helpful way of organizing data to assist ecological interpretation of events in the vegetation it reflects. Many diagrams have now broken with tradition and arrange taxa in ecologically meaningful groups (e.g. Moore 1973; Singh & Geissler 1985; Janssen & Braber 1987). One disadvantage of using this system is the difficulty experienced in placing certain broad taxa (like Gramineae) within a specific ecological group. The stratigraphic data may help in this, especially if detailed identifications of macroscopic fossils has been undertaken (Rybnickova & Rybnicek 1971).

Some pollen diagrams, particularly those emanating from Continental Europe, have several pollen types, usually trees, superimposed on one another, using standardized symbols for each taxon (see Faegri & Iversen 1975). This is economical on space, but often at the expense of clarity. This technique can be used to advantage, however, in summary diagrams.

6 *Other data.* It may be valuable to include data from other sources than pollen and spores on the pollen diagram. Other microfossils, such as algal cells or spores, fungal remains, rhizopods, etc. (see Chapter 5) may provide information that helps in the interpretation of features found in the pollen record, since they reflect local environmental conditions. They may be presented in a form similar to that of pollen taxa (see van Geel 1972). Chemical data, such as the loss on ignition or the concentration of individual elements (Bengtsson & Enell 1986) and the abundance of charcoal particles (Tolonen 1986a) are often included on pollen diagrams or on adjacent diagrams.

7 *Diversity.* An indication of the number of taxa recorded at each level may sometimes be of interest and value. In small sites within forest, for example, an opening of the canopy above the site of deposition can permit the entry of a wider variety of pollen taxa as more of the canopy and regional component pollen types enter the sedimentary environment (Perry & Moore 1987). Increasing numbers of pollen taxa may, therefore, indicate canopy opening and may lead to interpretive conclusions about forest catastrophe or human impact. Similarly, in larger lake and mire sites an increased number of taxa (richness) may indicate a diversification of habitats in the pollen catchment.

An improvement on the number of taxa is an index of diversity (see, for example, Krebs 1972), particularly an index which takes into account both richness and evenness (equitability) in the pollen data. One useful index is the Shannon–Wiener function, which has provided a very sensitive indication of canopy disturbance in the studies of woodland destruction and blanket bog formation in British uplands. Figure 7.6 shows an example where pollen indicators of forest disturbance and recovery (shown in the zonation scheme) are mirrored in the curve of pollen diversity (Moore 1973).

8 *Summary diagram*. It is helpful to incorporate into a pollen diagram a summary chart which displays graphically the way in which the total pollen sum is divided into tree, shrub, dwarf shrub and herbaceous components at each level (Figs 7.3 and 7.6). This allows one to see at a glance any major shifts in, for example, arboreal to non-arboreal pollen ratios, that may be of importance in the zonation and interpretation of a pollen diagram.

These criteria are set out as general guidelines for the construction of pollen diagrams. When a particular point needs emphasis for the sake of developing an argument then, as in the case of the reorientation of diagrams to express population changes in the course of time, these conventions should not form the basis of slavish adherence.

Pollen diagram zonation

It is both conventional and convenient to divide the pollen diagram into a series of zones on the basis of their pollen content. Horizontal lines can be drawn across the diagram, thus dividing it into a sequence of units or pollen zones. Such divisions should be selected simply on the criteria

of their pollen components, without reference to timescale, sediment stratigraphy, inferences concerning past climate, vegetation, archaeology, etc. In other words, the pollen zone should be a *biostratigraphic unit*, a unit defined purely upon its pollen content.

It was Cushing (1967b) who first brought some uniformity into the use of pollen zones, for the idea has been used in a range of different ways by palynologists. Early pollen studies were usually concerned with forest history in temperate regions and the implications of such work in the reconstruction of past climates. Subdivision of early pollen diagrams was therefore designed to simplify the information and to construct a scheme to link these two processes. Hence the early European pollen zonation schemes of Jessen (1935) and of Godwin (1940) were based on amalgums of climatic and vegetational changes that were assumed also to have temporal significance. Hence the Godwin scheme was used in conjunction with the climatic postulates of Blytt and Sernander as a time framework for the arrangements and classification of fossil finds at a national level (Godwin 1975). The pollen criteria used in the scheme have been shown to be asynchronous in Britain (Smith & Pilcher 1973) and it has now been largely rejected as a basis for comparative studies (Birks 1982), though it played a very important part in the evolution of modern pollen analysis. A similar evolution of zonation schemes has taken place elsewhere in Europe, as described by de Beaulieu (1982) for France.

If one adopts the empirical approach of Cushing, then zonation of a pollen diagram can be undertaken without reference to any information besides the pollen data. Other diagrams from the area and even the sedimentary information about the site under investigation are not needed in the initial zonation of the diagram. The main aim of zonation, therefore, is to divide the pollen diagram into a series of convenient units, each of which is as internally homogeneous as possible, and displays recognizable pollen characteristics upon which it may be differentiated from adjacent zones. It follows that the zone boundary lines will be placed at those points where change in the pollen spectra is most marked.

At this stage it need not be anticipated that any zone will be of more than immediately local significance, and it should be regarded as a *local pollen assemblage zone*. West (1970) defines an assemblage zone as 'a body of strata characterized by a certain assemblage of fossils without regard to their ranges'. It is entirely possible that at some later stage parallels will be found with other diagrams in the area under study, in which case *regional pollen assemblage zones* can be erected. This, however, is a secondary consideration that should not influence the initial establishment of a zonation scheme for a diagram.

The selection of zone boundaries is based on the determination of points of maximum change in the total pollen assemblage. Traditionally this has been achieved by subjective means and, since the pollen zone is essentially a unit of convenience and a subdivision of a continuum of variation in vertical space (assuming conformity in the sediments), a subjective approach is normally satisfactory. The one danger in the use of subjective methods is the bias and mental weighting given to certain chosen taxa, which undoubtedly colours one's selection of certain levels as particularly significant.

There are some basic points to bear in mind when zoning a pollen diagram.

1 A zone boundary should pass between samples, never through a sample. Thus every sample must belong to a particular assemblage zone.

2 It may be helpful to begin zonation by selecting the most distinctive point of change in the diagram and then subdivide the remaining portions.

3 Zone boundaries should not normally be selected on the basis of a change in a single taxon. The more taxa exhibiting changes in abundance at a given horizon, the stronger the case for erecting a boundary at that point.

4 Zones consisting of a very small number of samples (say less than three) should be avoided.

There are occasions, however, when this is indicated; then it may be worth while considering more analytical work at such a level in order to clarify the indications of distinctive pollen assemblages in the diagram.

5 Given that some zone boundaries will be more marked than others, it may be sensible to indicate the lesser importance of the latter by regarding them as merely subzones.

Some workers have stressed that the vegetation reflected by the changing pollen spectra itself represents a collection of species, each varying in abundance independently in space and time. Unless there is some catastrophic change in the environment, the changes in species composition within the vegetation may not be synchronized between species; each will vary according to its own requirements (see Fig 7.5). This being so, it could be argued that the pollen profile of each taxon needs to be zoned independently of all others. The theoretical arguments for this approach are strong and there may be occasions when this approach is helpful, but this use of zonation does not fulfil one of its chief potentials, namely the simplification of the data set. For this reason it has not been very widely used.

The recognition and selection of appropriate pollen zone boundaries can be assisted by the use of statistical methods (Gordon & Birks 1972), and this has become a more widespread practice as computer facilities have become more generally available. Many methods are now available to the palynologist and are discussed in general terms by Birks and Birks (1980) and Birks (1986; 1987b), and in much greater detail by Birks and Gordon (1985). These methods provide the research worker with a tool by which personal bias and interpretive preconceptions can be avoided, and those who are inexperienced can reach the same, repeatable conclusions as the experienced. However, all zonation systems should be viewed with some degree of caution; they are mere aids to interpretation. Even an objective system of zonation could be misleading if accepted without critical appraisal. Profound changes in some local taxa, for example, resulting from successional or other developments in the vicinity of the site, may well be selected as a basis for zonation while regional changes of considerable interest but low numerical impact may be ignored. This is not an argument against the use of numerical methods, but a caution regarding complete reliance upon them.

The numerical methods in most general use for the establishment of diagram zonation are those of Gordon and Birks (1972) and they consist of two major techniques, one agglomerative and the other divisive. The agglomerative method (CONSLINK) compares samples with their stratigraphic neighbours and pairs those most similar in their total pollen assemblages. In fact, the actual technique used is a measurement of dissimilarity, so the fused pairs are those with lowest dissimilarity coefficients. Repeated pairing of the most similar adjacent samples results in the establishment of a hierarchy of clusters; higher order clusters can then form the basis of zones.

The alternative approach is the subdivision of the total data set into progressively smaller units until eventually only individual samples remain. Two methods are most frequently used for this divisive process; one (SPLITINF) is based upon information theory. The total information (heterogeneity) of the data set is estimated and subdivided into two subsets which result in the greatest fall in their information content. In practice this means the two groups with the greatest degree of internal homogeneity. As in CONSLINK, there is a stratigraphic constraint, so the division must take place between adjacent samples. The first division will be the most significant and subsequent divisions will be of progressively lesser value in the erection of zone boundaries.

The alternative, frequently used method is SPLITSQ, which is also a divisive technique, but employs the sum of least squares deviations. Details of all these methods and worked examples should be sought in Birks and Gordon (1985). A very readable and well illustrated introduction to the techniques is given by Birks (1987b).

8

INTERPRETING POLLEN DATA

The value of pollen grains and spores in palaeo-ecological work lies not only in their distinctive structure, allowing them to be identified with some precision, but also in their abundance and their wide properties of dispersal in nature. But the quantities of pollen produced varies considerably between species, and the distances over which they may be dispersed varies not only between taxa, but also with such factors as position of release and geographical location. The interpretation of the data presented in pollen diagrams depends upon an understanding of these variables.

POLLEN PRODUCTIVITY

Pollen grains provide the means of genetic transfer between individual plants in a population and this is usually of advantage to the species in evolutionary terms. Investment of energy in pollen grains is a drain upon the resources of the plant and natural selection tends to ensure that this investment should not greatly exceed the optimum level. In other words, those individuals that are unduly wasteful with pollen grains will suffer as a consequence, since the energy expended upon their manufacture cannot be diverted to other uses. But what constitutes the optimum production of pollen grains depends on the efficiency of the pollination mechanism employed. Wind pollination is very unpredictable and those species using this method must invest more energy in their pollen grain production than those relying on insect pollination, which permits a more efficient targeting of the pollen

produced. Pollination biology has an extensive literature of its own and those requiring more information on the complexities of this area of ecology are referred to such texts as Faegri and van der Pilj (1971), Proctor and Yeo (1973) and Real (1983).

The differences that exist between species in their pollen productivity is of consequence to the palaeopalynologist for it will affect the chances of a particular taxon being represented in a fossil pollen assemblage. There are two possible approaches to the study of differential pollen production:

1 one can study the numbers of pollen grains produced per anther, per flower or per individual plant and make comparisons between species, or

2 one can study pollen fallout in the immediate vicinity of different species and compare them in that way.

Direct estimates of pollen productivity are relatively few, since their establishment is laborious and of limited reliability. Table 8.1 shows some data presented by Erdtman (1969) in which a number of plants are compared in terms of their pollen productivity. It can be seen from this data that comparisons are difficult to conduct in any detail, but there is a clear tendency for wind-pollinated species to produce more pollen. Thus *Rumex acetosa* (wind-pollinated) produces almost 150 times as many pollen grains per anther as *Trifolium pratense*; and *Betula pubescens* produces about 30 times as many pollen grains per inflorescence as *Fagus sylvatica*. But such figures are of limited value in trying to interpret a fossil pollen assemblage in terms of its vegetation of origin.

TABLE 8.1 Pollen production of various species expressed in different ways. The index of relative pollen production in the final column is based upon estimates of the pollen production of an individual over a period of 50 years and is expressed relative to beech (estimated production 2.45×10^{10} pollen grains). This value is taken as unity (after Erdtman 1969).

Species	No. of pollen grains per anther	No. of pollen grains per flower	No. of pollen grains per catkin	Index of relative pollen production (*Fagus* — 1.0)
Trifolium pratense	220	—	—	—
Acer platanoides	1000	8000	—	—
Malus sylvestris	1400–6250	—	—	—
Calluna vulgaris	2000 tetrads	—	—	—
Fraxinus excelsior	12 500	—	—	—
Secale cereale	19 000	57 000	—	—
Rumex acetosa	30 000	180 000	—	—
Juniperus communis	—	400 000	—	—
Pinus sylvestris	—	160 000	—	15.8
Picea abies	—	600 000	—	13.4
Betula pubescens	—	—	6 000 000	—
Alnus glutinosa	—	—	4 500 000	17.7
Quercus robur	—	1 250 000	—	—
Fagus sylvatica	—	—	175 000	1.0
Quercus petraea	—	—	—	1.6
Carpinus betulus	—	—	—	7.7
Betula pendula	—	—	—	13.6
Corylus avellana	—	—	—	13.7
Tilia cordata	—	—	—	13.7

An alternative, and much more promising, approach has been used by Andersen (1970) in Denmark. He investigated pollen fallout in a forested area by examining the pollen content of moss polsters. He regarded the pollen catchment for such samples as no larger than a 30 m radius, so he was able to calculate the crown area of different tree species within that 30 m radius and compare it with the pollen percentages obtained. He expressed his data in relative terms, choosing beech (*Fagus sylvatica*) as a reference species (Table 8.2).

Further work by Bradshaw (1981) in the south of England has shown some variation from these figures, mainly in the lower production of pollen by *Quercus*, but the general rank order of the trees is comparable to Andersen's data. It has proved informative to use these correction factors for the adjustment of pollen frequencies in sites that are comparable to those where Andersen carried out his research, i.e. sites beneath a forest canopy (e.g. Baker *et al.* 1978; Rackham 1980). Great caution should be applied to the use of these figures for other types of site.

POLLEN DISPERSAL

Andersen's data, quoted above, relate in part to pollen productivity and in part to dispersal from the parent plant. Andersen (1974) conducted some trap experiments beneath the canopy and concluded that the pollen in these sites fell vertically, or was washed by rain, hence moved very little distance. The rain-wash itself has been found to redistribute pollen to some extent, since the stemflow down tree trunks will tend to concentrate pollen at tree bases. Experiments in which anemophilous herbaceous pollen in a

TABLE 8.2 Pollen deposition rates beneath a forest canopy. Values are expressed in relative terms, with *Fagus sylvatica* as the reference species (from Andersen 1970).

Quercus	×1.0
Betula	×1.0
Alnus	×1.0
Carpinus	×1.3
Ulmus	×2.0
Tilia	×8.0
Fraxinus	×8.0
Fagus	×4.0
Pinus	×1.0
Acer	×8.0
Picea	×2.0
Abies	×4.0
Corylus	
canopy	×1.0
understorey	×4.0

TABLE 8.3 Volume, weight and settling velocity in still air for a number of pollen types (from Erdtman 1969).

Pollen type	Volume, (μm^3)	Weight, ($g \times 10^{-9}$)	Settling velocity (cm/s)
Picea	132000	72.8	6.0
Fagus	51770	37.0	5.0
Pinus	47030	18.4	3.0
Corylus	10150	10.2	2.3
Alnus	9070	6.8	1.6
Betula	7540	6.1	1.5
Taxus	7130	4.1	1.0
Juniperus	9460	3.8	0.9

woodland understorey has been labelled and tracked (Handel 1976) has similarly shown that pollen in the forest understorey is not likely to move more than a few metres.

How far a pollen grain travels is therefore dependent in part upon the microclimate of the site in which it is released, but it is also related to the aerodynamics of the pollen grain itself. This is best expressed in the terminal velocity of a pollen grain falling in still air. Table 8.3 gives some figures for settling velocities as well as some dimensions of a selection of pollen grains.

The efficiency of pollen dispersal by wind is dependent on a number of factors. The higher the windspeed, the further the pollen is likely to be carried. Air movements within the vegetation layer are usually less than 10 m/s. Settling velocities are related to size and density, so most wind pollinated grains are small, smooth and thin-walled. Some larger types, however, are saccate and employ the air sacs to reduce density and hence settling velocity (Whitehead 1983).

Both of the factors, air turbulence at the point of release and pollen aerodynamics, affect pollen dispersal potential. There is now a considerable volume of literature on the capacity of some

pollen grains to be carried long distances from their source. Studies on rain-deposited pollen at Albany, New York, by Raynor *et al.* (1983), suggest that much of the pollen brought to a site by rainfall has travelled between 18 and 42 h in the atmosphere and in this time it may well have been transported over long distances. A similar conclusion was reached by Christie and Ritchie (1969), who studied dry deposition of long distance transported pollen at Churchill, Manitoba. They found pollen grains from the aspen parkland region of Canada, almost 1000 km to the south, which must have been in the atmosphere about 72 h.

The presence of substantial quantities of pollen in hailstones (Mandrioli *et al.* 1973) indicates that pollen grains can be carried to great heights by atmospheric turbulence, which gives them yet greater potential for long distance transport. This influence will inevitably be felt in any pollen diagram, and single grains of unlikely local occurrence will always be difficult to interpret because of the possibilities of distant origins. The magnitude of this influence, however, depends on the relative contribution made to the pollen assemblage by the rain component of incoming pollen (see Chapter 2). In sites with low local pollen production, the pollen washed from the atmosphere by rain, and therefore the quantities of long-distance pollen, will be proportionally greater. This can be observed particularly in tundra environments (for example, Ritchie *et al.* 1987) and in desert sites (for example, Ritchie 1986a), where tree pollen may figure prominently in the pollen rain despite the total absence of trees from the surrounding vegetation.

Genetic studies on pollen movements as a source of gene flow have concentrated on such techniques as marking pollen grains or tracking pollinators using coloured dusts and dyes to establish their ranges (Waser 1988). Such work has shown that mating frequency between plants drops off very rapidly within a few metres of the pollen source and is insignificant beyond about 100 m. In cases where such studies have concentrated on the arrival of the pollen grains at their destination and the clues given by the cytology and genetic constitution of the resulting seed, some very different results have been obtained. Ellstrand (1988), for example, has worked with wild radish (*Raphanus sativus*) populations and concludes that between 4.5 and 18% of pollen arriving at a flower and effecting pollination has travelled between 100 and 1000 m. The palynologist must therefore be cautious in the interpretation of isolated fossil pollen grains, even if insect transported, as in the case of the radish.

In pollen stratigraphic studies, the megafossil content of the sediment matrix may provide important information concerning the local occurrence of some taxa, which can greatly assist in the interpretation of pollen transport distances (Gray 1985).

SURFACE STUDIES

Perhaps the most promising method of approaching the interpretation of past assemblages of pollen grains in terms of the vegetation which gave rise to them, is the study of surface samples from modern vegetation types. Some of the practical aspects of this type of work are discussed in Chapter 3. An example of the results which can be obtained from transects of surface samples is given in Fig. 8.1, taken from the work of Janssen (1966).

This provides a graphical representation of pollen rain beneath a range of vegetation types and, even without further analysis, is a useful source of information upon which fossil data could be interpreted. The kind of feature which is evident from this display is the very local dispersal of some pollen types, such as *Tilia*, *Acer* and *Rhamnus*, in comparison to the blanket rain of *Pinus*, *Ambrosia* and the grasses.

Even more information is available from such data, however, if it is subjected to more detailed analysis. Davis (1963) proposed the use of a correction factor (R) based upon the ratio of pollen in a surface assemblage to the abundance of that taxon in the surrounding vegetation.

$$R = \frac{\text{Percentage of pollen of taxon in surface sample}}{\text{Percentage of taxon in surrounding vegetation}}.$$

An improvement on the use of percentages based on surface samples is the establishment of absolute pollen influx using pollen traps (see Chapter 3).

The concept of the *R-value* has its limitations. The value obtained depends upon the radius (catchment) from which pollen is assumed to come. This may well differ from one site to another. It will also depend upon the vegetation structure and the local topography, which influence pollen release and transport. The application of this type of approach has, however, been quite successful in forest environments (e.g. Bradshaw 1988; Bradshaw & Miller 1988) where the pollen catchment is very small and easily defined, as described above. The original concept of the *R-value* has undergone considerable modification and has been subjected to statistical treatments (e.g. Parsons & Prentice 1981).

Much more sophisticated models of pollen sources, movement and accumulation have been developed which can take into account spatial variations in vegetation. These models have been reviewed by Birks and Gordon (1985) and Prentice (1988a).

ORDINATION OF SURFACE AND FOSSIL POLLEN SAMPLES

An alternative approach to the numerical analysis of surface pollen samples and their comparison with fossil samples is to use multivariate systems of analysis similar to those used for vegetation analysis (Gauch 1982). This is discussed in considerable detail in Birks and Gordon (1985). Just as with vegetation analysis, a very wide range of techniques has been employed in the construction of spatial models for the representation of complex data sets. Most operate on the principle that axes of variation can be extracted from the data sets which account for a portion of the total variation in the set. Some axes account for more of the variation than others and these can be selected in order to represent the data as a scatter diagram in which each sample has a coordinate on any selected axis, hence its position can be shown in two dimensions. A helpful introduction to the concepts involved is given by Birks (1987a).

These techniques have a number of possible applications.

1 They allow large sets of surface sample data to be presented in a simple, graphic form.
2 They make possible the arrangement of data into groups based upon clusters in the diagram, which are in turn a consequence of similarities between samples.
3 One can determine which pollen taxa are most influential in each of the axes.
4 It is possible to ordinate fossil data in the same way as modern data and hence observe the way in which vegetation over the course of time follows a particular track or sequence.
5 The track of past vegetation (as represented by fossil pollen assemblages) can be interpreted in the light of modern patterns of surface samples.

ENVIRONMENTAL RECONSTRUCTION

Pollen stratigraphy provides a record of the changing vegetation of the past which, although difficult to interpret and reconstruct, itself supplies information on past climates and land use history. There are thus two steps of interpretation leading from fossil pollen assemblage to vegetation and then from vegetation to the palaeoenvironment (see Moore & Webb 1978), Obviously, such reconstruction is best achieved using all available botanical information including: pollen and spores, diatoms (Battarbee 1986), and other algae (Cronberg 1986; van Geel 1986); fungal remains (van Geel 1986); plant megafossils together with sediment lithology, chemistry, charcoal (Tolonen 1986a); and animal fossils such as rhizopods (Evans 1972; Tolonen 1986b), and beetles (Coope 1986). All possible sources of information about past conditions should be exploited.

There are two main approaches to environmental reconstruction from fossil pollen data. Certain individual taxa may provide information because they have very precise environmental requirements. Such taxa may serve as *indicator species*, whose very presence permits certain definitive statements concerning past conditions. The use of indicator species in this way does assume that the physiological requirements of the taxon have not changed during the period that has elapsed since the fossil was deposited and that one is dealing with a similar physiological race of the taxon. It is thus an interpretive method that is best suited to stable species with few ecotypes. The approach is discussed in some detail by Birks and Birks (1980).

An alternative to the indicator species approach is the use of entire assemblages. On the argument that the full spectrum of vegetation is a better reflection of environmental conditions, such as climate and land use, than is any individual species, the full pollen assemblage should provide a better guide to past environmental conditions than individual indicator species taken in isolation. The relationship between the assemblage and the environment can be reduced to a numerical function, termed a *transfer function* (see Howe & Webb 1983). The calcu-

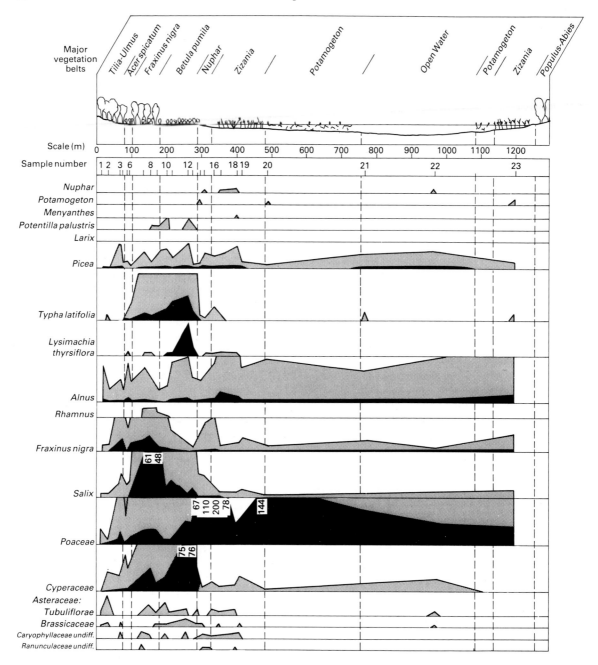

FIG. 8.1. Pollen analysis of surface samples along a transect through several vegetation types in Minnesota (redrawn with permission from Janssen 1966). Only selected pollen types are given. Dark shaded curves show the proportional representation of each pollen type and light shading shows these values exaggerated ×10.

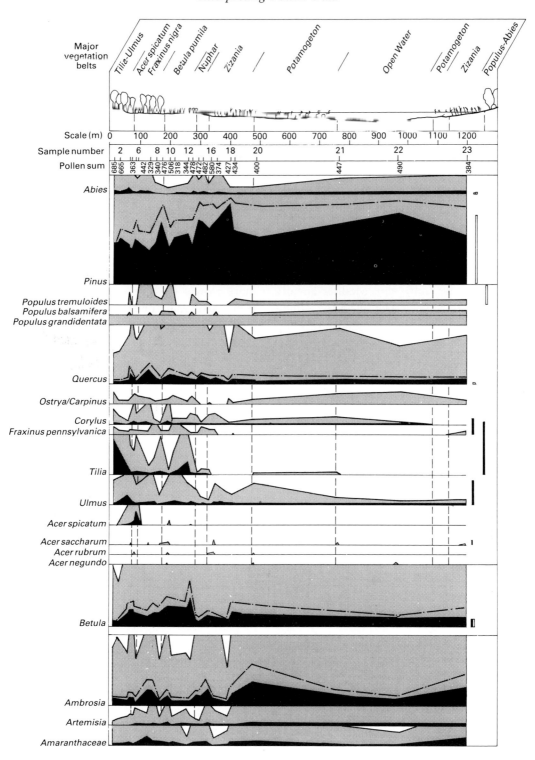

lation of these functions involves multiple linear regression, and the details of the techniques are discussed by Birks and Gordon (1985) and Arigio *et al*. (1986). A very readable introduction to the concepts involved is provided by Webb (1980).

The use of transfer functions has proved to be of considerable value in detailed climatic reconstruction over wide areas. For example, Huntley and Prentice (1988) have been able to map the mean July temperatures across Europe 6000 years ago using a transfer function approach. They were able to show that central and northern Europe were 2°C warmer than present at that time. From this they were able to speculate that stronger winds would have resulted from the elevated continental summer temperatures, bringing more rain into the Mediterranean region. Guiot *et al*. (1989) have used the same technique to extend the palynological study of Europe's climate back 140 000 years. Similar studies have been conducted by Webb (1985) in North America.

The interpretation of pollen stratigraphy simply in terms of climate, however, is not possible, for other factors need to be constantly taken into consideration. The sequence of change in vegetation over long periods of time (centuries and millennia) is affected not only by climatic changes but also by such variables as the invasion rates of different species (which influences the speed of equilibration of vegetation to the climatic template); competitive interactions between plant species; the influence of grazers and plant pathogens; and the possibility, even in early prehistoric times, of human modification of vegetation by management practices such as the use of fire (see Prentice 1986a; Ritchie 1986b).

Broad scale patterns of vegetation change using pollen data have been effectively reconstructed by mapping techniques (e.g. Huntley & Birks 1983; Dexter *et al*. 1987; Webb 1987). In these studies, well dated pollen profiles scattered over wide areas have been selected and the status of a given pollen taxon at any point in time is recorded by its pollen proportion in the assemblage at that time. In this way, contours of equal pollen proportions can be drawn ('isopolls') enclosing all sites that have achieved a particular level of representation for the taxon. The resulting maps provide a graphic representation of the spatial movements of plant species in the course of time. The results show that many trees achieved considerable rates of spread during the early part of the present interglacial, often reaching between 300 and 500 m/year. It is also apparent that trees with widely differing dispersal mechanisms (such as birch and hazel) nevertheless achieved similar rates of spread. In many cases, the rate of spread varied at different times. For example, lime (*Tilia*) spread rapidly (up to 500 m/year) in the early part of this interglacial (from 10 000 to 8000 BP) but spread much more slowly (between 50 and 130 m/year) between 7000 and 5000 BP. Evidently it is not the dispersal potential which limited the spread of the species but some other factor, either climatic or one related to the competitive resistance of established forest to further invasion.

The general conclusion from such mapping exercises is that, with the exception of the very early stages in forest succession, climatic and competitive factors are more influential in determining vegetation composition than the potential rate of spread of individual species.

Competitive interactions are much more difficult to study. The potential niche, and therefore the potential geographic range, of almost all species is greater than the realized niche and range. The difference is due to the greater efficiency of competitor organisms in parts of the potential range, particularly at the extreme edges. The study of the competitive interactions of plants in the past has only become possible with the development of absolute techniques in pollen analysis: prior to that the percentage method of expression obscured any study of individual species in isolation from others (Bennett & Lamb 1988). It may even be possible to interpret such data in terms of the population dynamics of the species involved (Bennett 1986; Prentice 1988a).

Diagrams of pollen accumulation rates provide

a means of tracing the individual histories of tree taxa and clearly show the sharp decline of some species coincident with the invasion of another. For example, birch in Ontario declines with pine invasion, while birch in eastern England declines with the arrival of hazel. The shape of the birch curves and the persistence of the taxon at lower frequencies suggest that in the absence of the new arrival the former high levels could have been maintained. Competitive forces, probably in the form of greater shade, have determined the sequence of events. What is still difficult to demonstrate is the suppression of the spread of a species as a result of the competitive strength and consequent inertia of a vegetation type. The occasional rapid expansion of a species, such as that of alder in the British Isles in the mid-Holocene, may suggest that some competitive constraint has been withdrawn. But the interaction of climatic and possibly human factors render the explanation of such events very difficult. The application of absolute techniques of analysis has, however, permitted palaeoecologists to think in terms of plant population dynamics (Bennett 1983; Prentice 1988b).

Sudden declines in pollen taxa in the stratigraphic record may be a consequence of climatic or competitive changes, but they can sometimes be associated with outbreaks of pathogens or environmental manipulation by mankind. Allison *et al.* (1986), for example, have interpreted the decline of *Tsuga* in the pollen record of the northeastern United States as a consequence of a pathogen outbreak. It also seems very probable that the European elm decline of 5000 years ago was caused by the spread of a fungal pathogen, carried by a beetle, and possibly also connected with human movements and changing forest management (Girling & Grieg 1985; Perry & Moore 1987).

The case of the elm decline demonstrates the problems of attempting to extricate single causes for an event in vegetation history. Many factors may contribute to such an event and these factors may themselves interact. The complexity of such interaction becomes yet greater when human

beings are involved. The multiplicity of ways in which prehistoric peoples may have modified their environment, and the various climatic, pedological and sociological processes underlying their motivation provide a considerable challenge to the palynologist.

HUMAN INFLUENCE ON VEGETATION

Reconstructing the history of human vegetation management has proved just as difficult a task as that of reconstructing climatic history. It is less easy to define the elements which constitute a human managed (or anthropogenic) environment than is the case with climatic variables, and determining the response of plant species to human impact is even more difficult. Interpretative processes are therefore dependent on the use of indicator species whose ecology can be linked to man-induced aspects of the environment, such as fire, disturbed soils, open canopies, nitrogen and phosphorus flushing, etc. The plant species associated with such habitats can be regarded as anthropogenic indicators and their pollen provides a basis for the recognition of human activity in pollen diagrams.

In Europe, palynologists have been fortunate in that certain wind pollinated species of weeds have recognizable pollen types and have proved relatively sensitive indicators of human activity. *Plantago lanceolata*, for example, is determinable to species level and is a regular constituent of pollen samples from sites disturbed by human activity. Many other such species are available from different parts of the temperate zone (see Behre 1986; Fig. 8.2), and Turner (1964) was able to devise an estimate of the relative importance of pastoral and arable agriculture on the basis of the different proportions of selected indicator pollen types in a sample. Vuorela (1973) has been able to supplement this work by studies of pollen rain in the neighbourhood of cultivated fields. This approach has been developed further by Behre (1981) who has devised detailed tabulations of the association between key indicator pollen types and the form of agriculture or disturbance being practised (Fig. 8.2). The major limitation of such a systematic approach is the geographical one; weeds behaving in a distinctive way at one site may not do so in a different geographical setting. But the method needs to be extended in the light of wider geographical experience.

In North America a similar approach has proved helpful and both native weeds (e.g. *Ambrosia*) and those brought to the continent by European settlers have been used in detecting past land use. Pre-European disturbance is more

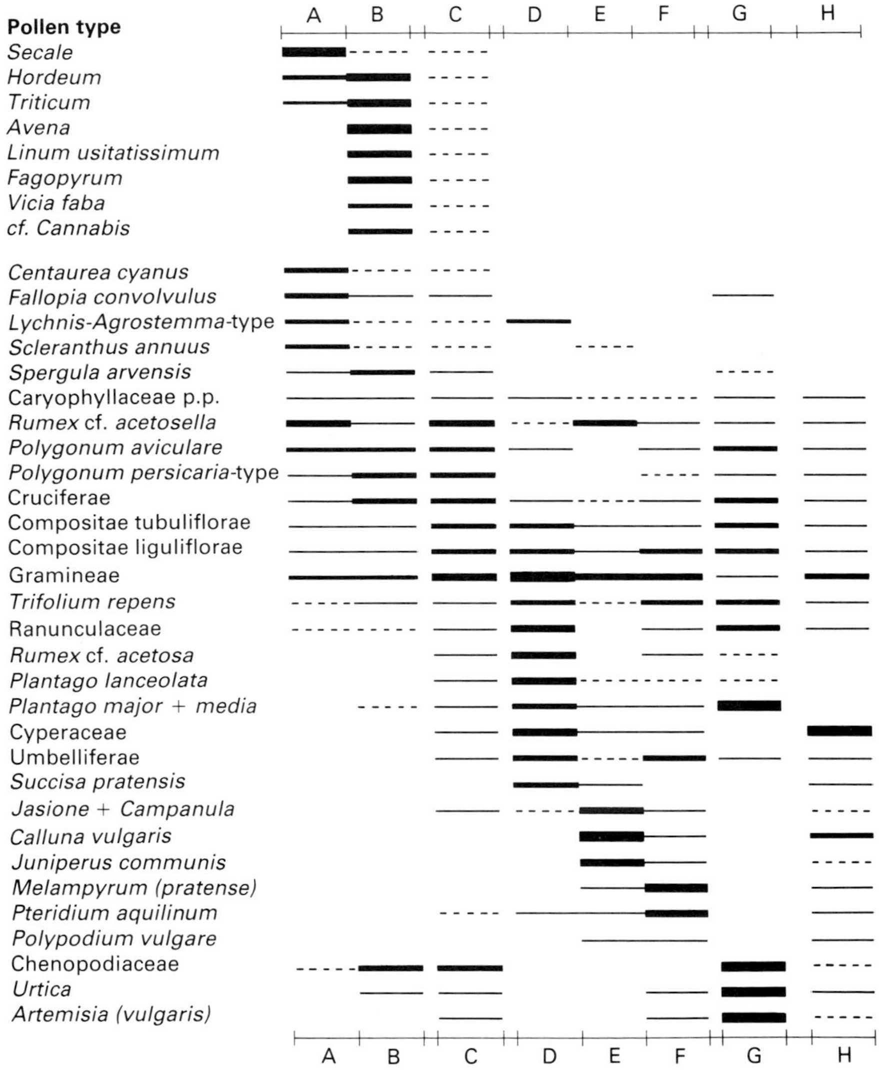

FIG. 8.2. The value of various pollen types as indicators of specific, man-managed habitats (A–H) in northern Europe (redrawn with permission from Behre 1986). The scheme can only be taken as a rough guide to interpretation as geographical variation is to be expected. Key to habitats: (A) winter cereals; (B) summer cereals and root crops; (C) fallow land; (D) wet meadows and pasture; (E) dry pastures (heaths, etc.); (F) grazed forest; (G) footpath and ruderal communities; (H) natural communities (e.g. mires).

difficult to study, but considerable advances are being made in this area (Delcourt 1987).

An increasing number of studies are now being carried out with the aim of tracing human impact in other areas of the world. In Australia, for example, it is possible to detect recent increases in fire tolerant plants, such as *Eucalyptus* species, which are associated in the fossil record with greater densities of charcoal particles (Singh & Geissler 1985). In this setting it may well prove that association with high charcoal density is the most effective way of identifying anthropogenic indicators. But, as in Europe, geographic variations are inevitable. Thus, although *Eucalyptus* expands at the expense of *Casuarina* in Singh and Geissler's dry sites in Australia, the picture in South East Asia is quite different. In Papua New Guinea, Flenley (1988) has found that *Casuarina* expands (together with such taxa as Urticaceae and *Trema*) at times when the overall forest cover is decreasing. The first stage in tracing the impact of past human cultures on vegetation using the technique of pollen analysis is thus the establishment of which pollen taxa will prove the most valuable indicators and this will vary from one locality to another.

In this area of research pollen analysis still has a long way to go, but ultimately will be able to contribute much to studies in tropical archaeology and anthropology.

SUCCESSIONAL STUDIES

Finally, mention must be made of those pollen stratigraphic studies that are concerned with short-term changes (of the order of decades or centuries) rather than longer term ones. Here, the sequential stratification of pollen reflects successional changes within vegetation rather than long-term shifts in the vegetation/climate equilibrium. Delcourt *et al.* (1983) have reviewed the problems in isolating processes of differing scale (both temporal and spatial) in pollen diagrams. When pollen catchments are small, as in peaty hollows within forest stands, the pollen stratigraphy provides a direct record of very local changes in vegetation and is an ideal approach to the study of microsuccessions or 'patch dynamics'. Work on mor humus profiles, such as those of Bradshaw (1988) in England, and Bradshaw and Miller 1988) in Ireland, and analyses of small peaty deposits in forest, such as the work of Heide (1984) in Wisconsin, illustrate the great potential of such sites for the elucidation of the history of forest mosaics. It is in this area that palynology may be able to make a significant contribution to problems of conservation and habitat management.

BIBLIOGRAPHY

Aaby, B. (1983) Forest development, soil genesis and human activity illustrated by pollen and hypha analysis of two neighbouring podzols in Draved Forest, Denmark. *Danm. Geol. Unders. IIR*, **114**, 1–114.

Aaby, B. (1986) Palaeoecological studies of mires. In *Handbook of Holocene Palaeoecology and Palaeohydrology*, ed. B.E. Berglund, John Wiley, Chichester, pp. 145–164.

Aaby, B. & Berglund, B.E. (1986) Characterization of peat and lake deposits. In *Handbook of Holocene Palaeoecology and Palaeohydrology*, ed. B.E. Berglund, John Wiley, Chichester, pp. 231–246.

Aaby, B. & Digerfeldt, G. (1986) Sampling techniques for lakes and bogs. In *Handbook of Holocene Palaeoecology and Palaeohydrology*, ed. B.E. Berglund, John Wiley, Chichester, pp. 181–194.

Adam, D.P. & Mehringer, P.J. (1975) Modern pollen surface samples — an analysis of subsamples. *J. Res. US Geol. Surv.* **3**, 733–736.

Adams, R.J. & Morton, J.K. (1972) *An Atlas of Pollen of the Trees of Eastern Canada and the Adjacent United States. I. Gymnospermae to Fagaceae.* University of Waterloo, Biological Series No. 8.

Adams, R.J. & Smith, M.V. (1981) Seasonal pollen analysis of nectar from the hive and of extracted honey. *J. Apicultural Res.* **20**, 243–248.

Adams, R.J., Smith, M.V. & Townsend, G.F. (1979) Identification of honey sources by pollen analysis of nectar from the hive. *J. Apicultural Res.* **18**, 292–297.

Afzelius, B.M. (1956) Electron microscope investigations into exine stratification. *Grana palynol.* **2**, 22–37.

Allison, T.D., Moeller, R.E. & Davis, M.B. (1986) Pollen in laminated sediments provides evidence for a mid-Holocene forest pathogen outbreak. *Ecology* **67**, 1101–1105.

Ambach, W. Bortenschlager, S. & Eisner, H. (1966) Pollen analysis investigations of a 20 m firn pit on the Kesselwandfener (Ötzal alps). *J. Glaciol.* **6**, 233–236.

Andersen, S.T. (1960) Silicone oil as a mounting medium for pollen grains. *Danm. Geol. Unders. IV*, **4**(1), 1–24.

Andersen, S.T. (1961) Vegetation and its environment in Denmark in the Early Weichselian Glacial (last glacial). *Danm. Geol. Unders. II*, **75**, 1–175.

Andersen, S.T. (1965) Mounting media and mounting techniques. In *Handbook of Palaeontological Techniques*, ed. B.G. Kummel & D.M. Raup. Freeman, San Francisco, pp. 587–598.

Andersen, S.T. (1967) Tree pollen rain in a mixed deciduous forest in south Jutland (Denmark). *Rev. Palaeobotan. Palynol.* **3**, 267–275.

Andersen, S.T. (1970) The relative pollen productivity and pollen representation of North European trees, and correction factors for tree pollen spectra. *Danm. Geol. Unders. II*, **96**, 1–99.

Andersen, S.T. (1974) Wind conditions and pollen deposition in a mixed deciduous forest. II. Seasonal and annual pollen deposition 1967–1972. *Grana* **14**, 64–77.

Andersen, S.T. (1978) On the size of *Corylus avellana* L. pollen mounted in silicone oil. *Grana* **17**, 5–13.

Andersen, S.T. (1979) Identification of wild grass and cereal pollen. *Danm. Geol. Unders. Årbog 1978*, 69–92.

Andersen, S.T. (1986) Palaeoecological studies of terrestrial soils. In *Handbook of Holocene Palaeoecology and Palaeohydrology*, ed. B.E. Berglund, John Wiley, Chichester, pp. 165–177.

Andrew, R. (1984) *A Practical Pollen Guide to the British Flora.* Technical Guide One. Quaternary Research Association, Cambridge.

Andrews, J. (1985) *Modern pollen studies in Scords Wood, Brasted, Kent.* Unpublished PhD thesis, University of London, London.

Arigio, R., Howe, S.E. & Webb, T. III (1986) Climatic calibration of pollen data: an example and annotated computing instructions. In *Handbook of Holocene Palaeoecology and Palaeohydrology*, ed. B.E. Berglund, John Wiley, Chichester, pp. 817–849.

Baker, C.A., Moxey, P.A. & Oxford, P.M. (1978) Woodland continuity and change in Epping Forest. *Field Studies* **4**, 645–669.

Bakker, M. & van Smeerdijk, D.G. (1982) A palaeo-ecological study of a late Holocene section from 'Het Ilperveld', western Netherlands. *Rev. Palaeobotan. Palynol.* **36**, 95–163.

Ball, D.F. (1986) Site and soils. In *Methods in Plant Ecology* 2nd edn, eds. P.D. Moore & S.B. Chapman, Blackwell Scientific Publications, Oxford, pp. 215–284.

Banerjee, U.C. & Barghoorn, E.S. (1971) The tapetal membranes in grasses and Ubisch body control of mature exine pattern. In *Sporopollenin*, ed. J.B. Brooks, P.R. Grant, M. Muir, P. Van Gijzel & G. Shaw. Academic Press, London, p. 127.

Barber, K.E. (1976) History of vegetation. In *Methods in Plant Ecology*, 1st edn, ed. S.B. Chapman. Blackwell Scientific Publications, Oxford, pp. 5–83.

Barber, K.E. (1981) *Peat Stratigraphy and Climatic Change*. A.A. Balkema, Rotterdam.

Barnes, S.H. & Blackmore, S. (1986) Some functional features in pollen development. In *Pollen and Spores: Form and Function*, ed. S. Blackmore & I.K. Ferguson, Academic Press, London, pp. 71–80.

Bassett, I.J. & Crompton, C.W. (1968) Pollen morphology and chromosome numbers of the family Plantaginaceae in North America. *Can. J. Bot.* **46**, 349–361.

Bates, C.D., Coxon, P. & Gibbard, P.L. (1978) A new method for the preparation of clay-rich sediment samples for palynological investigation. *New Phytol.* **81**, 459–463.

Battarbee, R.W. (1986) Diatom analysis. In *Handbook of Holocene Palaeoecology and Palaeohydrology*, ed. B.E. Berglund, John Wiley, Chichester, pp. 527–570.

Batten, D.J. & Morrison, L. (1983) Methods of palynological preparation for palaeo-environmental, source potential and organic maturation studies. *Bull. Norw. Pet. Direct.* **2**, 35–53.

Beckett, S.C. (1979) Pollen influx in peat deposits: values from raised bogs in the Somerset Levels, southwestern England. *New Phytol.* **83**, 839–847.

Beer, R. (1911) Studies in spore development. *Ann. Bot.* **25**, 199–214.

Behre, K-E. (1981) The interpretation of anthropogenic indicators in pollen diagrams. *Pollen Spores* **23**, 225–245.

Behre, K-E. (ed.) (1986) *Anthropogenic Indicators in Pollen Diagrams*. A.A. Balkema, Rotterdam.

Bengtsson, L. & Enell, M. (1986) Chemical analysis. In *Handbook of Holocene Palaeoecology and Palaeohydrology*, ed. B.E. Berglund, John Wiley, Chichester, pp. 527–570.

Bennett, K.D. (1983) Postglacial population expansion of forest trees in Norfolk, UK. *Nature* **303**, 164–167.

Bennett, K.D. (1986) Competitive interactions among forest tree populations in Norfolk, England, during the last 10 000 years. *New Phytol.* **103**, 603–620.

Bennett, K.D. (1986) Population dynamics. *Phil. Trans. R. Soc. Lond. B* **314**, 523–531.

Bennett, K.D. (1987) Holocene history of forest trees in southern Ontario. *Can. J. Bot.* **65**, 1792–1801.

Bennett, K.D. & Lamb, H.F. (1988) Holocene pollen sequences as a record of competitive interactions among tree populations. *Trends Ecol. Evolut.* **3**(6), 141–144.

Benninghoff, W.S. (1962) Calculation of pollen and spore density in sediments by addition of exotic pollen in known quantities. *Pollen Spores* **4**, 332–333.

Berglund, B.E., Erdtman, G. & Praglowski, J. (1960) On the index of refraction of embedding media and its importance in palynological investigations. *Svensk. Bot. Tidskr.* **53**, 452–468.

Berglund, B.E. & Ralska-Jasiewiczowa, M. (1986) Pollen analysis and pollen diagrams. In *Handbook of Holocene Palaeoecology and Palaeohydrology*, ed. B.E. Berglund, John Wiley, Chichester, pp. 455–484.

Beug, H-J. (1961) *Leitfaden der Pollenbestimmung für Mitteleuropa und angrenzende Gebiete*. Fischer, Stuttgart.

Bhadresa, R. (1986) Faecal analysis and exclosure studies. In *Methods in Plant Ecology* 2nd edn, ed. P.D. Moore & S.B. Chapman, Blackwell Scientific Publications, Oxford, pp. 61–71.

Biesboer, D.D. (1975) Pollen morphology of the Aceraceae. *Grana* **15**, 19–27.

Birks, H.H. (1970) Studies in the vegetational history of Scotland. I. A pollen diagram from Abernethy Forest, Inverness-shire. *J. Ecol.* **58**, 827–846.

Birks, H.J.B. (1968) The identification of *Betula nana* pollen. *New Phytol.* **67**, 309–314.

Birks, H.J.B. (1970) Inwashed pollen spectra at Loch Fada, Isle of Skye. *New Phytol.* **69**, 807–820.

Birks, H.J.B. (1973) *Past and Present Vegetation of the Isle of Skye: a Palaeoecological Study*. Cambridge University Press, Cambridge.

Birks, H.J.B. (1978) Geographic variation of *Picea abies* (L.) Karsten pollen in Europe. *Grana* **17**, 149–160.

Birks, H.J.B. (1982) Holocene (Flandrian) chronostratigraphy of the British Isles: a review. *Striae* **16**, 99–105.

Birks, H.J.B. (1986) Numerical zonation, comparison and correlation of Quaternary pollen-stratigraphical data. In *Handbook of Holocene Palaeoecology and Palaeohydrology*, ed. B.E. Berglund, John Wiley, Chichester, pp. 743–774.

Birks, H.J.B. (1987a) Multivariate analysis in geology and geochemistry: an introduction. *Chemomet. and Int. Lab. Sys.* **2**, 15–28.

Birks, H.J.B. (1987b) Multivariate analysis of stratigraphic data in geology: a review. *Chemomet. and Int. Lab. Sys.* **2**, 109–126.

Birks, H.J.B. & Birks, H.H. (1980) *Quaternary Palaeoecology*. Edward Arnold, London.

Birks, H.J.B. & Gordon, A.D. (1985) *Numerical Methods in Quaternary Pollen Analysis.* Academic Press, London.

Blackmore, S. (1984) The Northwest European pollen flora, 32: Compositae — Lactuceae. *Rev. Palaeobotan. Palynol.* **42**, 45–85.

Blackmore, S. & Barnes, S.H. (1986) Harmomegathic mechanisms in pollen grains. In *Pollen and Spores: Form and Function*, ed. S. Blackmore & I.K. Ferguson, Academic Press, London, pp. 137–149.

Blackmore, S. & Ferguson, I.K. (eds) (1986) *Pollen and Spores: Form and Function.* Academic Press, London.

Blackmore, S. & Heath, G.L.A. (1984) The Northwest European pollen flora, 30: Berberidaceae. *Rev. Palaeobotan. Palynol.* **42**, 7–21.

Blyth, A.W. (1984) A mechanical aid for use with peat samplers. *Proc. 7th Int. Peat Congr., Dublin* **1**, 39–44.

Bocher, T.W. & Larson, K. (1955) Chromosome studies in some flowering plants. *Svensk. Bot. Tidskr.* **52**, 125–132.

Bonny, A.P. (1972) A method for determining absolute pollen frequencies in lake sediments. *New Phytol.* **71**, 393–405.

Bonny, A.P. (1976) Recruitment of pollen to the seston and sediment of some Lake District lakes. *J. Ecol.* **64**, 859–887.

Bonny, A.P. (1978) The effect of pollen recruitment processes on pollen distribution over the sediment surface of a small lake in Cumbria. *J. Ecol.* **66**, 385–416.

Bonny, A.P. (1980) Seasonal and annual variation over 5 years in contemporary airborne polled trapped at a Cumbrian lake. *J. Ecol.* **68**, 421–441.

Bonny, A.P. & Allen, P.V. (1983) Comparison of pollen data from Tauber traps paired in the field with simple cylindrical collectors. *Grana* **22**, 51–58.

Bonny, A.P. & Allen, P.V. (1984) Pollen recruitment to the sediments of an enclosed lake in Shropshire, England. In *Lake Sediments and Environmental History*, ed. E.Y. Haworth & J.W.G. Lund, Leicester University Press, Leicester. pp. 231–259.

Bortenschlager, S. (1969) Pollen analyse des Gletschereises — Grundlegende Fragen der Pollen-analyse überhaupt. *Bericht deutsche botanische Geschichte* **81**, 491–497.

Boyd, W.E. & Dickson, J.H. (1987a) The pollen morphology of four *Sorbus* species with special reference to two Scottish endemic species *S. arranensis* Hedl. and *S. pseudofennica* E.F. Warb. *Pollen Spores* **29**, 59–72.

Boyd, W.E. & Dickson, J.H. (1987b) A post-glacial pollen sequence from Loch a 'Mhuilinn, North Arran: a record of vegetation history with special reference to the history of endemic *Sorbus* species. *New Phytol.* **107**, 221–244.

Bradshaw, R.H.W. (1981) Modern pollen representation factors for woods in south-east England. *J. Ecol.* **69**, 45–70.

Bradshaw, R.H.W. (1988) Spatially-precise studies of forest dynamics. In *Vegetation History*, eds B. Huntley & T. Webb III. Kluwer Academic Publishers, pp. 725–751.

Bradshaw, R.H.W. & Miller, N.G. (1988) Recent successional processes investigated by pollen analysis of closed-canopy forest sites. *Vegetatio* **76**, 45–54.

Bredenkamp, C.L. & Hamilton-Attwell, V.L. (1988) A filter technique for preparing pollen for scanning electron microscopy. *Pollen Spores* **30**, 89–94.

Brooks, D. & Thomas, K.W. (1967) The distribution of pollen grains on microscope slides. I. The non-randomness of the distribution. *Pollen Spores* **9**, 621–629.

Brooks, J. & Shaw, G. (1968) Identity of sporopollenin with older kerogen and new evidence for the possible biological source of chemicals in sedimentary rocks. *Nature* **220**, 678–679.

Brooks, J. & Shaw, G. (1971) Recent developments in the chemistry, biochemistry, geochemistry and post-tetrad ontogeny of sporopollenins derived from pollen and spore exines. In *Pollen: Development and Physiology*, ed. J. Heslop-Harrison, Butterworth, London, pp. 99–114.

Bryant, V.M. & Holloway, R.G. (eds) (1985) *Pollen records of Late Quaternary North American sediments.* American Association of Stratigraphic Palynologists Foundation, Austin, Texas.

Burrin, P.J. & Scaife, R.G. (1984) Aspects of Holocene valley sedimentation and floodplain development in southern England. *Proc. Geol. Assoc.* **95**, 81–96.

Bush, M.B. & Flenley, J.R. (1987) The age of the British chalk grassland. *Nature* **329**, 434–436.

Chaloner, W.G. (1976) The evolution of adaptive features in fossil exines. In *The Evolutionary Significance of the Exine*, eds I.K. Ferguson & J. Muller. Academic Press, London, pp. 1–14.

Chaloner, W.G. (1986) Electrostatic forces in insect pollination and their significance in exine ornament. In *Pollen and Spores: Form and Function*, ed. S. Blackmore & I.K. Ferguson, Academic Press, London, pp. 103–108.

Chamberlain, A.C. & Chadwick, R.C. (1972) Deposition of spores and other particles on vegetation and soil. *Ann. Appl. Biol.* **71**, 141–158.

Chanda, S. (1962) On the pollen morphology of some Scandinavian Caryophyllaceae. *Grana Palynol.* **3**, 67–69.

Chapman, J.L. (1985) Preservation and durability of stored palynological specimens. *Pollen Spores* **28**, 113–120.

Chissoe, W.F., Vezey, E.L. & Skvarla, J.J. (1990) Drying

of pollen with Peldri II (proprietary fluorocarbon) for scanning electron microscopy. *Rev. Palaeobotan. Palynol.* **63**, 29—34.

Christensen, P.B. & Blackmore, S. (1988) The Northwest European pollen flora, 40: Tiliaceae. *Rev. Palaeobotan. Palynol.* **57**, 33—43.

Christie, A.D. & Ritchie, J.C. (1969) On the use of isentropic trajectories in the study of pollen transports. *Naturaliste Can.* **96**, 531—549.

Clapham, A.R., Tutin, T.G. & Moore, D.M. (1987) *Flora of the British Isles*, 3rd edn. Cambridge University Press, Cambridge.

Clarke, G.C.S. (1976) The Northwest European pollen flora. 7: Guttiferae. *Rev. Palaeobotan. Palynol.* **21**, 125—142.

Clarke, G.C.S. (1977) The Northwest European pollen flora, 10: Boraginaceae. *Rev. Palaeobotan. Palynol.* **24**, 59—101.

Clarke, G.C.S. & Jones, M.R. (1977a) The Northwest European pollen flora, 15: Plantaginaceae. *Rev. Palaeobotan. Palynol.* **24**, 129—154.

Clarke, G.C.S. & Jones, M.R. (1977b) The Northwest European pollen flora, 16: Valerianaceae, *Rev. Palaeobotan. Palynol.* **24**, 155—179.

Clarke, G.C.S. & Jones, M.R. (1978) The Northwest European pollen flora, 26: Aceraceae. *Rev. Palaeobotan. Palynol.* **26**, 181—193.

Clarke, G.C.S. & Jones, M.R. (1981a) The Northwest European pollen flora, 21: Dipsacaceae. *Rev. Palaeobotan. Palynol.* **33**, 1—25.

Clarke, G.C.S. & Jones, M.R. (1981b) The Northwest European pollen flora, 23: Dioscoreaceae. *Rev. Palaeobotan. Palynol.* **33**, 45—50.

Clarke, G.C.S. & Jones, M.R. (1981c) The Northwest European pollen flora, 24: Cabombaceae. *Rev. Palaeobotan. Palynol.* **33**, 51—55.

Cloutman, E.W. (1987) A mini-monolith cutter for absolute pollen analysis and fine sectioning of peats and sediments. *New Phytol.* **107**, 245—248.

Clymo, R.S. (1965) Experiments on the breakdown of *Sphagnum* in two bogs. *J. Ecol.* **53**, 747—758.

Clymo, R.S. (1973) The growth of *Sphagnum*: some effects of environment. *J. Ecol.* **61**, 849—869.

Clymo, R.S. (1988) A high resolution sampler of surface peat. *Functional Ecol.* **2**, 425—431.

Clymo, R.S. & Mackay, D. (1987) Upwash and down-wash of pollen and spores in the unsaturated surface layer of *Sphagnum*-dominated peat. *New Phytol.* **105**, 175—183.

Colinvaux, P.A. (1964) Sampling stiff sediments of an ice-covered lake. *Limnol. Oceanogr.* **9**, 262—264.

Collinson, M.E. (1987) Special problems in the conservation of palaeobotanical material. *Geological Curator* **4**, 439—445.

Coope, G.R. (1986) Coleoptera analysis. In *Handbook of Holocene Palaeoecology and Palaeohydrology*, ed. B.E. Berglund, John Wiley, Chichester, pp. 703—713.

Corbet, S.A. (1973) An illustrated introduction to the testate rhizopods in *Sphagnum*, with special reference to the area around Malham Tarn, Yorkshire. *Field Studies* **3**, 801—838.

Corbet, S.A., Beament, J. & Eisikowitch, D. (1982) Are electrostatic forces involved in pollen transfer? *Plant Cell Env.* **5**, 125—129.

Couteaux, M. & de Beaulieu, J.-L. (1976) L'analyse pollinique des 'argiles d'Eybens' prouve un age glaciaire. *Comptes Rendus Acad. Sci. Paris D* **282**, 277—280.

Cowan, J.W. (1988) Pollen in honey. *Plants Today* **1**, 95—99.

Cox, C.B. & Moore, P.D. (1985) *Biogeography; An Ecological and Evolutionary Approach*, 4th edn. Blackwell Scientific Publications, Oxford.

Cox, F.E.G. (1970) Separation of parasites in sucrose gradients. *Nature* **227**, 192.

Craig, A. (1972) Pollen influx to laminated sediments: a pollen diagram from northeastern Minnesota. *Ecology* **53**, 46—57.

Cronberg, G. (1986) Blue-green algae, green algae and Chrysophyceae in sediments. In *Handbook of Holocene Palaeoecology and Palaeohydrology*, ed. B.E. Berglund, John Wiley, Chichester, pp. 507—526.

Cronk, Q.C.B. & Clarke, G.C.S. (1981) The Northwest European pollen flora, 28: Convolvulaceae. *Rev. Palaeobotan. Palynol.* **33**, 117—135.

Culhane, K.J. & Blackmore, S. (1988) The Northwest European pollen flora, 41: Malvaceae. *Rev. Palaeobotan. Palynol.* **57**, 45—74.

Cundill, P.R. (1986) A new design of pollen trap for modern pollen studies. *J. Biogeog.* **13**, 83—98.

Cushing, E.J. (1961) Size increase in pollen grains mounted in thin slides. *Pollen Spores* **3**, 265—274.

Cushing, E.J. (1967a) Evidence for differential pollen preservation in late Quaternary sediments in Minnesota. *Rev. Palaeobotan. Palynol.* **4**, 87—101.

Cushing, E.J. (1967b) Late-Wisconsin pollen stratigraphy and the glacial sequence in Minnesota. In *Quaternary Palaeoecology*, eds E.J. Cushing & H.E. Wright. Yale University Press, New Haven, Connecticut, pp. 59—88.

Cushing, E.J. & Wright, H.E. (1965) Hand-operated piston corers for lake sediments. *Ecology* **46**, 380—384.

Cwynar, L.C. (1978) Recent history of fire and vegetation from laminated sediment of Greenleaf Lake, Algonquin Park, Ontario. *Can. J. Bot.* **56**, 10—21.

Cwynar, L.C., Burden, E. & McAndrews, J.H. (1979) An inexpensive sieving method for concentrating pollen and spores from fine-grained sediments. *Can. J. Earth Sci.* **16**, 1115—1120.

Damon, P.E., Donahue, D.J., Gore, B.H., Hatheway, A.L., Jull, A.J.T., Linick, T.W., Sercel, P.J.,

Toolin, L.J., Bronk, C.R., Hall, E.T., Hedges, R.E.M., Housley, R., Law, I.A., Perry, C., Bonani, G., Trumbore, S., Woelfli, W., Ambers, J.C., Bowman, S.G.E., Leese, M.N. & Tite, M.S. (1989) Radiocarbon dating of the Shroud of Turin. *Nature* **337**, 611–615.

Davies, R.I., Coulson, C.B. & Lewis, D.A. (1964) Polyphenols in plant, humus and soil. *J. Soil Sci.* **15**, 299–317.

Davis, M.B. (1963) On the theory of pollen analysis. *Amer. J. Sci.* **261**, 897–912.

Davis, M.B. (1965) A method for determination of absolute pollen frequency. In *Handbook of Palaeontological Techniques*, eds B.G. Kummel & D.M. Raup, Freeman, San Francisco, pp. 647–686.

Davis, M.B. (1966) Determination of absolute pollen frequency. *Ecology* **47**, 310–311.

Davis, M.B. (1967) Pollen deposition in lakes as measured by sediment traps. *Geol. Soc. Amer. Bull.* **78**, 849–858.

Davis, M.B. (1968) Pollen grains in lake sediments: redeposition caused by seasonal water circulation. *Science* **162**, 796–799.

Davis, M.B. (1969) Climatic changes in southern Connecticut recorded by pollen deposition at Rogers Lake. *Ecology* **50**, 409–422.

Davis, M.B., Brubaker, L.B. & Beiswenger, J.M. (1971) Pollen grains in lake sediments: pollen percentages in surface sediments from Southern Michigan. *Quat. Res.* **1**, 450–467.

Davis, M.B. & Deevey, E.S. (1964) Pollen accumulation rates: estimates from late-glacial sediment of Rogers Lake. *Science* **145**, 1293–1295.

Davis, M.B. & Ford, M.S. (1982) Sediment focusing in Mirror Lake, New Hampshire. *Limnol. Oceanogr.* **27**, 137–150.

De Beaulieu, J-L. (1982) Palynological subdivision of the Holocene in France. *Striae* **16**, 106–110.

Deevey, E.S. & Potzger, J.E. (1951) Peat samples for radiocarbon analysis: problems in pollen statistics. *Am. J. Sci.* **249**, 473–511.

Delcourt, H.R. (1987) The impact of prehistoric agriculture and land occupation on natural vegetation. *Trends Ecol. Evolut.* **2**, 39–44.

Delcourt, P.A. & Delcourt H.R. (1980) Pollen preservation and Quaternary environmental history in the southeastern United States. *Palynology*, **4**, 215–231.

Delcourt, H.R., Delcourt, P.A. & Webb, T. (1983) Dynamic plant ecology: the spectrum of vegetational change in space and time. *Quat. Sci. Revs.* **1**, 153–175.

De Leeuw, J. & Largeau, C. (in press) A review of macromolecular organic compounds that comprise living organisms and their role in kerogen, coal and petroleum formation. In *Organic Geochemistry*, eds M. Engel & S. Macko.

Delwiche, C.F., Graham, L.E. & Thomson, N. (1989) Lignin-like compounds and sporopollenin in *Coleochaete*, an algal model for land plant ancestry. *Science* **245**, 399–401.

Dexter, F., Banks, H.T. & Webb, T. (1987) Modeling Holocene changes in the location and abundance of beech populations in eastern North America. *Rev. Palaeobotan. Palynol.* **50**, 273–292.

Dickson, J.H. (1973) *Bryophytes of the Pleistocene*. Cambridge University Press, Cambridge.

Dickson, J.H. (1978) Bronze age mead. *Antiquity* **52**, 108–113.

Dickson, J.H. (1986) Bryophyte analysis. In *Handbook of Holocene Palaeoecology and Palaeohydrology*, ed. B.E. Berglund, John Wiley, Chichester, pp. 627–643.

Dimbleby, G.W. (1957) Pollen analysis of terrestrial soils. *New Phytol.* **56**, 12–28.

Dimbleby, G.W. (1985) *The Palynology of Archaeological Sites*. Academic Press, London.

Dodson, J.R. (1983) Pollen recovery from organic lake clays: a comparison of two techniques. *Pollen Spores* **25**, 131–138.

Echlin, P. (1968) Pollen. *Scientific American* **218**, 80–90.

Echlin, P. & Godwin, H. (1968) The ultrastructure and ontogeny of pollen of *Helleborus foetidus* L. I. The development of the tapetum and Ubisch bodies. *J. Cell Sci.* **3**, 175–186.

Edwards, K.J. (1983) Quaternary palynology: multiple profile studies and pollen variability. *Progress Phys. Geogr.* **7**, 587–609.

Edwards, K.J. & Gunson, A.R. (1978) A procedure for the determination of exotic pollen concentrations with a coulter counter. *Pollen Spores* **20**, 303–309.

Eide, F. (1981) Key for the Northwest European Rosaceae pollen. *Grana* **20**, 101–118.

Ellstrand, N.C. (1988) Pollen as a vehicle for the escape of engineered genes? *Trends Ecol. Evolut.* **3**(4), S30–S32.

Ellis, A.C. & van Geel, B. (1978) Fossil zygospores of *Debarya glyptosperma* (De Bary) Wittr. (Zygnemataceae) in Holocene sandy soils. *Acta Bot. Neerl.* **27**, 389–396.

Engel, M.S. (1978) The Northwest European pollen flora, 19: Haloragaceae. *Rev. Palaeobotan. Palynol.* **26**, 199–207.

Engstrom, D.R. & Maher, L.J. (1972) A new technique for volumetric sampling of sediment cores for concentrations of pollen and other microfossils. *Rev. Palaeobotan. Palynol.* **14**, 353–357.

Erdtman, G. (1943a) *An Introduction to Pollen Analysis*. Chronica Botanica, Waltham, Massachusetts.

Erdtman, G. (1943b) Polenspektra fran svenska vaxtsamhallen hamte pollenanalytiska markstudier i sodra Lappland. *Geol. Foren Stockh. Forhandl.* **65**, 37–66.

Erdtman, G. (1952) *An Introduction to Palynology I.*

Pollen morphology and plant taxonomy. Angiosperms. Almqvist & Wiksell, Stockholm.

Erdtman, G. (1956) 'LO analysis' and 'Welcker's rule'. A centenary. *Svensk. Bot. Tidskr.* **54**, 135–141.

Erdtman, G. (1957) *An Introduction to Palynology II. Pollen and spore morphology/plant taxonomy. Gymnospermae, Pteridophyta, Bryophyta.* Almqvist & Wiksell, Stockholm.

Erdtman, G. (1960) The acetolysis method. *Svensk. Bot. Tidskr.* **54**, 561–564.

Erdtman, G. (1966) *Sporoderm morphology and morphogensis.* A collection of data and suppositions. *Grana palynol.* **6**, 318–323.

Erdtman, G. (1969) *Handbook of Palynology.* Munksgaard, Copenhagen.

Erdtman, G., Berglund, B.E. & Praglowski, J. (1961) *An Introduction to a Scandinavian Pollen Flora. I.* Almqvist & Wicksell, Stockholm.

Erdtman, G. & Erdtman, H. (1933) The improvement of pollen analysis technique. *Svensk. Bot. Tidskr.* **27**, 347–357.

Erdtman, G. & Nordborg, G. (1961) Über die Möglichkeiten die Geschichte verschiedener Chromosomenzahlenrassen von *Sanguisorba minor* und *S. officinalis* pollenanalytisch zu beleuchten. *Bot. Notiser* **114**, 19–21.

Erdtman, G., Praglowski, J. & Nilsson, S. (1963) *An Introduction to a Scandinavian Pollen Flora. II.* Almqvist & Wicksell, Stockholm.

Erdtman, G. & Sorsa, P. (1971) *An Introduction to Palynology IV. Pollen and spore morphology/plant taxonomy. Pteridophyta.* Almqvist & Wiksell, Stockholm.

Etherington, J. (1983) *Wetland Ecology.* Studies in Biology No. 154. Edward Arnold, London.

Evans, A.T. & Moore, P.D. (1985) Surface pollen studies of *Calluna vulgaris* (L.) Jull and their relevance to the interpretation of bog and moorland pollen diagrams. *Circaea* **3**, 173–178.

Evans, J.G. (1972) *Land Snails in Archaeology.* Academic Press, London.

Faegri, K. (1956) Recent trends in palynology. *Bot. Rev.* **22**, 639–664.

Faegri, K. & Deuse, P. (1960) Size variation in pollen grains with different treatments. *Pollen Spores* **2**, 293–298.

Faegri, K. & Iversen, J. (1964) *Textbook of Pollen Analysis*, 2nd edn. Blackwell Scientific Publications, Oxford.

Faegri, K. & Iversen, J. (1975) *Textbook of Pollen Analysis*, 3rd edn. Blackwell Scientific Publications, Oxford.

Faegri, K. & Iversen, J. (1989) *Textbook of Pollen Analysis*, 4th edn, (revised by K. Faegri, P.E. Kaland & K. Krzywinski). John Wiley & Sons, Chichester.

Faegri, K. & van der Pilj, L. (1971) *The Principles of Pollination Ecology*, 2nd edn. Pergamon, Oxford.

Fagerlind, F. (1952) The real signification of pollen diagrams. *Bot. Notiser* **105**, 185–224.

Fast, A.W. & Wetzel, R.G. (1974) A close-interval fractionator for sediment cores. *Ecology* **55**, 202–204.

Feinsinger, P. (1987) Effects of species on each other's pollination: is community structure influenced? *Trends Ecol. Evolut.* **2**, 123–126.

Ferguson, I.K. & Webb, D.A. (1970) Pollen morphology in the genus *Saxifraga* and its taxonomic significance. *Bot. J. Linn. Soc. Lond.* **63**, 295–312.

Fink, J. & Kukla, G.J. (1977) Pleistocene climate in central Europe: at least 17 interglacials after the Olduvai event. *Quat. Res*, **7**, 363–371.

Flenley, J.R. (1988) Palynological evidence for land use changes in South East Asia. *J. Biogeogr.* **15**, 185–197.

Ford, T.D. (1989) Tufa: a freshwater limestone. *Geology Today* **5**, 60–63.

Foss, P.J. (1988) A palynological study of the Irish Ericaceae and *Empetrum*. *Pollen Spores* **30**, 151–178.

Francis, E. & Hall, V. (1985) Preliminary investigations into the causes of 'clumping' during standard pretreatments using *Lycopodium* spore tablets in absolute pollen analysis. *Circaea* **3**, 151–152.

Fredskild, B. (1967) Palaeobotanical investigations at Semermuit, Jakobshavn, West Greenland. *Meddr. Grønland.* **178**, 1–54.

Fredskild, B. & Wagner, P. (1974) Pollen and fragments of plant tissue in core samples from the Greenland Ice Cap. *Boreas* **3**, 105–108.

Free, J.B. (1968) Dandelion as a competitor to fruit trees for bee visits. *J. Appl. Ecol.* **5**, 169–178.

Frei, M. (1979) Plant species of pollen samples from the Shroud. Appendix to *The Turin Shroud*, I. Wilson. Penguin Books, London, Appendix E.

French, C.N. & Moore, P.D. (1986) Deforestation, *Cannabis* cultivation and schwingmor formation at Cors Llyn (Llyn Mire), central Wales. *New Phytol.* **102**, 469–482.

Frey, D.G. (1955) A differential flotation technique for recovering microfossils from inorganic sediments. *New Phytol.* **54**, 257–258.

Fries, M. & Hafsten, U. (1965) Asbjornsen's peat sampler the prototype of the Hiller sampler. *Geol. Foren. Stockh. Forh.* **87**, 307–313.

Gauch, H.G. (1982) *Multivariate Analysis in Community Ecology.* Cambridge University Press, Cambridge.

Girling, M.A. & Greig, J. (1985) A first fossil record for *Scolytus scolytus* (F.) (elm bark beetle): its occurrence in elm decline deposits from London and the implications for Neolithic elm disease. *J. Archaeol. Sci.* **12**, 347–351.

Glew, J.R. (1988) A portable extruding device for close interval sectioning of unconsolidated core samples.

J. Paleolimnol. **1**, 235–239.

Godwin, H. (1940) Pollen analysis and forest history of England and Wales. *New Phytol.* **33**, 278–305.

Godwin, H. (1949) Pollen analysis of glaciers in special relation to the formation of various types of glacier bands. *J. Glaciol.* **1**, 325–333.

Godwin, H. (1967) Pollen-analytical evidence for the cultivation of *Cannabis* in England. *Rev. Palaeobotan. Palynol.* **4**, 71–80.

Godwin, H. (1968) The origin of the exine. *New Phytol.* **67**, 667–676.

Godwin, H. (1975) *History of The British Flora*, 2nd edn. Cambridge University Press.

Godwin, H. & Andrew, R. (1951) A fungal fruit body common in postglacial peat deposits. *New Phytol.* **50**, 179–183.

Gordon, A.D. & Birks, H.J.B. (1972) Numerical methods in Quaternary palaeoecology. I. Zonation of pollen diagrams. *New Phytol.* **71**, 961–967.

Gordon, A.D. & Prentice, I.C. (1977) Numerical methods in Quaternary palaeoecology. IV. Separating mixtures of morphologically similar pollen taxa. *Rev. Palaeobotan. Palynol.* **23**, 359–372.

Gore, A.J.P. (ed.) (1983) *Ecosystems of the World. 4. Mires: Swamp, Bog, Fen and Moor*. Elsevier, Oxford.

Gray, J. (1965) Extraction techniques. In *Handbook of Palaeontological Techniques*, ed. B.G. Kummel & D.M. Raup, Freeman, San Francisco, pp. 530–587.

Gray, J. (1985) Interpretation of co-occurring megafossils and pollen: a comparative study with Clarkia as an example. In *Late Cenozoic History of the Pacific Northwest*, ed. C.J. Smiley, Pacific Division of the American Association for the Advancement of Science, San Francisco, pp. 185–239.

Grayum, M.H. (1986) Correlations between pollination biology and pollen morphology in the Araceae, with some implications for angiosperm evolution. In *Pollen and Spores: Form and Function*, ed. S. Blackmore & I.K. Ferguson, Academic Press, London, pp. 313–327.

Green, D.G. (1983) The ecological interpretation of fine-resolution pollen records. *New Phytol.* **94**, 459–477.

Green, D., Singh, G., Polach, H., Moss, D., Banks, J. & Geissler, E.A. (1988) A fine-resolution palaeoecology and palaeoclimatology from southeastern Australia. *J. Ecol.* **76**, 790–806.

Gregory, K.J. (ed.) (1983) *Background to Paleohydrology*. John Wiley, Chichester.

Greig, J. (1981) The investigations of a medieval barrel-latrine from Worcester. *J. Archaeol. Sci.* **8**, 265–282.

Greig, J. (1986) The archaeobotany of the Cowick medieval moat and some thoughts on moat studies. *Circaea* **4**, 43–50.

Grohne, U. (1957) Die Bedeutung des Phasenkontrastverfahrens für die Pollenanalyse, dargelegt am Beispiel der Gramineenpollen vom Getreidetyp. *Photogr. Forsch.* **7**, 237–248.

Grosse-Brauckmann, G. (1986) Analysis of vegetative plant macrofossils. In *Handbook of Holocene Palaeoecology and Palaeohydrology*, ed. B.E. Berglund, John Wiley, Chichester, pp. 591–618.

Guildford, W.J., Schneider, D.M., Larowitz, J. & Opella, S.J. (1988) High resolution solid state BCNMR spectroscopy of sporopollenin from different plant taxa. *Plant Physiol.* **86**, 134–136.

Guillet, B. & Planchais, N. (1969) Note sur une technique d'extraction des pollens des sols par une solution dense. *Pollen Spores* **11**, 141–145.

Guinet, Ph. (1986) Geographic patterns of the main pollen characters in genus *Acacia* (Leguminosae), with particular reference to subgenus *Phyllodineae*. In *Pollen and Spores: Form and Function*, eds S. Blackmore & I.K. Ferguson, Academic Press, London, pp. 297–311.

Guiot, J., Pons, A., de Beaulieu, J-L. & Reille, M. (1989) A 140 000 year continental climate reconstruction from two European pollen records. *Nature* **338**, 309–313.

Hall, V. (1989) A comparison of grass foliage, moss polsters and soil surfaces as pollen traps in modern pollen studies. *Circaea* **6**, 63–69.

Handel, S.N. (1976) Restricted pollen flow of two woodland herbs determined by neutron-activation analysis. *Nature* **260**, 422–423.

Handel, S.N. (1983) Pollination ecology, plant population structure, and gene flow. In *Pollination Biology*, ed. L. Real, Academic Press, Orlando, pp. 163–211.

Hansen, B.S. & Cushing, E.J. (1973) Identification of *Pinus* pollen of Quaternary age from the Chuska Mountains, New Mexico. *Geol. Soc. Amer. Bull.* **84**(2), 1181–1200.

Havinga, A.J. (1964) Investigation into the differential corrosion susceptibility of pollen and spores. *Pollen Spores* **6**, 621–635.

Havinga, A.J. (1971) An experimental investigation into the decay of pollen and spores in various soil types. In *Sporopollenin*, eds J. Brooks, P.R. Grant, M.D. Muir, P. van Gijzel & G. Shaw, Academic Press, London.

Havinga, A.J. (1974) Problems in the interpretation of pollen diagrams of mineral soils. *Geol. Mijnb.* **53**, 449–453.

Havinga, A.J. (1985) A 20-year experimental investigation into the differential corrosion susceptibility of pollen and spores in various soil types. *Pollen Spores* **26**, 541–558.

Heath, G.W., Arnold, M.K. & Edwards, C.A. (1965) Studies in leaf litter breakdown. I. Breakdown rates of leaves of different species. *Pedobiologia* **6**, 1–12.

Hebda, R.J., Chinnappa, C.C. & Smith, B.M. (1988) Pollen morphology of the Rosaceae of western Canada.

I. *Agrimonia* to *Crataegus*. *Grana* **27**, 95–113.

Heide, K. (1984) Holocene pollen stratigraphy from a lake and small hollow in north-central Wisconsin, USA. *Palynology* **8**, 3–20

Heslop-Harrison, J. (1968) Pollen wall development. *Science* **161**, 230–238.

Heslop-Harrison, J. (1971a) The pollen wall: structure and development. In *Pollen: Development and Physiology*, ed. J. Heslop-Harrison, Butterworth, London, pp. 75–98.

Heslop-Harrison, J. (1971b) Sporopollenin in the biological context. In *Sporopollenin*, eds J.B. Brooks, P.R. Grant, M. Muir, P. van Gijsel & G. Shaw, Academic Press, London, pp. 1–30.

Heslop-Harrison, J. (1979) An interpretation of the hydrodynamics of pollen. *Am. J. Bot.* **66**, 737–743.

Hill, C.R. (1983) Glycerine jelly mounts. *A.A.S.P. Newsletter* **16**, 3.

Hirst, J.M. (1952) An automatic volumetric spore trap. *Ann. Appl. Biol.* **39**, 257–265.

Hirst, J.M., Stedman, O.J. & Hogg, W.H. (1967) Long-distance spore transport: methods of measurement, vertical spore profiles and the detection of immigrant spores. *J. Gen. Microbiol.* **48**, 357–377.

Hodges, D. (1974) *The Pollen Loads of the Honey Bee*. Bee Research Association, London.

Holmes, P.L. (1990) Differential transport of spores and pollen: a laboratory study. *Rev. Palaeobotan. Palynol.* **64**, 289–296.

Horner, J.T. & Pearson, C.B. (1978) Pollen wall and aperture development in *Helianthus annuus* (Compositae: Heliantheae). *Am. J. Bot.* **65**, 293–309.

Howe, S. & Webb, T. (1983) Calibrating pollen data in climatic terms: improving the methods. *Quatern. Sci. Revs.* **2**, 17–51.

Hulme, P.D. & Shirriffs, J. (1985) Pollen analysis of a radiocarbon-dated core from North Mains, Strathallan, Perthshire. *Proc. Soc. Antiq. Scot.* **115**, 105–113.

Hunt, C.O. & Gale, S.J. (1986) Palynology: a neglected tool in British cave studies. In *New Directions in Karst*, eds K. Paterson & M.M. Sweeting. Geo-Abstracts, Norwich, pp. 323–332.

Huntley, B. & Birks, H.J.B. (1983) *An Atlas of Past and Present Pollen Maps for Europe: 0–13 000 years ago*. Cambridge University Press, Cambridge.

Huntley, B. & Prentice, I.C. (1988) July temperatures in Europe from pollen data, 6000 years before present. *Science* **241**, 687–690.

Huttenen, P. (1980) Early land-use, especially the slash and burn cultivation in the commune of Lammi, southern Finland, interpreted mainly using pollen and charcoal analysis. *Acta Bot. Fennica* **113**, 1–45.

Hyde, H.A. (1950) Studies in atmospheric pollen 4. Pollen deposition in Great Britain, 1943. Part 1, The influence of situation and weather. *New Phytol.* **49**, 398–406.

Hyde, H.A. (1969) Aeropalynology in Britain — an outline. *New Phytol.* **68**, 579–590.

Ingram, H.A.P. (1985) Hydrology. In *Ecosystems of the World. Mires, Swamp, Bog, Fen and Moor. 1. General Studies*, ed. A.J.P. Gore. Elsevier, Amsterdam, pp. 67–158.

Ivanov, K.E. (1981) *Water Movement in Mirelands* (translated by A. Thompson & H.A.P. Ingram). Academic Press, London.

Iversen, J. (1964) Retrogressive vegetational succession in the post-glacial. *J. Ecol.* **52** (Jubilee Symp. Suppl.), 59–70.

Jacobson, G.L. & Bradshaw, R.H.W. (1981) The selection of sites for paleovegetational studies. *Quatern. Res.* **16**, 80–96.

Janssen, C.R. (1959) *Alnus* as a disturbing factor in pollen diagrams. *Acta Bot. Neerland.* **8**, 55–58.

Janssen, C.R. (1966) Recent pollen spectra from the deciduous and coniferous-deciduous forests of northeastern Minnesota: a study in pollen dispersal. *Ecology* **47**, 804–825.

Janssen, C.R. (1974) *Verkenningen in de Palynologie*. Oosthoek, Scheltema & Holkema, Utrecht.

Janssen, C.R. & Braber, F.I. (1987) The present and past grassland vegetation of the Chajoux and Moselotte valleys (Vosges, France). *Proc. Konink. Nederl. Akad. Wetens. C* **90**, 115–138.

Jemmett, G. & Owen, J.A.K. (1990) Where has all the pollen gone? *Rev. Palaeobotan. Palynol.* **64**, 205–211.

Jessen, K. (1935) Archaeological dating in the history of North Jutland's vegetation. *Acta Archaeol.* **5**, 185–214.

Jones, M.R. & Blackmore, S. (1988) The Northwest European pollen flora, 38: Lycopodiaceae. *Rev. Palaeobotan. Palynol.* **57**, 1–25.

Jones, M.R. & Clarke, G.C.S. (1981) The Northwest European pollen flora, 25: Nymphaceae. *Rev. Palaeobotan. Palynol.* **33**, 57–67.

Jonsell, B. (1968) Studies in the Northwest European species of *Rorippa s. str. Symbolae Bot. Upsal.* **19**(2), 1–222.

Jorgensen, S. (1967) A new method of absolute pollen counting. *New Phytol.* **66**, 489–493.

Jowsey, P.C. (1966) An improved peat sampler. *New Phytol.* **65**, 245–248.

Kalis, A.J. (1979) The Northwest European pollen flora, 20: Papaveraceae. *Rev. Palaeobotan. Palynol.* **28**, 209–260.

Kapp, R.O. (1969) *How to Know Pollen and Spores*. Brown & Co, Dubuque, W.C.

Keatinge, T.H. (1983a) Development of pollen assemblage zones in soil profiles in southeastern England. *Boreas* **12**, 1–12.

Keatinge, T.H. (1983b) Influence of stem-flow on the representation of pollen of *Tilia* in soils. *Grana* **21**,

171−174.

Kerp, H. (1991) The study of fossil gymnosperms by means of cuticular analysis. *Palaios*, **5**, 548−569.

Kerney, M.P., Brown, E.H. & Chandler, T.J. (1964) The late-glacial and post-glacial history of the chalk escarpment near Brook, Kent. *Phil. Trans. R. Soc. Lond.* B **248**, 135−204.

Kerney, M.P., Preece, R.C. & Turner, C. (1980) Molluscan and plant biostratigraphy of some late Devensian and Flandrian deposits in Kent. *Phil. Trans. R. Soc. Lond.* B **291**, 1−43.

Klaus, W. (1978) On the taphonomic significance of tectum sculpture characters in alpine *Pinus* species. *Grana* **17**, 161−166.

Knox, A.S. (1942) The use of bromoform in the separation of non-calcareous microfossils. *Science* **95**, 307.

Knox, E.M. (1951) Spore morphology in British ferns. *Trans. Bot. Soc. Edinb.* **35**, 437−439.

Knox, R.B. (1979) *Pollen and Allergy*. Studies in Biology No. 107. Edward Arnold, London.

Krebs, C.J. (1972) *Ecology: The Experimental Analysis of Distribution and Abundance*. Harper & Row, New York.

Kurmann, M.H. (1985) An opal phytolith and palynomorph study of extant and fossil soils in Kansas (USA). *Palaeogeogr. Palaeoclimatol. Palaeoecol.* **49**, 217−235.

Langford, M., Taylor, G.E. & Flenley, J.R. (1990) Computerized identification of pollen grains by texture analysis. *Rev. Palaeobotan. Palynol.* **64**, 197−203.

Leroi-Gourhan, A. (1969) Pollen grains of Gramineae and Cerealia from Shanidar and Zawi Chemi. In *Domestication and Exploitation of Plants and Animals*, eds P.J. Ucko & G.W. Dimbleby, Duckworth, London, pp. 143−148.

Lewis, D. (1979) *Sexual Incompatibility in Plants*. Studies in Biology No. 110. Edward Arnold, London.

Lewis, D.M. & Ogden, E.C. (1965) Trapping methods for modern pollen rain studies. In *Handbook of Paleontological Techniques*, eds B.G. Kummel & D.M. Raup, Freeman, San Francisco, pp. 613−626.

Lieux, M.H. (1972) A melissopalynological study of 54 Louisiana (USA) honeys. *Rev. Palaeobotan Palynol.* **13**, 95−124.

Lieux, M.H. (1983) An atlas of pollen of trees, shrubs and other woody vines of Louisiana and other south eastern states, part V. Lythraceae to Euphorbiaceae. *Pollen Spores* **25**, 321−350.

Louveaux, J., Maurizio, A. & Vorwohl, G. (1978) Methods of melissopalynology. *Bee World* **59**, 139−157.

Lowe, J.J. and Walker, M.J.C. (1984) *Reconstructing Quaternary Environments*. Longmans, London.

Mackereth, F.J.H. (1958) A portable core-sampler for lake deposits. *Limnol. Oceanogr.* **3**, 181−191.

Macvicar, S.M. (1971) *The Student's Handbook of British Hepatics*, 2nd edn. Wheldon & Wesley, Hitchin.

Maher, L.J. (1972) Nomograms for computing 0.95 confidence limits of pollen data. *Rev. Paleobotan. Palynol.* **13**, 95−124.

Mandrioli, P., Puppi, G.L., Bagni, N. & Prodi, F. (1973) Distribution of microorganisms in hailstones. *Nature* **246**, 416−417.

Marceau, L. (1969) Effets, sur le pollen, des ultra-sons de base frequence. *Pollen Spores* **11**, 147−164.

Martin, H.A. (1973) Palynology and historical ecology of some cave excavations in the Australian Nullabor. *Aust. J. Bot.* **21**, 283−316.

Martin, P.S. & Mehringer, P.J. (1965) Pleistocene pollen analysis. In *Quaternary of the United States*, eds H.E. Wright & P.G. Frey, Princeton University Press, Princeton, 433−451.

Matthews, J. (1969) The assessment of a method for the determination of absolute pollen frequencies. *New Phytol.* **68**, 161−166.

McAndrews, J.H., Berti, A.A. & Norris, G. (1973) *Key to the Quaternary Pollen and Spores of the Great Lakes Region*. Toronto Royal University Museum Miscellaneous Publications.

McAndrews, J.H. & King, J.E. (1976) Pollen of the North American Quaternary: the top twenty. *Geosci. and Man* **15**, 41−49.

McDonald, J.E. (1962) Collection and washout of airborne pollens and spores by raindrops. *Science* **135**, 435−436.

Mehringer, P.J., Blinman, E. & Petersen, K.L. (1977) Pollen influx and volcanic ash. *Science* **198**, 257−261.

Middeldorp, A.A. (1982) Pollen concentration as a basis for indirect dating and quantifying net organic and fungal production in a peat bog ecosystem. *Rev. Palaeobotan. Palynol.* **37**, 225−282.

Mildenhall, D.C. (1990) Forensic palynology in New Zealand. *Rev. Palaeobotan. Palynol.* **64**, 227−234.

Moe, D. (1974) Identification key for trilete microspores of Fennoscandian pteridophyta. *Grana* **14**, 132−142.

Moe, D. (1983) Palynology of sheep's faeces: relationship between pollen content, diet and local pollen rain. *Grana* **22**, 105−113.

Moore, P.D. (1973) The influence of prehistoric cultures upon the initiation and spread of blanket bog in upland Wales. *Nature* **241**, 350−353.

Moore, P.D. (1977) Stratigraphy and pollen analysis of Claish Moss, northwest Scotland: significance for the origin of surface pools and forest history. *J. Ecol.* **65**, 375−397.

Moore, P.D. (1981) Neolithic land-use in mid-Wales. *Proc. 4th. Int. Conf. Palynol., Lucknow* **3**, 279−290.

Moore, P.D. (ed.) (1984a) *European Mires*. Academic Press, London.

Moore, P.D. (1984b) Clues to past climate in river sediments. *Nature* **308**, 316.

Moore, P.D. (1986) Site history. In *Methods in Plant Ecology*, 2nd edn, eds P.D. Moore & S.B. Chapman. Blackwell Scientific Publications, Oxford, pp. 525–556.

Moore, P.D. (1987) Ecological and hydrological aspects of peat formation. In *Coal and Coal-bearing Strata: Recent Advances*, ed. A.C. Scott. Geological Society Special Publications No. 32. Blackwell Scientific Publications, Oxford, pp. 7–15.

Moore, P.D. (1990) Vegetation's place in history. *Nature* 347, 710.

Moore, P.D. & Bellamy, D.J. (1974) *Peatlands*. Elek Science, London.

Moore, P.D. & Stevenson, A.C. (1982) Pollen studies in arid environments with particular reference to Turan. In *Desertification and Development: Dry Land Ecology in Social Perspective*, eds B. Spooner & H.S. Mani, Academic Press, London, pp. 249–267.

Moore, P.D. and Webb, J.A. (1978) *An Illustrated Guide to Pollen Analysis*. Hodder & Stoughton, London.

Nilsson, S., Praglowski, J. & Nilsson, L. (1977) *Atlas of Airborne Pollen Grains and Spores in Northern Europe*. Natur och Kultur, Stockholm.

Nowicke, J.W. & Meselson, M. (1984) Yellow rain — a palynological analysis. *Nature* 309, 205–206.

Ogden, E.C., Hayes, J.V. & Raynor, G.S. (1969) Diurnal patterns of pollen emission in *Ambrosia*, *Phleum* and *Ricinus*. *Am. J. Bot.* 56, 16–21.

Oldfield, F. (1959) The pollen morphology of some of the west European Ericales. *Pollen Spores* 1, 19–48.

Oldfield, F., Brown, A. & Thompson, R. (1979) The effect of microtopography and vegetation on the catchment of airborne particles measured by remanent magnetism. *Quat. Res.* 12, 326–332.

O'Sullivan, P.E. (1983) Annually laminated lake sediments and the study of Quaternary environmental changes — a review. *Quat. Sci. Revs.* 1, 245–313.

Palmer, P.G. (1976) Grass cuticles: a new paleoecological tool for East African lake sediments. *Can. J. Bot.* 54, 1725–1734.

Pals, J.P., van Geel, B. & Delfos, A. (1980) Paleo-ecological studies in the Klokkeweel Bog near Hoogkarspel (Province of Noord-Holland). *Rev. Palaeobotan. Palynol.* 30, 371–418.

Parsons, R.W. & Prentice, I.C. (1981) Statistical approaches to *R*-values and the pollen–vegetation relationship. *Revs. Palaeobotan. Palynol.* 32, 127–152.

Pearsall, D.M. (1989) *Paleoethnobotany — A Handbook of Procedures*. Academic Press, San Diego.

Peck, R.M. (1972) Efficiency test on the Tauber trap used as a pollen sampler in turbulent water flow. *New Phytol.* 71, 187–198.

Peck, R.M. (1974) A comparison of four absolute pollen preparation techniques. *New Phytol.* 73, 567–587.

Peglar, S.M., Fritz, S.C., Alapieti, T., Saarnisto, M. &

Birks, H.J.B. (1984) Composition and formation of laminated sediments in Diss Mere, Norfolk, England. *Boreas* 13, 13–28.

Pennington, W. (1979) The origin of pollen in lake sediments: an enclosed lake compared with one receiving inflow streams. *New Phytol.* 83, 189–213.

Pennington, W. (1980) Modern pollen samples from west Greenland and the interpretation of pollen data from the British late-glacial (Late Devensian). *New Phytol.* 84, 171–201.

Pennington, W. (1981a) Records of a lake's life in time. *Hydrobiologia* 79, 197–219.

Pennington, W. (1981b) Sediment composition in relation to the interpretation of pollen data. *Proc. 4th Int. Palynol. Conf., Lucknow* 3, 188–213.

Pennington, W. & Bonny, A.P. (1970) Absolute pollen diagrams from the British late-glacial. *Nature* 226, 871–872.

Pentecost, A. (1978) Blue-green algae and freshwater carbonate deposits. *Proc. R. Soc. Lond. B* 200, 141–152.

Perry, I. & Moore, P.D. (1987) Dutch elm disease as an analogue of Neolithic elm decline. *Nature* 326, 72–73.

Phillips, L. (1972) An application of fluorescence microscopy to the problem of derived pollen in British Pleistocene deposits. *New Phytol.* 71, 755–762.

Piperno, D.R. (1988) *Phytolith Analysis. An Archaeological and Geological Perspective*. Academic Press, London.

Piperno, D.R. (1989) The occurrence of phytoliths in the reproductive structures of selected tropical angiosperms and their significance in tropical paleoecology, paleoethnobotany and systematics. *Rev. Palaeobotan. Palynol.* 61, 147–173.

Pitkin, P.H. (1975) Variability and seasonality of the growth of some corticolous pleurocarpous mosses. *J. Bryology* 8, 337–356.

Planchais, N. (1987) Impact de l'homme lors du remplissage de l'estuaire du Lez (Palavas Herault) mis en evidence par l'analyse pollinique. *Pollen Spores* 29, 73–88.

Postek, M.T., Howard, K.S., Johnson, A.H. & McMichael, K.L. (1980) *Scanning Electron Microscopy — A Student's Handbook*. M.T. Postek & Ladd Research Industries.

Pott, R. (1986) Der pollenanalytische Nachweis extensiver Waldbewirtschaftungen in den Haubergen des Siegerlandes. In *Anthropogenic Indicators in Pollen Diagrams*, ed. K-E. Behre, A.A. Balkema, Rotterdam, pp. 125–134.

Potter, L.D. (1967) Differential pollen accumulation in water-tank sediments and adjacent soils. *Ecology* 48, 1041–1043.

Praglowski, J. (1970) The effects of pre-treatment and the embedding media on the shape of pollen grains.

Rev. Palaeobotan. Palynol. **10**, 203–208.

Preece, R.C. (1980) The biostratigraphy and dating of the tufa deposit at the Mesolithic site at Blashenwell, Dorset, England. *J. Archaeol. Sci.* **7**, 345–362.

Preece, R.C., Coxon, P. & Robinson, J.E. (1986) New biostratigraphic evidence of the post-glacial colonization of Ireland and for Mesolithic forest disturbance. *J. Biogeog.* **13**, 487–509.

Prentice, I.C. (1981) Quantitative birch (*Betula* L.) pollen separation by analysis of size frequency data. *New Phytol.* **89**, 145–157.

Prentice, I.C. (1985) Pollen representation, source area, and basin size: toward a unified theory of pollen analysis. *Quatern. Res.* **23**, 76–86.

Prentice, I.C. (1986a) Vegetation responses to past climatic variation. *Vegetatio* **67**, 131–141.

Prentice, I.C. (1986b) Forest-composition calibration of pollen data. In *Handbook of Holocene Palaeoecology and Palaeohydrology*, ed. B.E. Berglund, John Wiley, Chichester, pp. 799–816.

Prentice, I.C. (1986c) Multivariate methods for data analysis. In *Handbook of Holocene Palaeoecology and Palaeohydrology*, ed. B.E. Berglund, John Wiley, Chichester pp. 775–797.

Prentice, I.C. (1988a) Records of vegetation in time and space. In *Vegetation History*, eds B. Huntley & T. Webb III, Kluwer Academic Publishers, pp. 17–42.

Prentice, I.C. (1988b) Palaeoecology and plant population dynamics. *Trends Ecol. Evolut.* **3**(12), 343–345.

Prentice, I.C. & Webb, T. (1986) Pollen percentages, tree abundances and the Fagerlind effect. *J. Quatern. Sci.* **1**, 35–43.

Price, M.D.R. & Moore, P.D. (1984) Pollen dispersion in the hills of Wales: a pollen shed hypothesis. *Pollen Spores* **26**, 127–136.

Proctor, M.C.F. & Yeo, P.F. (1973) *The Pollination of Flowers*. Collins, London.

Punt, W. (1975) The Northwest European pollen flora, 5: Sparganiaceae and Typhaceae. *Rev. Palaeobotan. Palynol.* **19**, 75–88.

Punt, W. (1984) The Northwest European pollen flora, 37: Umbelliferae. *Rev. Palaeobotan. Palynol.* **42**, 155–363.

Punt, W. & den Breejen, P. (1981) The Northwest European pollen flora, 27: Linaceae. *Rev. Palaeobotan. Palynol.* **33**, 75–115.

Punt, W. & Langewis, E.A. (1988) The Northwest European pollen flora, 42: Verbenaceae. *Rev. Palaeobotan. Palynol.* **57**, 75–79.

Punt, W., de Leeuw van Weenen, J.S. & van Oostrum, W.A.P. (1974) The Northwest European pollen flora, 3: Primulaceae. *Rev. Palaeobotan. Palynol.* **17**, 31–70.

Punt, W. & Neinhuis, W. (1976) The Northwest European pollen flora, 6: Gentianaceae. *Rev. Palaeo-botan. Palynol.* **21**, 89–123.

Punt, W. & Malotaux, M. (1984) The Northwest European pollen flora, 31: Cannabaceae, Moraceae and Urticaceae. *Rev. Palaeobotan. Palynol.* **42**, 23–44.

Punt, W. & Monna-Brands, M. (1977) The Northwest European pollen flora, 8: Solanaceae. *Rev. Palaeobotan. Palynol.* **23**, 1–30.

Punt, W., Reitsma, T. & Reuvers, A.A.M.L. (1974) The Northwest European pollen flora, 2. Caprifoliaceae. *Rev. Palaeobotan. Palynol.* **17**, 5–29.

Punt, W. & Reumer, J.W. (1981) The Northwest European pollen flora, 22: Alismataceae. *Rev. Palaeobotan. Palynol.* **33**, 27–44.

Rackhan, O. (1980) *Ancient Woodland: Its History, Vegetation and Uses in England*. Edward Arnold, London.

Ranson, J.F. & Leopold, E.B. (1962) The standard rain gauge as an efficient sampler of air-borne pollen and spores. *Pollen Spores* **4**, 373.

Rathcke, B. (1983) Competition and facilitation among plants for pollination. In *Pollination Biology*, ed. L. Real, Academic Press, Orlando, Florida, pp. 305–329.

Raynor, G.S., Hayes, J.V. & Lewis, D.M. (1983) Testing of the air resources laboratories trajectory model on cases of pollen wet deposition after long-distance transport from known source regions. *Atmos. Environ.* **17**, 213–220.

Real, L. (ed.) (1983) *Pollination Biology*, Academic Press, Orlando, Florida.

Reitsma, T. (1966) Pollen morphology of some European Rosaceae. *Acta Bot. Neerl.* **15**, 290–307.

Reitsma, T. (1969) Size modification of pollen grains under different treatments. *Rev. Palaeobotan. Palynol.* **9**, 175–202.

Reitsma, T. (1970) Suggestions towards unification of descriptive terminology of angiosperm pollen grains. *Rev. Palaeobotan. Palynol.* **10**, 39–60.

Reitsma, T. & Reuvers, A.A.M.L. (1975) The Northwest European pollen flora, 4: Adoxaceae. *Rev. Palaeobotan. Palynol.* **19**, 71–73.

Ritchie, J.C. (1984) *Past and Present Vegetation of the Far Northwest of Canada*. University of Toronto Press, Toronto.

Ritchie, J.C. (1986a) Modern pollen spectra from Dakleh Oasis, Western Egyptian Desert. *Grana* **25**, 177–182.

Ritchie, J.C. (1986b) Climate change and vegetation response. *Vegetatio* **67**, 65–74.

Ritchie, J.C., Hadden, K.A. & Gajewski, K. (1987) Modern pollen spectra from lakes in arctic western Canada. *Can. J. Bot.* **65**, 1605–1613.

Rovner, I. (1986) *Plant Opal Phytolith Analysis in Archaeology and Paleoecology*. North Carolina State University, Raleigh, North Carolina.

Rowley, J.R. (1960) The exine structure of 'cereal' and

'wild' type grass pollen. *Grana Palynol.* **2**, 9—15.

Rowley, J.R. (1963) Non-homogeneous sporopollenin in microspores of *Poa annua*. *Grana palynol.* **3**, 3—20.

Rowley, J.R. & Dahl, A.O. (1956) Modifications in design and use of the Livingstone sampler. *Ecology* **37**, 849—851.

Rowley, J.R. Dahl, A.O., Sungupta, S. & Rowley, J.S. (1981) A model of exine substructure based on dissection of pollen and spore exines. *Palynology* **5**, 107—152.

Rybnickova, E. (1973) Pollenanalytische Unterlagen für die Rekonstruktion der ursprunglichen Waldvegetation im mittleren Teil des Otava-Bohmerwladvorgebirges. *Folia Geobot. Phytotax.*, *Praha* **8**, 117—142.

Rybnickova, E. & Rybnicek, K. (1971) The determination and elimination of local elements in pollen spectra from different sediments. *Rev. Palaeobotan. Palynol.* **11**, 165—176.

Saarnisto, M. (1979) Applications of annually laminated lake sediments; a review. *Acta Univ. Ouluensis A* **82**; *Geology* **3**, 97—108.

Saarnisto, M. (1986) Annually laminated lake sediments. In *Handbook of Holocene Palaeoecology and Palaeohydrology*, ed. B.E. Berglund, John Wiley, Chichester, pp. 343—370.

Salmi, M. (1962) Investigations on the distribution of pollens in an extensive raised bog. *Comp. Rend. Soc. Geol. Finland* **34**, 159—193.

Santisuk, T. (1979) A palynological study of the tribe Ranunculaceae. *Opera Botanica* **48**, 1—74.

Sawyer, R. (1985) Misrepresentation — a simplified procedure. *Bee Craft* **67**, 167—169.

Sawyer, R. (1988) *Honey Identification*, University College Cardiff Press, Cardiff.

Schoenwetter, J. (1974) Pollen analysis of human paleofeces from Upper Salts Cave. In *Archeology of the Mammoth Cave Area*, Academic Press, New York, pp. 49—58.

Shapiro, J. (1958) The core freezer — a new sampler for lake sediments. *Ecology* **39**, 748.

Simola, H.L.K., Coard, M.A. & O'Sullivan, P.E. (1981) Annual laminations in the sediments of Loe Pool, Cornwall. *Nature* **290**, 238—241.

Singh, G. & Geissler, E.A. (1985) Late Cainozoic history of vegetation, fire, lake levels and climate, at Lake George, New South Wales, Australia. *Phil. Trans. R. Soc. Lond. B* **311**, 379—447.

Singh, G., Joshi, R.R., Chopra, S.K. & Singh, A.B. (1974) Late-Quaternary history of vegetation and climate of the Rajasthan Desert, India. *Phil. Trans. R. Soc. Lond. B* **267**, 467—501.

Skvarla, J.J. & Larson, D.A. (1966) Fine structural studies of *Zea mays* pollen. I. Cell membranes and exine ontogeny. *Amer. J. Bot.* **53**, 1112—1125.

Smith, A.G. & Pilcher, J.R. (1973) Radiocarbon dates and vegetational history of the British Isles. *New Phytol.* **72**, 903—914.

Smith, A.G., Pilcher, J.R. & Singh, G. (1968) A large capacity hand-operated peat sampler. *New Phytol.* **67**, 119—124.

Solomon, A.M., Blasing, T.J. & Solomon, J.A. (1982) Interpretation of floodplain pollen in alluvial sediments from an arid region. *Quat. Res.* **18**, 52—71.

Sorsa, P. (1964) Studies on the spore morphology & Fennoscandian fern species. *Ann. Bot. Fenn.* **1**, 179—201.

Southworth, D. (1974) Solubility of pollen exines. *Am. J. Bot.* **61**, 36—44.

Stevenson, A.C. (1981) *Pollen Studies in Semi-arid Areas: North-east Iran and South-west Spain*. Unpublished PhD thesis, University of London, London.

Stevenson, A.C. (1985) Studies in the vegetational history of S.W. Spain. I. Modern pollen rain in the Doñana National Park, Huelva. *J. Biogeog.* **12**, 243—268.

Stewart, J.M. & Durno, S.E. (1969) Structural variations in peat. *New Phytol.* **69**, 875—883.

Stockmarr, J. (1970) Species identification of *Ulmus* pollen. *Danm. Geol. Unders. IVR* **4**(11), 1—19.

Stockmarr, J. (1971) Tablets with spores used in absolute pollen analysis. *Pollen Spores* **13**, 614—621.

Stockmarr, J. (1973) Determination of spore concentration with an electronic particle counter. *Danm. Geol. Unders. Arbog* **1972**, 87—89.

Stockmarr, J. (1975) Retrogressive forest development as reflected in a mor pollen diagram from Mantingerbos, Drenthe, the Netherlands. *Palaeohistoria* **17**, 37—51.

Sturludottir, S.A. & Turner, J. (1985) The elm decline at Pawley mire: an anthropogenic interpretation. *New Phytol.* **99**, 323—329.

Tallis, J.H. (1962) The identification of *Sphagnum* spores. *Trans. Br. Bryol. Soc.* **4**, 209—213.

Tappan, H. (1980) *The Paleobiology of Plant Protists*. Freeman & Co, San Francisco.

Tauber, H. (1965) Differential pollen dispersion and the interpretation of pollen diagrams. *Danm. Geol. Unders. IIR* **89**, 1—69.

Tauber, H. (1974) A static non-overload pollen collector. *New Phytol.* **73**, 359—369.

Thomas, K.W. (1964) A new design for a peat sampler. *New Phytol.* **63**, 422—425.

Ting, W.S. (1965) The saccate grains of Pinaceae mainly of California. *Grana* **6**, 270—289.

Ting, W.S. (1966) Determination of *Pinus* pollen species by pollen statistics. *Univ. Calif. Publ. Geol. Sci.* **58**, 1—182.

Tinsley, H.M. & Smith, R.T. (1974) Surface pollen studies across a woodland heath transition and their application to the interpretation of pollen diagrams. *New Phytol.* **73**, 547—565.

Tipping, R. (1985) A problem with pollen concentration procedures. *Pollen Spores* **27**, 121–129.

Tolonen, K. (1986a) Charred particle analysis. In *Handbook of Holocene Palaeoecology and Palaeohydrology*, ed. B.E. Berglund, John Wiley, Chichester, pp. 485–496.

Tolonen, K. (1986b) Rhizopod analysis. In *Handbook of Holocene Palaeoecology and Palaeohydrology*, ed. B.E. Berglund, John Wiley, Chichester, pp. 645–666.

Tolonen, M. (1980a) Identification of fossil *Ulmus* pollen in sediments of Lake Lamminjarvi, S. Finland. *Ann. Bot. Fennici* **17**, 7–10.

Tolonen, M. (1980b) Degradation analysis of pollen in sediments of Lake Lamminjarvi, S. Finland. *Ann. Bot. Fennici* **17**, 11–14.

Tolonen, K. & Tolonen, M. (1988) Synchronous pollen changes and traditional land use in South Finland, studied from three adjacent sites: a lake, a bog and a forest soil. In *Lake, Mire and River Environments*, eds G. Lang & C. Schluchter. A.A. Balkema, Rotterdam, pp. 83–97.

Tomlinson, P. (1984) Ultrasonic filtration as an aid in pollen analysis of archaeological deposits. *Circaea* **2**, 139–140.

Trusswell, E.M. & Owen, J.A. (1990) The proceedings of the 7th International Palynological Congress (Part I). *Rev. Palaeobotan. Palynol.* **64**, 1–366.

Traverse, A. (1988) *Paleopalynology*. Unwin Hyman, Boston.

Troels-Smith, J. (1955) Characterization of unconsolidated sediments. *Danm. Geol. Unders. IVR* **3**(101), 1–73.

Tryton, A.F. (1986) Stasis, diversity and function in spores based on an electron microscope survey of the Pteridophyta. In *Pollen and Spores: Form and Function*, eds S. Blackmore & I.K. Ferguson, Academic Press, London, pp. 233–249.

Turner, J. (1964) The anthropogenic factor in vegetational history. I. Tregaron and Whixall Mosses. *New Phytol.* **63**, 73–90.

Turner, S.C. & Blackmore, S. (1984) The Northwest European pollen flora, 36: Plumbaginaceae. *Rev. Palaeobotan. Palynol.* **42**, 133–154.

Ueno, J. (1958) Some palynological observations of Pinaceae. *J. Biol. Osaka City Univ.* **9**, 163–186.

Van den Assem (1971) Airborne pollen in relation to pollinosis. In *Scandinavian Aerobiology*, ed. S. Nilsson, Swedish Natural Science Research Council, Stockholm, pp. 181–196.

Van der Knaap, (1989) Past vegetation and reindeer on Edgeoya (Spitsbergen) between *c.* 7900 and *c.* 3800 BP, studied by means of peat layers and reindeer faecal pellets. *J. Biogeogr.* **16**, 379–394.

Van Geel, B. (1972) Palynology of a section from the raised peat bog 'Wietmarscher Moor', with special reference to fungal remains. *Acta Bot. Neerl.* **21**, 261–284.

Van Geel, B. (1976) Fossil spores of Zygnemataceae in ditches of a prehistoric settlement in Hoogkarspel (The Netherlands). *Rev. Palaeobotan. Palynol.* **22**, 337–344.

Van Geel, B. (1978) A palaeoecological study of Holocene peat bog sections in Germany and the Netherlands, based on the analysis of pollen, spores and macro- and microscopic remains of fungi, algae, cormophytes and animals. *Rev. Palaeobotan. Palynol.* **25**, 1–120.

Van Geel, B. (1979) Preliminary report on the history of Zygnemataceae and the use of their spores as ecological markers. *Proc. 4th Int. Palynol. Conf., Lucknow*, **1**, 467–469.

Van Geel, B. (1986) Application of fungal and algal remains and other microfossils in palynological analyses. In *Handbook of Holocene Palaeoecology and Palaeohydrology*, ed. B.E. Berglund, John Wiley, Chichester, pp. 497–505.

Van Geel, B., Bohncke, S.J.P. & Dee, H. (1981) A palaeoecological study of an upper Late Glacial and Holocene sequence from 'De Borchert', The Netherlands. *Rev. Palaeobotan. Palynol.* **31**, 367–448.

Van Geel, B., Coope, G.R. & van der Hammen, T. (1989) Palaeoecology and stratigraphy of the Late Glacial type section at Usselo (The Netherlands). *Rev. Palaeobot. Palynol.* **60**, 25–129.

Van Geel, B., Hallewas, D.P. & Pals, J.P. (1983) A late Holocene deposit under the Westfriese Zeedijk near Enkhuizen (Prov. of Noord-Holland, The Netherlands): palaeoecological and archaeological aspects. *Rev. Palaeobotan. Palynol.* **38**, 269–335.

Van Geel, B. & van der Hammen, T. (1978) Zygnemataceae in Quaternary Colombian sediments. *Rev. Palaeobotan. Palynol.* **25**, 377–392.

Van Helvoort, H.A.M. & Punt, W. (1984) The Northwest European pollen flora, 29: Araliaceae. *Rev. Palaeobotan. Palynol.* **42**, 1–5.

Van Leeuwen, P., Punt, W. & Hoen, P.P. (1988) The Northwest European pollen flora, 43: Polygonaceae. *Rev. Palaeobotan. Palynol.* **57**, 81–151.

Vareschi, V. (1934) Pollen analysen aus Gletschereis. *Bericht über das Geobotanische Forschungsinstitut Rübel in Zürich für das Jahr 1934*, 81–89.

Verbeek-Reuvers, A.A.L.M. (1977a) The Northwest European pollen flora, 9: Saxifragaceae. *Rev. Palaeobotan. Palynol.* **24**, 31–58.

Verbeek-Reuvers, A.A.L.M. (1977b) The Northwest European pollen flora, 12: Grossulariaceae. *Rev. Palaeobotan. Palynol.* **24**, 107–116.

Verbeek-Reuvers, A.A.L.M. (1977c) The Northwest European pollen flora, 14: Parnassiaceae. *Rev. Palaeobotan. Palynol.* **24**, 123–128.

Vitt, D.H. & Hamilton, C.D. (1974) A scanning electron microscope study of the spores and selected peristomes of the North American Encalyptaceae (Musci). *Can. J. Bot.* **52**, 1973–1981.

Von Post, L. (1916) Om skogstradspollen i sydsvenska torfmosselagerfoljder (foredragsreferat). *Geol. For. Stock. Forh.* **38**, 384–394.

Vuorela, I. (1973) Relative pollen rain around cultivated fields. *Acta Bot. Fennica* **102**, 1–27.

Walch, K.M., Rowley, J.R. & Norton, N.J. (1970) Displacement of pollen grains by earthworms. *Pollen et Spores* **12**, 39–44.

Waser, N.M. (1988) Comparative pollen and dye transfer by pollinators of *Delphinium nelsonii. Functional Ecol.* **2**, 41–48.

Wasylikowa, K. (1986) Analysis of fossil fruits and seeds. In *Handbook of Holocene Palaeoecology and Palaeohydrology*, ed. B.E. Berglund, John Wiley, Chichester, pp. 571–590.

Watson, E.V. (1981) *British mosses and liverworts*, 3rd edn. Cambridge University Press, Cambridge.

Watt, I.M. (1985) *The Principles and Practice of Electron Microscopy*. Cambridge University Press, Cambridge.

Webb, J.A. (1977) *Studies on the late-Devensian vegetation of the Whitlaw Mosses, south-east Scotland*. Unpublished PhD thesis, University of London, London.

Webb, J.A. & Moore, P.D. (1982) The Late Devensian vegetational history of the Whitlaw Mosses, southeast Scotland. *New Phytol.* **91**, 341–398.

Webb, T. III (1980) The reconstruction of climatic sequences from botanical data. *J. Interdisc. Hist.* **10**, 749–772.

Webb, T. III (1985) Holocene palynology and climate. In *Paleoclimate Analysis and Modeling*, ed. A.D. Hecht, John Wiley, New York, pp. 163–195.

Webb, T. III (1987) The appearance and disappearance of major vegetational assemblages: Long-term vegetational dynamics in eastern North America. *Vegetatio* **69**, 177–187.

Webb, T. III, Howe, S.E., Bradshaw, R.H.E. & Heide, K.M. (1981) Estimating plant abundances from pollen percentages: the use of regression analysis. *Rev. Palaeobotan. Palynol.* **34**, 269–300.

West, R.G. (1970) Pollen zones in the Pleistocene of Great Britain and their correlation. *New Phytol.* **69**, 1179–1183.

West, R.G. (1977) *Pleistocene Geology and Biology*, 2nd edn. Longman, London.

Whitehead, D.R. (1983) Wind pollination: some ecological and evolutionary perspectives. In *Pollination Biology*, ed. L. Real, Academic Press, Orlando, California, pp. 97–108.

Whittington, G. & Gordon, A.D. (1987) The differentiation of the pollen of *Cannabis sativa* L. from that of *Humulus lupulus* L. *Pollen Spores* **29**, 111–120.

Wilson, M.F. & Burley, N. (1983) *Mate Choice in Plants*, Princeton University Press, Princeton.

Wiltshire, P.E.J. (1988) A simple device for obtaining contiguous peat samples of small volume for pollen analysis. *Circaea* **5**(2), 97–99.

Wodehouse, P.P. (1935) *Pollen Grains*. McGraw-Hill, New York.

Wolfe, J.A. & Upchurch, G.R. (1987) Leaf assemblages across the Cretaceous/Tertiary boundary in the Raton Basin, New Mexico and Colorado. *Proc. Nat. Acad. Sci.* **84**, 5096–5100.

Woodhead, N. & Hodgson, L.M. (1935) Preliminary study of some Snowdonian peats. *New Phytol.* **34**, 263–282.

Woosley, A. (1978) Pollen extraction for arid land sediments. *J. Field Archaeol.* **5**, 349–355.

Wright, H.E. (1967) The use of surface samples in Quaternary pollen research. *Rev. Palaeobotan. Palynol.* **2**, 321–330.

Wright, H.E. (1980) Cores of soft lake sediments. *Boreas* **9**, 107–114.

Wright, H.E., Livingstone, D.A. & Cushing, E.J. (1965) Coring devices for lake sediments. In *Handbook of Palaeontological Techniques*, eds B. Kummel & D.M. Raup, Freeman, San Francisco, pp. 494–520.

Wright, H.E., Mann, D.H. & Glaser P.H. (1984) Piston corers for peat and lake sediments. *Ecology*, **65**, 657–659.

Zetzsche, F. (1932) Kork und cuticularsubstanzen. In *Handbuch der Pflanzenanalyse*, ed. G. Klein. Springer-Verlag, Berlin, pp. 205–215.

INDEX

Page references in *italics* refer to figures.
References in **bold** refer to tables, and plates are denoted by pl.